C0-APH-378

24. Colloquium
der Gesellschaft für Biologische Chemie
26.—28. April 1973 in Mosbach/Baden

Regulation of Transcription and Translation in Eukaryotes

Edited by

E. K. F. Bautz, P. Karlson, and H. Kersten

With 131 Figures

Springer-Verlag Berlin · Heidelberg · New York 1973

Professor Dr. E. K. F. BAUTZ, Institut für Molekulare Genetik, Zoologisches Institut der Universität, D-6900 Heidelberg, Neuenheimer Feld/Federal Republic of Germany

Professor Dr. P. KARLSON, Physiologisch-Chemisches Institut der Universität, D-3550 Marburg/Federal Republic of Germany

Professor Dr. H. KERSTEN, Physiologisch-Chemisches Institut der Universität, D-8520 Erlangen/Federal Republic of Germany

ISBN 3-540-06472-9 Springer-Verlag Berlin · Heidelberg · New York
ISBN 0-387-06472-9 Springer-Verlag New York · Heidelberg · Berlin

Typesetting, Printing and Binding: Brühlsche Universitätsdruckerei, Gießen

Preface

This volume represents the proceedings of the 24th Mosbach Colloquium on "Regulation of Transcription and Translation in Eukaryotes" which was held April 26—28, 1973, in Mosbach, Germany, under the auspices of the Gesellschaft für Biologische Chemie. To the three of us (H. KERSTEN, P. KARLSON and myself) who were commissioned with the invitation of speakers, it was a difficult decision as to whether we should attempt to cover with some twenty contributions as many aspects of this broad topic as possible, or to sacrifice the intellectually perhaps more pleasing but more speculative concepts and to concentrate on a few aspects of gene expression in reasonable detail. We unanimously decided on the latter course, leaving such important and timely topics as for example, hormone action, cyclic AMP and reverse transcription to the proceedings of other symposia, and concentrating on the four questions which are most basic to an understanding of the mechanisms of transcription and translation and for which fragmentary but nonetheless reliable experimental results have become available within the last few years. These are the structure of chromatin, the synthesis of messenger RNA, the structure of the active ribosome, and the role of initiation factors in protein synthesis. Thus, even though restricting ourselves to these questions is liable to give the impression that the subject "regulation" is not adequately dealt with, we felt that progress in the field of gene expression is now at the point where it is worthwhile summarizing the biochemical evidence on the organization and function of the genetic material.

To me, as sort of an outsider to research on eukaryotes, this meeting has served a useful purpose in not only showing me the present state of wisdom and ignorance, but it also has given a fairly good idea of which problems will present us with the greatest difficulties and which questions are easier to be answered than expected. There is no doubt, in my mind at least, that the biochemistry of eukaryotic translation is progressing almost without difficulties as the structures of the translational machinery are very similar in pro- and eukaryotic systems. Thus, almost any information obtained with one system can be used to build upon for further work with the other.

The reverse seems to apply to the problem of transcription. While good progress has been made on messenger processing thanks to the discovery of poly A sequences at the 3' end of most mammalian messages, still very little is known about the chromosomal structures which are actually transcribed. How much of the genetic information is translatable into the primary sequence of proteins and how much of it represents regulatory signals remains an unsolved question. Furthermore, to find out how such signals are masked and unmasked during development requires a far greater knowledge of the fine structure of chromatin than is available today.

It is hoped that the present volume provides a necessarily short-lived but for the moment informative survey of what we know already and also of what we need to know in order to understand the molecular mechanisms by which gene expression is regulated.

I am very much indepted to Dr. Dorothy Fields for her editorial help. Her topical and linguistic competence has simplified the editors' task extraordinarily.

Heidelberg, October 1973 E. K. F. Bautz

Contents

Chromosome Structure and Function

Transcription I

Transcription II

Translation I

Translation II

Chromosome Structure and Function

The Organization of Sequences in Chromosomal DNA

C. A. Thomas, Jr.

Department of Biological Chemistry, Harvard Medical School, Boston, USA

With 8 Figures

Late in 1968 a simple experiment was done. A commercial sample of salmon sperm DNA was denatured, annealed and examined in the EM. The results startled us: there were many rings to be found among the denatured chains and unidentifiable material. Bacterial and phage DNA's did not produce these rings. Since that time we have produced these rings (called "slipped rings") from every species of eukaryotic DNA that has come into our hands, including *Drosophila* polytene chromosome DNA. Prokaryotic DNA's always failed to produce slipped rings. Such rings would be expected if chromosomal DNA contained a lot of tandem repetitions. The experiment was done to see if we could prove or disprove Callan's suggestion that chromomeres were largely organized into tandemly-repeating sequences (Callan and Lloyd, 1960; Callan, 1967).

These slipped rings encouraged us, but they proved difficult to quantitate since they represented only a portion of the DNA – most of which remained denatured and possibly uncountable. We turned to the method of "folded-rings". In these experiments, purified chromosomal DNA was sheared to various lengths, and the fragments "resected" or partly digested with an exonuclease (λ-exonuclease *or* exonuclease III) to produce single-chain terminals. Again many rings were formed after a brief period of annealing while prokaryotic DNA's produced extremely few rings. Since most of the DNA is still duplex, it is quite easy to count the number of rings and linear fragments and get reproducible ring frequencies. We published these findings (Thomas et al., 1970) pointing out that these results were expected for a DNA containing tandemly-repeating regions, in accord with Callan's model. At that time we suggested a competing model – the intermittent repetition model – and suggested that in order to account for the ring frequency, the intermittent repeats would have to be very densely clustered along the chromatid.

During the past three years we have been studying how the ring frequency, R, depends on the species of DNA, its initial mode of fragmentation, the degree of resection, and the fragment length. The thermal stability profiles of rings have been determined, and the closure regions have been examined in the presence of increasing concentrations of destabilizing formamide. We have mathematically analyzed various models for the eukaryotic chromosome, including intermittent repetition and tandem repetition models. Our results lead us to a very simple view of the chromosome. For those of you who want to hurry ahead, a diagram is shown in Fig. 2. For those of you who want to see where we were led astray, the following account is offered. More detailed papers have now appeared (Lee and Thomas; Thomas and Dancis; Pyeritz and Thomas; Bick, Huang, and Thomas; Thomas, Zimm, and Dancis, 1973).

Folded Rings

The minimum requirement for the formation of a folded ring is the simultaneous exposure of two single chain regions on the same DNA fragment that are sufficiently complementary to unite to form a double helical region of sufficient stability. There are only two kinds of sequences that could be responsible for such complementary single chains: *ordinary* repetitions and *inverted* repetitions. This is shown in Fig. 1.

Ordinary repetitions can produce rings and lariats, when exonucleases operate normally – that is, exclusively from the 3′ *or* 5′ chain end but not both. Inverted repetitions will produce lariats only if the resection by exonuclease occurs abnormally. Since the specificity of these enzymes is well known (Richardson, 1969), we can be sure that the folded ring experiments signal the presence of *ordinary* repetitions. This is confirmed by the fact that most rings appear without a "handle", which our thermal stability experiments indicate would be long enough to see in the EM. Thus, we can be very sure that these are ordinary repetitions that are responsible for the majority – perhaps all – of the rings and lariats seen.

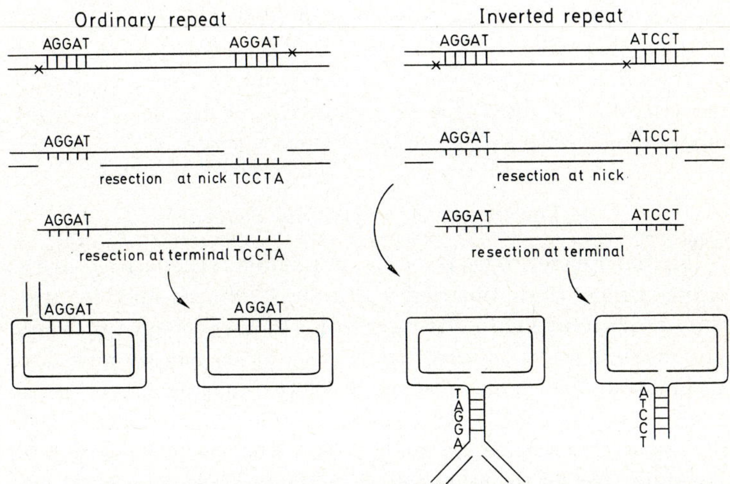

Fig. 1. Rings from ordinary and inverted repetitions. The double helix is depicted as two horizontal parallel lines. The 5 nucleotide pairs representing the repetitious sequence are depicted by "rungs" and labeled. The "×" locates the position of a single chain breakage

What determines which single-chain regions are exposed by the action of the exonuclease? Clearly, it is the location of the double-chain or single-chain breakage that determines the location of the resected region. In the case of λ-exonuclease, which will not operate at nicks, it is the molecular end only that determines the resected region. In many of our experiments the single-chain lengths of the DNA as isolated were much longer than the eventual fragment lengths: in this way nicked fragments can be infrequent. Therefore, even in the case of experiments with exo III, which can resect at nicks, the terminals are the prime site of resection. This is consistent with the finding that simple rings are the most abundant circular structure. Lariats are always less that 30% and quite rare with λ-exonuclease.

Thus we must ask what determines the location of the original double-chain breakage. Does it occur at a special weak point, thus allowing the selective exposure of complementary sequences? This does not appear to be correct. A variety of different shearing conditions, limited treatment with endo *I* or certain restricting endonucleases or with pancreatic DNAase in the presence of Mn^{++}, all produce about the same frequency of folded rings. One cannot imagine that these quite different modes of breakage will all hit at the same few hypothetical "special points". Therefore, we will proceed with the discussion assuming that the fragments were formed *randomly* with respect to nucleotide sequence.

1. The Density of Repetitious Sequences

If the resected regions occur at random with respect to the ordinary repetitions, then the repetitious sequences must be rather densely clustered in order to account for the frequency of the rings. For example let us assume that for ring formation we must have an initial double-chain breakage in a repetitious block. If the repetitious blocks amounted to, say, 10% of all the nucleotides, then the chance that a breakage would occur within such a block is 10%. The same goes for the other end, since the two events are stochastically independent (as implied by the assumption of random breakage). Thus, if *all* the DNA in the genome contained repeating blocks amounting to 10%, *and* if the efficiency of resection and ring formation were 100%, one would expect to see only 1% rings. We find 15–30% rings depending on the species of DNA, at optimal fragment lengths and degrees of resection. To produce 25% rings, this simple argument would require that 50% of all the DNA be in repetitious blocks – even if the efficiency of ring formation were perfect. Clearly, something is not right about this interpretation.

One of the things that is wrong is that the simple calculation is too simple. A more complete analysis of this problem gives

$$F = [1 - (1 - \alpha)^{\left\{\frac{r + g' - 2b_0 + 1}{g'}\right\}}]^2 \qquad (1)$$

for randomly located, homopolymeric repetitious blocks of length g', and

$$F = \left[\frac{r + g' - 2b_0 + 1}{s}\right]^2 \qquad (2)$$

if the repetitious blocks are located at intervals of s nucleotides. This is the "intermittent repetition" model that has proved to be the strongest competing model. This and other models are described more fully in Fig. 2. The quantity r is number of nucleotides resected, b_0 the minimum number of perfectly complementary nucleotides required to form a ring and α the fraction of nucleotides

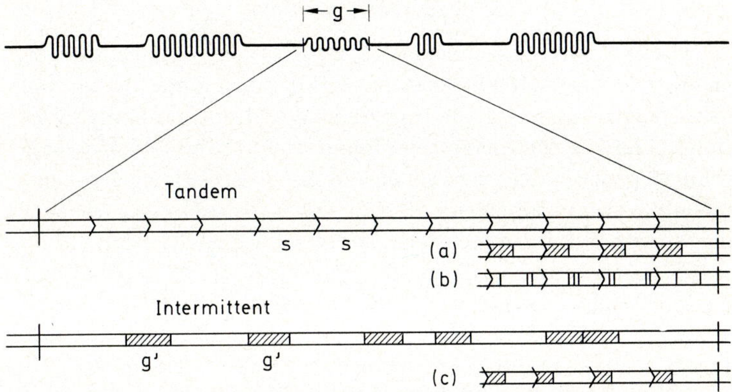

Fig. 2. The regionally repetitious chromatid. The single line bearing folded regions reminiscent of chromomeres represents the mononemic chromatid that is organized into distinctive regions, the contour length of which is g nucleotide pairs. The fraction of all the nucleotides organized into such regions is γ. We can picture two extreme models for the organization of sequences within these regions. At one extreme we have tandem repetition where each sequence s units long is repeated f times to make up the total region g ($g = fs$). This model may be degraded by: a) supposing that only a portion of s is repeating, the remainder being non-repeating. This might be called the "fractional tandem" model. Alternatively: b) one might picture the sequences as being only partially repeating — that is, containing occasional substitutions that differ from repetition to repetition. This might be called the "variegated tandem". The second model, called *intermittent repetition*, assumes that each region containing a total of g nucleotides is composed of non-repetitious DNA containing irregularly arranged sequences g' nucleotides long. The number of them, n', and the fraction, α, of the nucleotides of the region represented in intermittently-repetitious sequences is negotiable ($n'g' = \alpha g$). For mathematical convenience we assume all blocks of g' are copolymers — that is tandemly-repetitous with a very short s'. This clearly overestimates the fraction of cyclizable fragments. The intervening non-repetitious DNA is pictured to be similar to *E. coli* DNA. If the repetitious sequences are regularly spaced as shown in (c), we return to the fractional tandem-repetition model (a)

that are presumed to be located in the repeating blocks which are g' units in length ($\alpha = g'/s$). These equations assume that the efficiency of ring formation is 100 % and that *all* of the DNA contains frequently-spaced homopolymer-like repeating blocks. Before proceeding with this analysis we must examine these two extreme assumptions, which are incorrect.

2. Clustered Repetitions

The optimal ring frequency, ca. 20 %, is generally seen for fragments 1.5–2.0 μ (4500–6000 nucleotide pairs) long. As the fragment length increases, the ring frequency drops gradually. While rings as long as 20 μ have been found, it is rare to find any longer than 10 μ or so. The dependence of the observed ring frequency, including lariats, is shown in Fig. 3. This decrease in ring frequency could not be for kinetic reasons, because resected T7 DNA (12.5 μ) will form rings at a uniform 10 %/min until 80–90 % of all molecules are rings. Neither could this be reasonably attributed to less

---

Fig. 3a—e. Ring frequency as a function of fragment length. (a) Samples of mouse satellite, (b) main band DNA and (c) *Necturus* DNA were sheared at 0 °C in 0.5M NaCl, 1mN EDTA, 10mM Tris-HCl (pH 7.5) by passing the DNA solution (< 20 μg/ml) through a No. 28 syringe needle ten times at maximum thumb pressure. After shearing, the samples were concentrated and sedimented through sucrose gradients in order to select fractions of a narrow range of fragment length. Fractions were dialyzed against 0.1 M Tris-HCl (pH 8.0) and treated with exonuclease III. The extent of the reaction was determined by the fraction of acid soluble nucleotides. The average extent of resection is shown for each experiment on the drawing. Aliquots of the exonuclease III-treated fragments were brought to either 2X SSC or 0.5 M NaCl, 24 mM EDTA, 37.5 mM Tris-HCl (pH 7.4) (0.6 M Na^+) and annealed at 65 °C, generally for 2 hours. Electron microscope grids were prepared by the aqueous technique. Entire grid squares were systematically scanned and 250—500 molecules were scored as linear or circular. Simple rings, lariats and polyrings, generally found in the proportion 7 : 2 : 1, are all scored as circular structures. Lengths of fragments were determined by measuring EM photographs of linear and circular fragments. Under aqueous conditions, the single chains are not seen, thus the observed lengths should be equal to $l - 2r$, which is slightly shorter than l. The sedimentation fractions generally contained a distribution of lengths covering a range of ±25 % of the mean value. If length was measured by s before resection, then 2× the average resection is subtracted for consistency. The results of this procedure is shown by the solid ● ■ or open ⊙ ⊡ data points. The precision of these determinations is a few percentage units, and no error flags are shown. Similar experiments

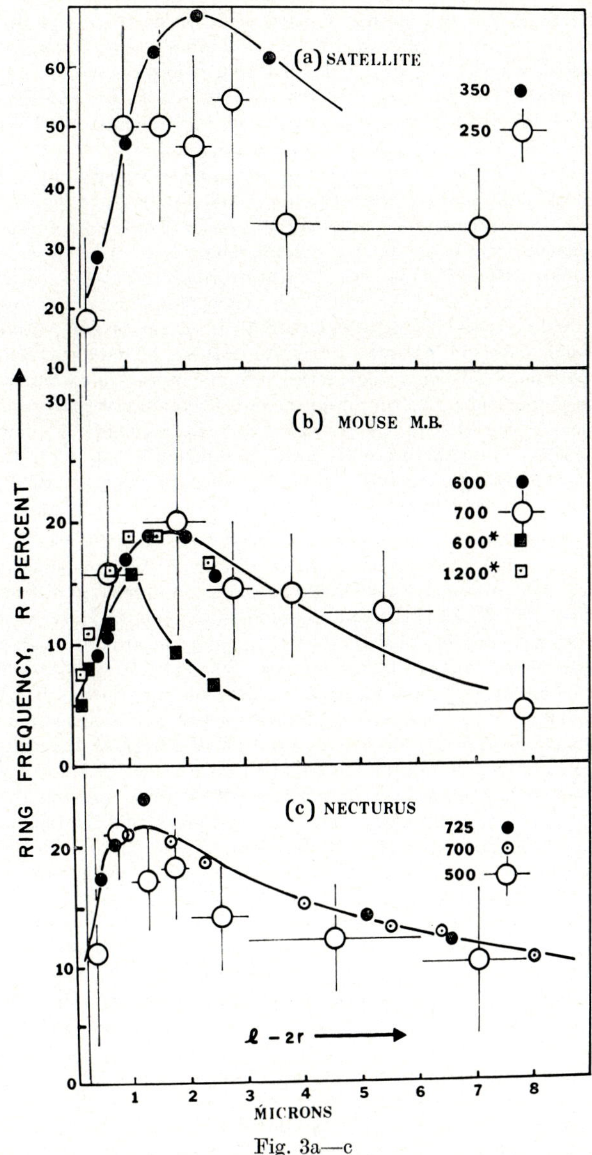
(a) SATELLITE
350
250
(b) MOUSE M.B.
600
700
600*
1200*
(c) NECTURUS
725
700
500
RING FREQUENCY, R - PERCENT
ℓ - 2r
MICRONS
60
50
40
30
20
10
30
20
10
0
20
10
0
1
2
3
4
5
6
7
8

Fig. 3a—c

Legend to Fig. 3a—e continued.

were performed with a wide distribution of fragment lengths produced by limited shear breakage of long DNA. For example with *Necturus* DNA, histograms displayed a mean length of 2.50 μ before resection, and 2.20 μ after resection. This shortening (0.30 μ) correlates with the average degree of resection (500 nucleotides) measured by the external ^{32}P-T7 method ($2 \times 500 = 1000$ nucleotides or 0.3 μ). Annealing in 2 × SSC, 65 °C, occurred for 1/2, 2 and 4 hours and the ring frequency data was pooled, as no change was evident. EM photographs were made and the contour lengths of all linear and circular structures seen were measured. These were grouped into length classes demarcated by the lateral bars (—○—) and the fraction of circular structures plotted. The vertical error bars indicate ± 1 s.d. calculated from the number of molecules. The same procedure was followed for mouse satellite (a) and the main band DNA (b). As can be seen,both experimental approaches give essentially the same results. All data shown here are from experiments done on a single series of fragment lengths all derived from the same preparation. The ring frequencies continue to descend at fragment lengths longer than shown. For example, *Necturus* fragments 27 μ long gave only 3% rings. Note the definitely lower profile in (b —■—■—). This collection of fragments was underresected. The resection values marked (*) are probably 2 × too high. In this same panel all unfractionated fragment points —♦— have been reduced by a factor of 0.200/0.353 in order to correct for the overestimation of ring frequency resulting from the preferential photography of rings. (d) *Drosophila* DNA fragments were prepared in two ways: by limited endo $R \cdot H$ treatment followed by sucrose gradient centrifugation to produce fractions of various lengths and by shearing through various gauge hypodermic needles. In the first case 300 μg of *D. virilis* adult DNA in 2 ml of an incubation medium containing 60 mM NaCl, 6.6 mM Tris (pH 7.5), 6.6 mM $MgCl_2$ and 6.6 mM 2 mercaptoethanol, was digested with a few μl of endo $R \cdot H$ at 37° for 30 min. A portion of this digest was sedimented in a sucrose gradient at 39 krpm for 5 h in an SW 41 rotor. Various fractions were collected, dialysed against 70 mM Tris (pH 8), then resected and folded as usual. For shearing, the samples containing 2—5 μg of DNA in more than 0.50 M NaCl were pressed through needles. Results were variable: for example, 1.3 μ fragments were produced by No. 27 and No. 28 gauge needles; 1.5 μ — No. 24 and No. 27; 1.7 μ — No. 21 and No. 24; 2.5 μ and 3.3 μ by No. 22 and No. 21½. All solutions, needles and syringes were at 0—4 °C. ● *D. virilis* adult DNA broken by endo $R \cdot H$ + sedimentation. ○ *D. virilis* salivary gland DNA — shear broken. □ *D. melanogaster* salivary gland DNA — shear broken. △ *D. hydei* salivary gland DNA (Laird) — shear broken. ▲ *D. hydei* salivary gland DNA (our prep.) — shear broken. —○— *D. virilis* adult DNA broken by endo I. (e) HeLa DNA fragments were sheared to 3 different mean lengths: squares 7.66 μ, triangles 2.90 μ, and circles 1.28 μ. These in turn were resected to different extents: □ 669, ■ 2327; ▼ 616, △ 1127, ▲ 2143; ⦰ 391, ○ 482, ● 969. EM photos were made and the linear and circular molecules classified according to length, and the fraction of perfect rings (excluding lariats, etc.) in each size class determined

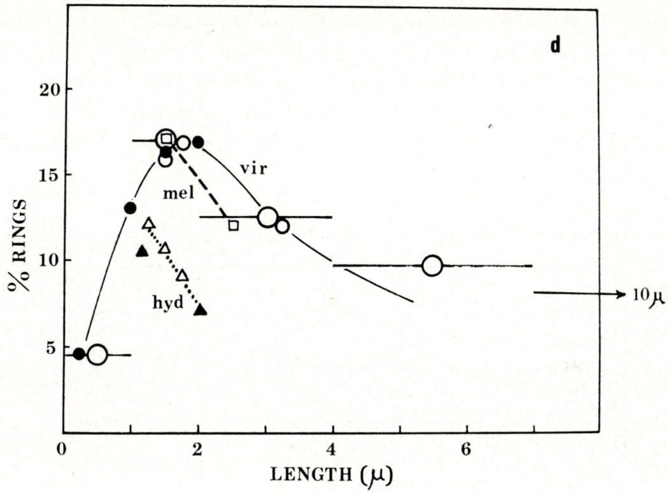

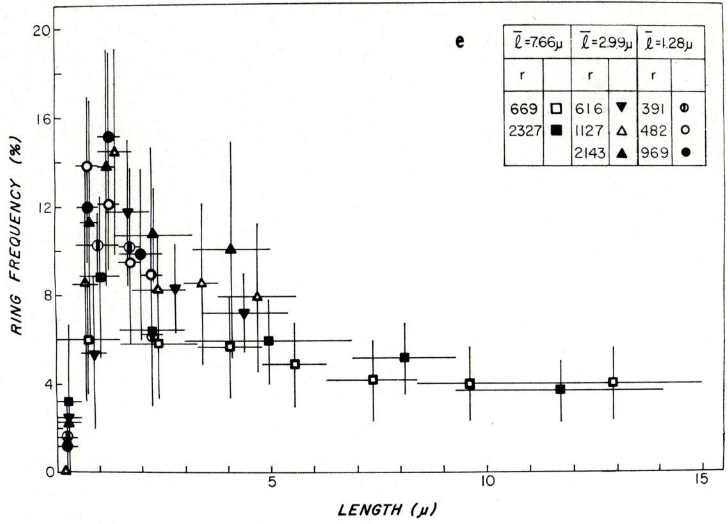

Fig. 3d and e. Legend, see opposite page

efficient resection of the terminals of long fragments. It must be due to the fact that a smaller fraction of long fragments have their terminals within the region containing the clustered repetitions of a given sort. This can be understood from the diagram shown in Fig. 4. Making the extreme (worst case) assumption that the region

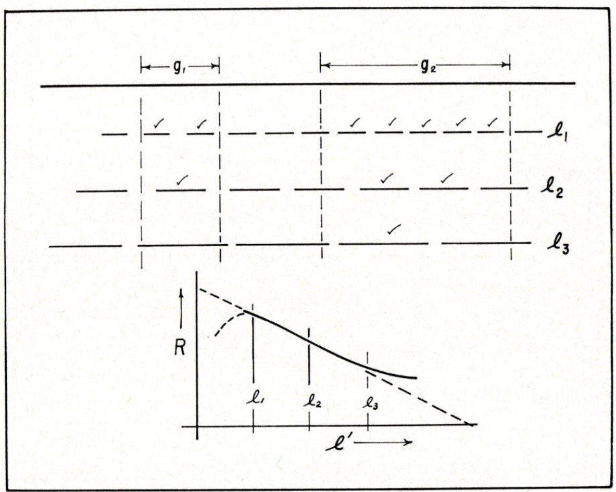

Fig. 4. Longer fragments produce fewer rings. Shorter fragments will produce more rings than long ones, if the repetitious sequences are clustered within regions (*g*-regions) that are comparable to the lengths of the fragments. In the diagram, those fragments that lie entirely within the region containing the clustered repetitions (a *g*-region) do have a chance of producing a ring. They are checked (✓). Longer fragments cannot form rings under these circumstances. In actuality it is not the length of the fragment but its resected length, l', that must lie in g: $l' = l - 2r + 2b_0 - 1$

containing the clustered repetitions is a pure homopolymer, the maximum frequency of rings is given by

$$R = \varepsilon\gamma^* [1 - l'/g^*], \tag{3}$$

where $l' = l - 2r + 2b_0 - 1$. The ring frequency, R, would drop to *zero* when the fragment length, l, and resection, r, were such that l' just equaled g, the length of the region containing the clustered repetitions. Because there is a distribution of g-region lengths, there will always be a few regions that will contribute ring-forming fragments, even though the fragment length were very long. This means that we can never get a true average value of g unless we get the slope of the R vs. l' curve at zero fragment length. This is not possible to do. The best that can be done is to select a value of l', draw the tangent and pick the intercept on the l' axis. This should be equal to the average value of g when that average is taken *only* for those g-regions that are longer than l'. This is the meaning of g^*. The data in Fig. 3 thus reflect the distribution of g-regions. The quantity γ^*, is the fraction of the chromosomal DNA that is found in all the g-regions having lengths equal to or longer than l'. The efficiency of ring formation, ε, is assumed to be independent of ring size or the nature of the repeats within the g-region. This assumption may eventually be revised.

By extrapolating the R vs. l' plots back to the ordinate, one can determine $\varepsilon\gamma^*$. For most of the DNA's that we have studied, this comes to 25%. This means that if ε were 100%, γ^* would be 25%. If ε were 25%, γ^* would be 100%. These represent the limits to ε and to γ^*. We think that ε is about 50% and γ^* is about 50% for the following reasons.

1. When we estimate g^* for *Drosophila* DNA, we get by extrapolation about 5 μ. This is about 1/2 of the average amount of DNA per chromomere. Thus it looks as though 1/2 of the average chromomere is equal to an average g-region. If $\gamma^* = 1/2$, and $\varepsilon\gamma^* = 1/4$, then $\varepsilon = 1/2$. Using this value of γ^* is appealing because we can calculate the number of g-regions in *Drosophila* from the following relation

$$n^* g^* = \gamma^* \Lambda, \tag{4}$$

where n^* is the number of g-regions (of length greater than l') and Λ is the total number of nucleotides in the total chromosomal DNA. When this is done,

$$n^* \cong 5000,$$

or about the same number of g-regions as bands in the polytene chromosome. This encourages one to think that there is one g-region per band.

But to return to ε: we have studied the maximum ring frequency from a sample of *Xenopus* 5s DNA, generously given us by Don Brown. This is thought to be a fairly pure sample of a tandemly-repeating DNA. The maximum ring frequency was 50% in repeated experiments. We feel that some fragment ends are damaged – perhaps 20% of them – so that proper resection does not take place. Thus only about $(80\%)^2 = 64\%$ are cyclizable, and these are cyclized at 80% efficiency, giving an over all efficiency of close to 50%.

Returning to the intermittent repetition model, and in particular Eq. (2), we calculate the ring frequency to be

$$R = \varepsilon\gamma^* [1 - l'/g^*] \left[\frac{r + g' - 2 b_0 + 1}{s}\right]^2, \tag{5}$$

provided the hypothetical intermittent repeats are more or less *regularly* spaced. Assigning $\varepsilon\gamma^*$ its observed value of 25% and assuming $1 - l'/g^* = 1$ (in an effort to tilt in favor of this model), we calculate from R vs. r data shown in Pyeritz and Thomas (1973) that the density of homopolymer-like repeats must be about *50*% in order to account for the ring frequency[1]. If they were not homopolymers, they must be even more densely distributed within the g-region. If they were randomly spaced, an even greater density would be required. This subject is developed more fully in the reference cited above.

What does this mean in therms of the g-region? It means that one can account for the ring frequency if homopolymer-like repeats were spaced at intervals equal to their own length: *100* 100 *100* 100 . . . , *200* 200 *200* 200 . . . , or *400* 400 *400* 400 . . . for about 15000 nucleotides. If the calculated density drops to 33% (for fragments of greater length) the region might be *100* 200 *100* 200 *100* 200 . . . to *400* 800 *400* 800 *400* 800 . . . . If such homopolymers existed one might have expected them to be detected by their reassociation rate. But there are also compelling reasons against such models that come from the folded ring experiments. These are described below.

3. Spaced – Yet Clustered Repetitions

As can be seen in Fig. 3, the ring frequency drops when the fragments become shorter than 1.5–2 μ. At first we thought that this might be due to the stiffness of the DNA molecule, and that

[1] For example taking Eq. (5) and substituting in a point from the R and r data from Fig. 7, for example $R = 15\%$ when $r = 200$, we calculate $\alpha = 46\%$ if $g' = 200$ and 58% if $g' = 400$. Taking the data for the longer fragment lengths $R = 20\%$ at $r = 600$ we may calculate 25% and 38% for the same values of g'.

shorter fragments had more difficulty forming rings. But physical chemists have demonstrated that rings only 0.2 μ around should form with no difficulty, a view confirmed by MERTZ and DAVIS (1972). The smallest rings that we see are about 0.2 μ. Thus, unless shorter fragments are less resectable or have a lower efficiency of joining their complementary terminals, this decrease means that short fragments are less likely to have complementary terminals. They must often fall between the repetitious sequences. The decrease in R as l' falls below 6000 nucleotides should reflect the distribution of spacings *between* the repetitious blocks. According to this, the mean spacing would be about 1500–3000 nucleotides, but could range down to 600 and up to 6000 or more. For example, if repeating blocks were 200 units long, spaced every 1800 nucleotides (*200* 1800 *200* 1800 *200*) then fragments shorter than $1800 + 2b_0$ would fail to cyclize as required by the observations, but the density of repetitious nucleotides would only be 10% (200/(1800 + 200)). Returning to Eq. (2), we can calculate the expected frequency of rings

$$R = (0.25)\,(1.0)\left[\frac{r + 200 - 66 + 1}{2000}\right]^2 = 0.25\left[\frac{r + 135}{2000}\right]^2.$$

In most experiments, maximal ring formation is seen when an average of 500 nucleotides have been resected per terminal.

$$R = 0.25\left[\frac{500 + 135}{2000}\right]^2 = 3.2\,\%.$$

In fact, one finds about 20% rings. Thus there is a conflict.

Somehow one must satisfy *simultaneously* the observations that require that repetitious sequences are clustered into g-regions, where they are spaced apart from each other by up to 6000 nucleotides, and yet be so densely spaced as to account for the observed abundance of rings. Bear in mind that the Eq. (2) and (5) are weighted as strongly as possible in favor of the intermittent repetition model. We conclude that no such model accounts for all these results, but, as we will see below, a model in which the g-regions are composed of tandem repeats fits the evidence very well.

4. How Ring Frequency Depends on the Degree of Resection

No rings are formed unless the fragments have been appropriately resected. Some typical results are seen in Fig. 5. Mouse main band fragments, 1.25 μ long, show an increase in ring frequency of 7% per 100 nucleotides resected, leveling off at 20% rings when 300 nucleotides per terminal have been removed. *Necturus* fragments of the same length show an increase of 4–5% per 100 nucleotides resected, leveling off at 20% when 600 nucleotides have been removed. If one tries to interpret these curves in terms of an intermittent repetition model [by differentiating Eq. (5)], one gain requires about 50% of the g-region to be in blocks of homopolymer repeats. Even the form of Eq. (5) is not appropriate. This expression predicts an r^2-dependence, while the results shown in Fig. 5 are clearly linear.

However, when interpreted on the basis of a tandemly-repeating g-region, one expects the ring frequency to vary directly as r:

$$R = \varepsilon\gamma^* [1 - l'/g^*] \frac{2(r - b_0)}{s_n}, \qquad (6)$$

where s_n is the number-average length of the tandemly-repeating sequence. This expression *assumes* that all values of s are shorter than all values of g. From these slopes one may calculate that $s_n = 700$ for *Drosophila* and mouse main band DNA. This value

---

Fig. 5a—e. The frequency of rings, R, observed after various extents of resection, r. Samples of mouse main band (a), *Necturus* (b), mouse satellite DNA (c), and T1 phage DNA (a) were sheared and fractionated according to fragment length by sedimentation. Resection was accomplished by exonuclease III in a reaction mixture containing 1.3mM $MgCl_2$, 10mM 2-mercaptoethanol or 10mM dithiothreitol, 67mM Tris-HCl (pH8.0) and 0.5—2 μg DNA. Enough enzyme was added to give an appropriate extent of digestion. A companion tube containing intact ^{32}P-T7 DNA (with or without added eukaryotic DNA fragments) was treated in an identical fashion and the fraction of acid soluble nucleotides determined. The average resection, r, was calculated assuming that T7 DNA is 38000 nucleotide pairs long and that all nucleotides are being removed from molecular ends (a fair assumption since less than 30% of the T7 single-chains were broken). (d) *D. virilis* polytene DNA fragments (1.52 μ) were digested with exo III. At various times an aliquot was removed, 20× SSC added to a final concentration of 2× SSC, and annealed at 65° for 2 h. The extent of resection was measured by shortening of the length of the double-helical portion of the fragments. (e) ^{3}H-HeLa DNA fragments of a wide distribution of lengths were resected to increasing extents, samples removed, annealed and photographed. Perfect rings and linears were grouped according to length and the ring frequency, R, calculated for each length group

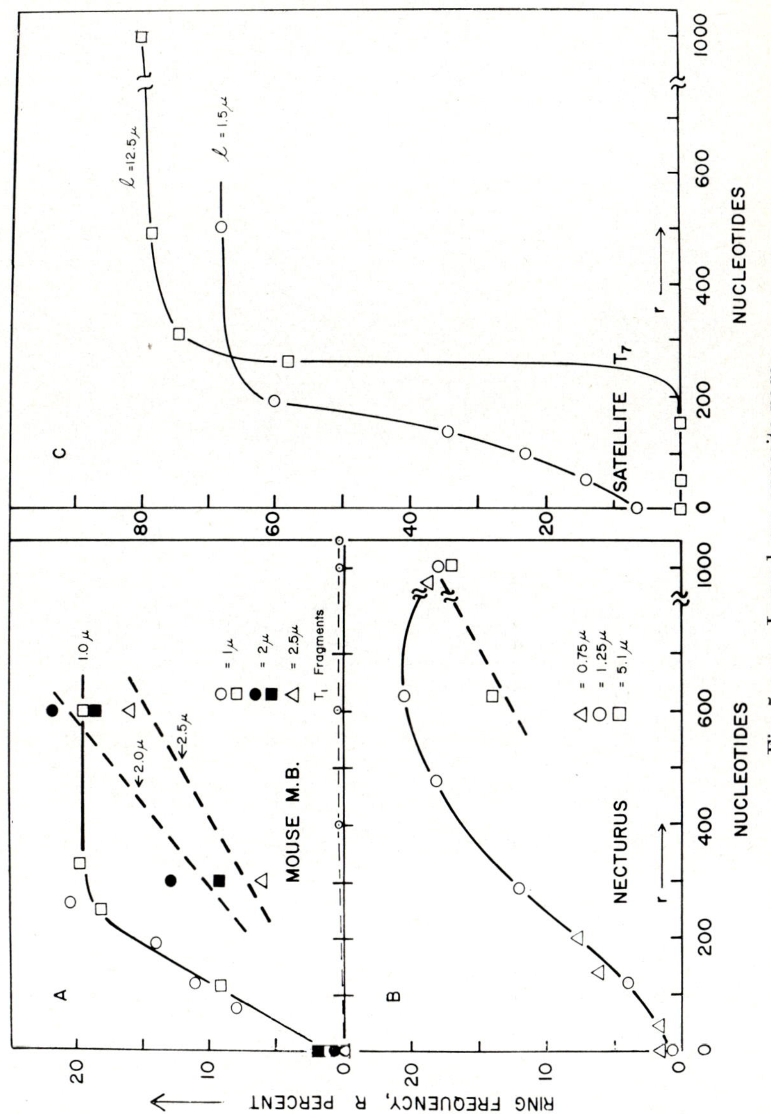

Fig. 5 a—c. Legend, see opposite page

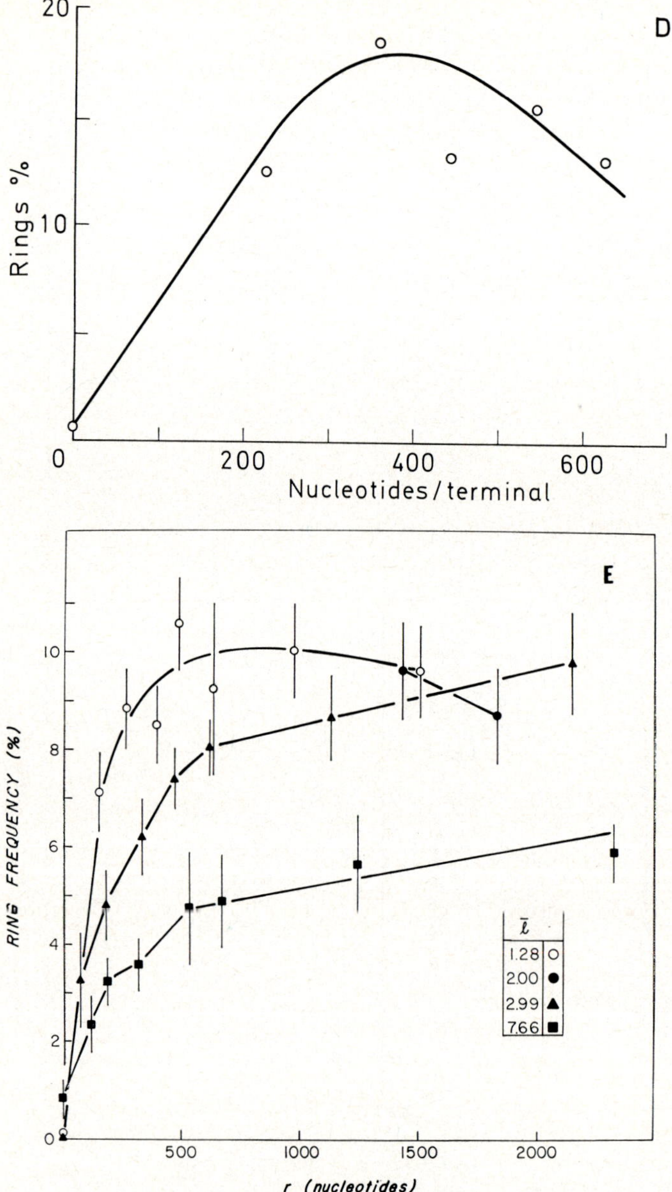

Fig. 5e and d. Legend, see p. 14

for the number average of s is in fair agreement with the distribution of s, ranging from 600–6000 nucleotide pairs that one obtains from the decreasing value of R at shorter fragment lengths. However, there are good reasons for thinking that this estimate of s_n is too low because all of those g-regions having tandem repeats longer than the fragment length, in this case 1–1.5 μ, will not be included in the average. This idea is confirmed by the finding that longer fragments show lower slopes giving larger values of s_n (Fig. 5).

It is very probable that many g-regions have values of s which are longer than the length of other g-regions. For example, the nucleolus organizer DNA of *Xenopus laevis* has a repeat length of 4.5 μ (13500 nucleotide pairs). Yet as seen in Fig. 3, fragments of this length do not cyclize optimally, presumably because the bulk of the g-regions are shorter than this. If this is true, the observed frequency of rings is significantly *lower* than it would be if all g-regions were longer than all values of s. Said another way, it is not possible to find a fragment length that will be longer than all values of s and much shorter than all values of g; thus the optimum frequency is not a true reflection of the fraction of DNA that is organized into g-regions.

5. The Thermal Stability and Closure Length of Folded Rings

The thermal stability of the folded rings has been measured in two ways: 1. by increasing the temperature of a solution containing folded rings and removing an aliquot after a brief incubation, mixing with cold cytochrome c and spreading for electron microscopy under aqueous conditions; 2. by mixing with increasing concentrations of formamide, and spreading for electron microscopy in formamide under the so called "isodenaturing" conditions of DAVIS and HYMAN (1971). Under these latter conditions any renaturation is largely prevented. The thermal stability profiles obtained by both methods are the same (Fig. 6a—f). Qualitatively the results are clear: the bulk of the folded rings are very stable, becoming linear only a few degrees below the melting temperature of DNA itself. T7 rings, which are closed by 260 nucleotide pairs, display a thermal stability that is approximately that expected for a duplex closure of this length.

Even if all rings were the result of pure tandem repeats, their stability should be *less* than that of DNA itself because the number

Fig. 6a—f. Thermal stability of folded rings. (a) Mouse main band DNA fragments (mean length 1μ) were resected an average of 400 nucleotides and the extent of resection was determined by carrying out the identical reaction in a separate tube containing ^{32}P-T7. Both samples were annealed at 60° in 2× SSC, 10 mM EDTA. Following the annealing treatment, the mouse and T7 samples were diluted with an equal volume of water and small volumes of these solutions (25 μl) were exposed for 10 min to various temperatures from 50—100 °C. After quickly chilling the heat-treated samples to 0 °C, 10 μl was mixed with cold spreading solution (25 μl 2.0 M NH_4OAc, 10 μl lmg/ml cytochrome c, 55 μl H_2O) and grids promptly prepared. Just after annealing and prior to the heat treatments, the mouse preparation contained 15.5% folded rings while the T7 DNA was 78% circular. The fraction of rings remaining of both types is plotted in Fig. 3a. T7 rings (●); mouse folded rings (+). The T_l of the T7 rings is approximately 82°, although we would expect it to be 4° higher in 1× SSC. The dashed line shows the theoretically expected profile for T7 with T_l of 86°. The solid line (————) is the theoretically expected transition calculated for folded rings formed by fragments resected by 400 nucleotides. The dotted line (··········) is the same transition displaced 8.2° to the left to suggest the transition that would be expected if the closure were limited to 100 nucleotide pairs. For this calculation we assume that mouse DNA is 41.5% $G + C$ and a standard deviation of melting temperatures of ±5°. This gives a median Tm of 86°, accidentally the same as the expected Tl of T7 rings. The solid curve to the extreme left is the calculated linearization profile for $\lambda b_2 b_5$ rings from Wang and Davidson (1968) in the present solvent. (b) *Necturus* DNA fragments averaging 1.25 μ in length were mixed with ^{32}P-T7 whole molecules prior to treatment with exonuclease III. Resection was stopped after an average of 700 nucleotides were removed and the sample annealed for 2 h at 65 °C in 0.5 M NaCl, 37.5 mM EDTA, 25 mM Tris-HCl (pH 8.0). Following the annealing period, the solution was dialysed at 4 °C against the above buffer (0.6 M Na^+). Aliquots were then given 10 min heat treatments as described in (a). Due to the tenfold difference in size between the T7 and *Necturus* molecules, ring frequencies for the two species could be scored simultaneously on each grid. T7 rings (●); *Necturus* folded rings (⫯). The initial frequency of T7 rings was 76%, of *Necturus* rings, 16.5%. The theoretical curves are calculated as described in (a) assuming *Necturus* DNA has 42% $G + C$, giving 94° for Tm of *Necturus* and T_l of T7 rings in this solvent. In this case the observed linearization profile of T7 rings agrees with expectation. (c) *D. virilis* polytene DNA fragments (1.6 μ) were resected by exo III to an average of 450 nucleotides (estimated by the added labeled T7 DNA method) extracted with phenol then annealed for 2 h, and mixed with a separately prepared solution of T7 rings. This solution was dialysed against 0.10 M NaCl, 0.001 M EDTA, 0.001 M Tris, pH 7, (0.103 M monovalent cation) and heated gradually to 100° in a water bath over a period of 60 min. After equilibration for 2—3 min, at the indicated temperatures, samples were removed and put into cytochrome c solution at 0° and EM grids were prepared immediately. Frequencies of both kinds of rings were scored. The starting ring frequencies in this experiment were: fragment rings 12.5%,

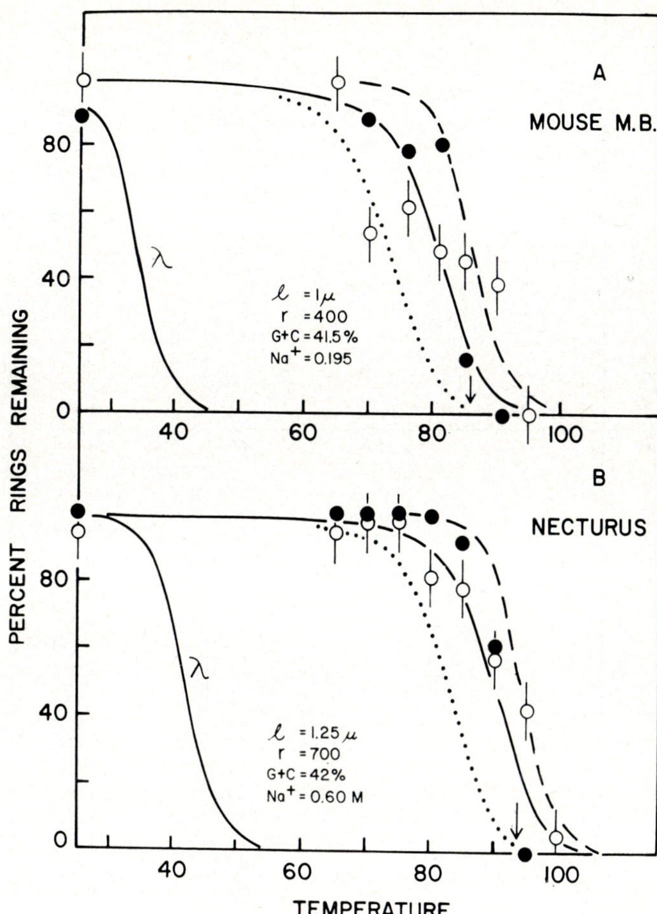

Fig. 6 a and b Legend continued

T7 rings 74%. Data points shown are normalized by these values. At 90 °C. all the T7 DNA was denatured and more than half to the polytene DNA was denatured (yet ~ 20% of polytene rings could be seen). ● Fraction of T7 rings remaining. ⦽ Fraction of fragment rings remaining. (d) *D. hydei* polytene fragment rings. Experiment same as above. Initial ring frequencies: polytene rings 12%, T7 rings 70%. The degree of resection of *D. hydei* fragments was estimated to be twice that of the *D. virilis* fragments by the same method. The error flags are ±1 standard deviation calculated from counting errors alone. Plotted on both graphs are the following curves: — — — (dashed line) the observed melting profile of *D. virilis* adult measured in the

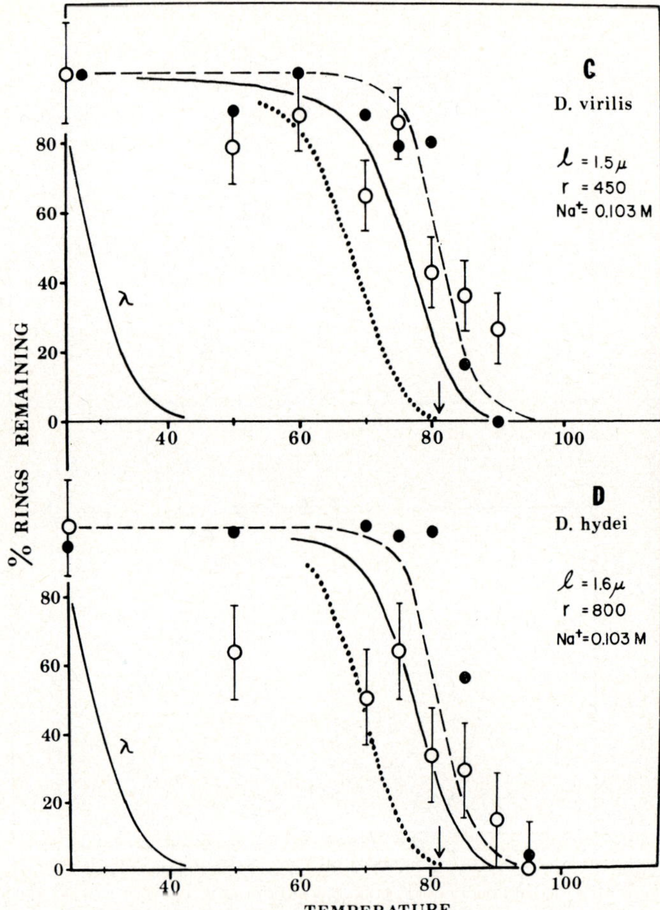

Fig. 6 c and d. Legend continued

same solvent used for the ring stability experiment. The Tm (81°) agrees with the value observed by Gall et al. (1971) of 85.8° for *D. melanogaster* in SSC after correction to our ionic strength (0.103 M). The Tm of *Drosophila* DNA was 4° below the Tm of *E. coli* DNA measured during the same experiment. This confirms that the %GC of *Drosophila* DNA is 10% below that of *E. coli* or 40%. Notice that this curve also conveniently describes the linearization profile of T7 rings. An expected stability profile for rings having complementary closure regions ranging from 33—450 nucleotides (or 33 to 800 in (d) is ——— (solid line). This represents the maximum stability

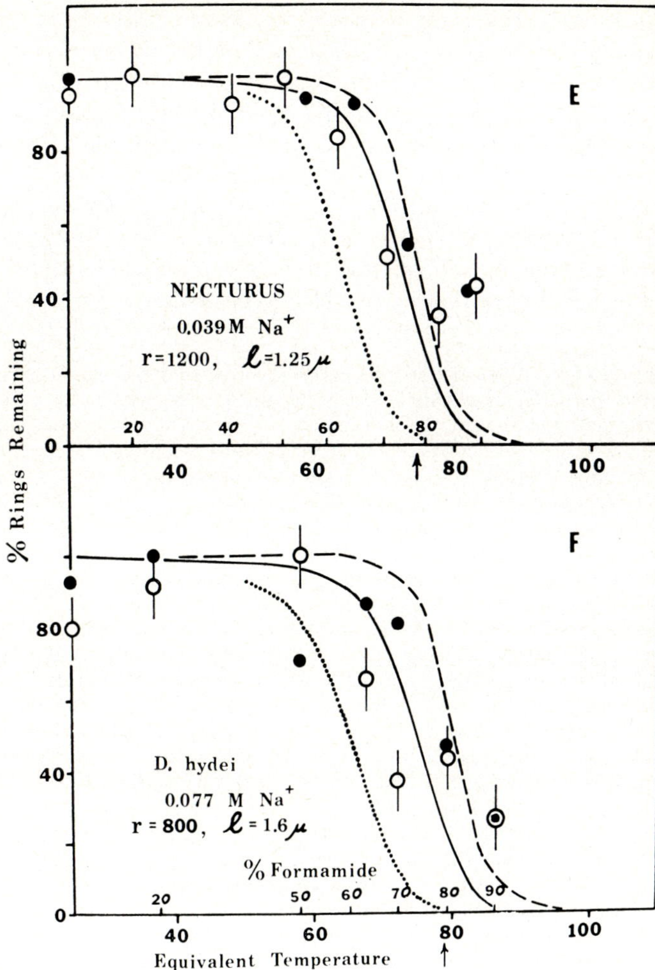

Fig. 6 e and f. Legend continued

that would ever be expected, presuming random fragmentation. Given the precision of these experiments, this theoretical curve is a reasonable representation of these data. .... (dotted line) is the same stability profile drawn above, now shifted 8.2° to the left. This curve should approximate the expected stability profile of the overlap were limited to 100 nucleotide pairs. Clearly, this is an unacceptable fit to the data; all (but one) points lie above or to the right of this curve. The curve to the left is the stability profile of of nucleotide pairs in the overlap is less than infinite. In fact the maximum closure would be r, the minimum b_0, and one would expect an even distribution between these two extremes. The solid lines in Fig. 6 are the theoretically expected transition profiles assuming that all fragments are resected by exactly the measured average value r, and that the lowering of the Tm of a short duplex segment b units long, in comparison with one of great length is $\Delta \mathrm{Tm} = 820/b$. (See appendix by THOMAS and DANCIS, 1973.) Generally speaking, the theoretical lines are a reasonable fit to the data points. The rings are as stable as one could expect given the

---

Legend Fig. 6 continued
rings taken from WANG and DAVIDSON (1968) corrected to the present solvent. (e) *Necturus* DNA fragments 3—4 μ in length were resected to an average of 1200 nucleotides and annealed to produce 19% circular structures, mixed with intact T7 DNA folded rings, then adjusted to increasing concentrations of formamide, held at 21 °C for 2—3 min and spread for EM at 21 °C under isodenaturing conditions. The normalized percentage of rings of both types is plotted against the percentage (v/v) formamide, or equivalent temperature assuming 7 °C for each 10% increase in formamide (MCCONAUGHY et al., 1969). ● T7 rings; ⫯ *Necturus* folded rings. The solid curve is the theoretical expectation, assuming that each terminal has been resected by 12000 nucleotides and that the Tm° of *Necturus* DNA is 75 °C (see arrow) as calculated from its presumed composition (42% CG) and the salt concentration (0.039 M). The solid curve is a reasonable fit to the data. When this curve is shifted 8.2° lower to indicate the stability profile of rings closed by 100 nucleotides (dotted curve ....) we see a poor fit, and conclude that the closure lengths must be longer than 100 nucleotides. The expected T_l of T7 rings is 76 °C which is 820/260 or 3° lower than the Tm of T7 DNA in this solvent, 79 °C. The dashed line (— — — —) is drawn for $\sigma = 4°$ with a midpoint of 76°. The data agree with this expectation. (f) *D. hydei* fragments 1.5 μ in length that were resected to an average of 800 nucleotides and annealed to produce 13% circular structures, were mixed with T7 DNA folded rings, then adjusted to increasing concentrations of formamide in 0.077 M salt, held at 23 °C for 2—3 minutes and spread for EM at 23 °C under isodenaturing conditions. The theoretical solid and dotted lines are constructed in the same way as above assuming a uniform resection of 800 nucleotides. The Tm of *D. hydei* DNA is assumed to be the same as that of *D. melanogaster* and *D. virilis* DNA's, and corrected to the present salt concentration (0.077 M). Again, we see a good fit assuming complementary sequences that are longer than the resected length and a poor fit assuming that closure lengths are limited to 100 nucleotides (dotted curve ....). The expected T_l of T7 rings in 80 °C which is 820/260 or 3° less than the Tm of T7 DNA in this solvent 83 °C. The dashed line (— — —) is drawn for $\sigma = 4°$ with a midpoint of 80°. The data are better represented by a Tm that is 3° lower than this

magnitude of the resection. What are the confidence limits? The dotted curves are the stability profiles expected if the closures were limited to 100 nucleotides. Clearly the rings are more stable than this. On this basis we conclude that the rings must be closed by more than 100 and probably more than 200 perfectly matched nucleotides of average composition.

By examining a large number of rings under formamide conditions, it is possible to find some rings that have two visible single-chain tails. By measuring the distance between these two tails,

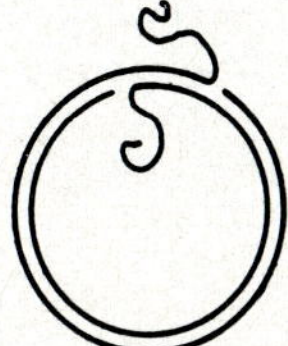

one should obtain a direct measure of the closure length. This was done for various kinds of rings, and the results plotted in histogram form Fig. 7. As can be seen about 96% are closed by more than 100 nucleotides, a value in accord with the thermal stability data.

These formamide studies also supply further evidence regarding the stability of the closure. It is possible to raise the concentration of formamide to the point that many circular and linear fragments become partially denatured, yet many rings remain intact, presumably being closed by regions richer in $G + C$ (Bick, Huang and Thomas, 1973).

The Search for Regions of Non-Homology in the Closure Region

We reasoned that if *g*-regions were intermittent repeats or tandem repeats containing blocks of non-repetitious sequences, one would occasionally find single-chain bubbles in the closure region as shown in the following diagram.

Folded rings from *Drosophila* and *Necturus* DNA were prepared and examined at low (sub-denaturing) levels of formamide. More than 1000 rings were studied, and some 200 photographed. Only 5 structures were found that could possibly be the result of non-homology. Under the conditions of resection, regions of non-homology greater than 1000 units in length may have been missed. Non-homologous regions less than 50 may not be resolved.

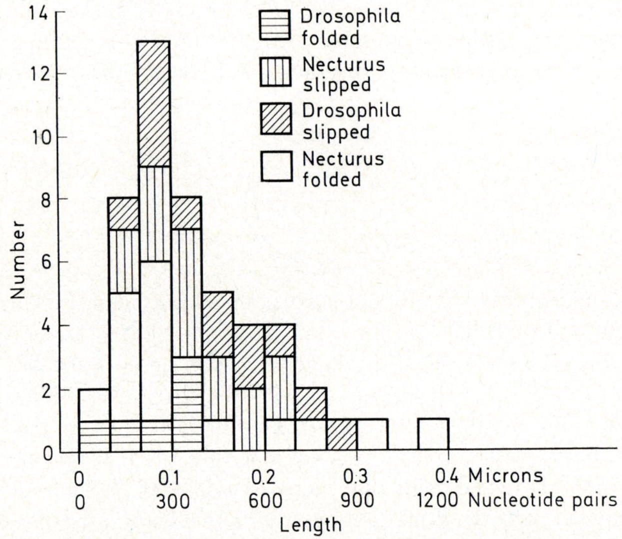

Fig. 7. Histogram of closure lengths. The shortest distance between the two single-chain "whiskers" on a ring is plotted as a histogram. Both folded and slipped rings formed from *Drosophila* and *Necturus* DNA fragments are included. As one can see, only 2/49 rings had closures of less than 100 nucleotides in length

In an effort to extend this kind of experiment, *Drosophila* fragments were cross-linked with reduced mitomycin (SYBALSKI and IYER, 1964), then digested as far as possible. The scheme is depicted in the following diagram.

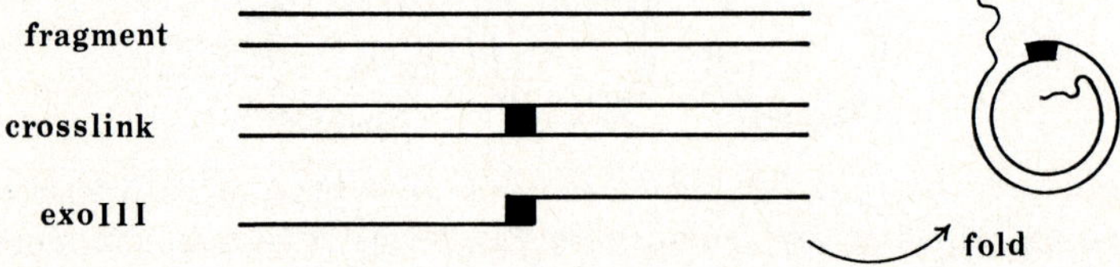

The rings that were formed were quite small, as expected, but again they appeared perfect with no obvious unpaired regions. Thus, in sum, these negative experiments indicate that if non-repetitive DNA exists within *g*-regions, we have not been able to find it.

Models for the Eukaryotic Chromatid

Figure 2 shows a diagram in which the eukaryotic chromatid is depicted as a single double helix. About 1/2 of the total DNA is organized into many thousands of *g*-regions – one for each chromomere in the case of *Drosophila*. The *g*-regions are thought to be composed of either pure tandem repeats or intermittent repeats. The evidence recounted above is in accord with a distribution of *g*-regions ranging in length from a micron or so up to 20 or more microns. The tandem repeats within these regions range from 600 to 6000 nucleotide pairs. There appears to be no detectable non-repetitious material within the *g*-region (<50 nucleotide pairs), and the fidelity of the repeat must be high ($<3-4\%$ nucleotide substitutions). Intermittent repetition models require a high density of homopolymers in order to account for the abundance of rings, and when this condition is satisfied, such models cannot account for the decrease in ring frequency when the fragment length falls much below 6000 nucleotide pairs.

Some workers have been under the impression that these results conflict with the rate of reassociation of single chain fragments. These studies have been interpreted to mean that 95% of the polytene chromosome DNA is composed of "single copy" DNA. In Fig. 8, the expected reassociation rate curves are plotted for a purely unique DNA, and a 50 : 50 mixture of unique DNA and the remainder present in 5, 10, 20 identical copies. One can see that 10 copies results in a difference of 3 in the value of $Cot_{1/2}$. It is our opinion that it would not be possible to detect this difference considering the uncertainties of determining $Cot_{1/2}$, and the precision with which the *c*-value of T_4, *E. coli* and *Drosophila* are known. It should be noted that the relationship between *c*-value and $Cot_{1/2}$ has never been calibrated with a tandemly-repeating DNA. Such sequences would be expected to form more compact intermediates of a significantly different structure than is the case with *E. coli* and T4 DNA. Thus tandemly-repeating sequences might be expected to associate more slowly than expected on the basis of a non-repeating DNA. Apparently, the formation of folded rings is a more sensitive measure of local repetition than is reassociation kinetics. Further comment on this subject is available (Laird et al., 1973).

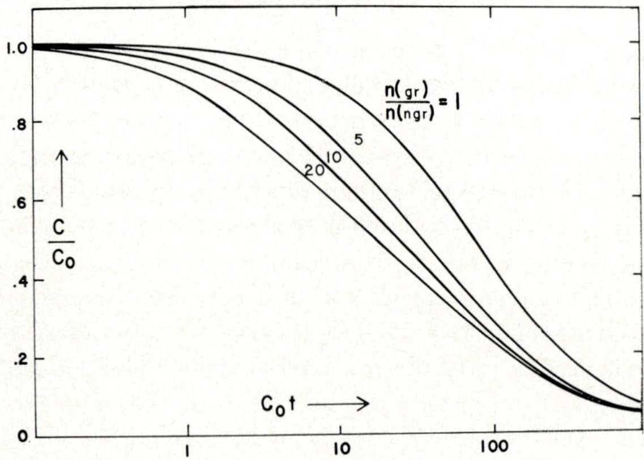

Fig. 8. Expected association kinetics. This graph shows the association kinetics expected for a hypothetical DNA containing 50% non-repetitious and 50% repetitious sequences. The repetitious DNA is pictured to be composed of 1, 5, 10, 20 copies. If the DNA within an average g-region of 15,000 nucleotide pairs were composed of 10 identical sequences of 1500 units, the value of Cot when 1/2 the nucleotides are associated is shifted a factor of 3 to lower values of Cot. Cot $_{1/2}$ was taken as 75 (Dickson et al. 1971)

Commentary

Some commentary and reinterpretation of these folded ring experiments has already appeared. For example, Bonner and Wu (1973) suggest a model in which the entire euchromatic portion of the *Drosophila* chromatid is composed of *125* 750 *125* 750 *125* ... where the *125* is thought to be an internally non-repetitious block that is identical with other blocks of 125 in any given g-region. All of the DNA in a chromomere is assumed to be in g-regions and the efficiency of ring formation is assumed to be perfect ($\varepsilon = 100\%$).

Bonner and Wu calculate

$$R = [1 - l/g]\left[\frac{r - g'}{s}\right]^2, \tag{7}$$

for $l > 2s + g'$. By substituting in $r = 450$, $g' + 125$ and $s = 875$, and assuming $l \ll g$, BONNER calculates $R = 13.8\%$. For $g' = 100$, $R = 16\%$. By the same token, Eq. (7) gives $R = 100\%$ when $r - g' = s$. This is contrary to the facts. The maximum ring frequency, extrapolating to $l' = 0$ is only 25%. Thus one or both of his assumptions are wrong: either the ring forming efficiency is less than 100%, or all the DNA does not contain clustered intermittent repetitions, or both are less than 100% as we have suggested.

There is a further difficulty with the *125* 750 *125* 750 model: it does not account for the short side reduction in ring frequency that occurs between 600–6000 nucleotide pairs. The data points of LEE and THOMAS (1973) were incorrectly plotted in Fig. 2 of the paper by BONNER and WU, a fact that may have misled them.

A fractional tandem model has been proposed by LAIRD et al. (1973). It calls for blocks of 150 nucleotides separated by blocks of 750 non-repeating nucleotides. The 150 unit blocks are thought to be of 5 different sequence types in any given g-region. The model is 750 $\mathit{150}_1$ 750 $\mathit{150}_2$ 750 $\mathit{150}_3$ 750 $\mathit{150}_4$ 750 $\mathit{150}_5$ 750 $\mathit{150}_1$ 750 $\mathit{150}_2$. Thus, a given g-region contains 150/900 or 16.7% repetitious sequences. Since they believe the maximum amount of "non-unique" DNA in the euchromatic portion is 7%, one must suppose that only 7.0/16.7 or 42% of the DNA is to be found in g-regions. Assuming that the efficiency of ring formation is 70% (rather than 50%, as we do) they make a calculation that gives a maximum of 10–15% rings.

Here the difference apparently lies in the calculation. From Eq. (5) assuming homopolymeric repeats we see

$$\begin{aligned} R &= (0.7)\,(0.42)\,[1 - l'/g^*]\left[\frac{500 + 150 - 2(33) + 1}{4500}\right]^2 \\ &= (0.294)\,(1.0)\left(\frac{585}{4500}\right)^2 \\ &= (0.294)\,(0.13)^2 = (0.294)\,(0.017) = 0.5\,\%. \end{aligned}$$

At the other extreme, let us suppose that the entire g-region were composed of tandem repeats 4500 units long (rather than the fractional tandem in which 1 – (0.167) of the DNA is non-repetitious). In this case Eq. (6) applies:

$$\begin{aligned} R &= (0.70)\,(0.42)\,[1 - l'/g^*]\,\frac{2\,(r - b_0)}{4500} \\ R &= 1.3 \times 10^{-4}(r - 33) \end{aligned}$$

When $r = 300$ to 500, $R = 3.5–6.1\%$ instead of the observed 17 % for *D. virilis* polytene chromosome DNA. The dependence of R on r, predicted by the above relation, is significantly different from that observed (see Fig. 5). This expression itself must be an exaggeration. At low degrees of resection we think it must be multiplied by the fraction of the g-region that is composed of repeating sequences, in this case 0.167. This lowers the predicted ring frequency to something comparable with the ring frequency predicted from Eq. (5) (THOMAS, ZIMM, and DANCIS, 1973).

The differences here derive from the different calculations and could be resolved. We have some confidence in Eq. (6) because it givens a value of s for *X. laevis* 5 s DNA that is in accord with the thermal denaturation maps. When we attempt the calculational route followed by LAIRD, our results give a *lower* value of R than given by Eq. 5 for any given density of intermittent repetition (see THOMAS, ZIMM, and DANCIS, 1973).

Finally, this fractional tandem model conflicts with the apparent absence of non-repetitious sequences in the overlap region (BICK, HUANG, and THOMAS, 1973)

In sum, no intermittent or fractional tandem model that we can think of will satisfy the magnitude of the ring frequency, its dependence on the degree of resection, and its generally observed, characteristic dependence on the fragment length.

Acknowledgements

I am indebted to my collaborators who are, in fact, responsible for this work: Drs. C.S. LEE, R.E. PYERITZ, B.M. DANCIS, M.D. BICK, and H.H. HUANG. We also thank Prof. B.H. ZIMM for his invaluable theoretical work. The names of all of these individuals appear on the primary publications in the Journal of Molecular Biology.

We also thank Dr. D.D. BROWN for his gift of *Xenopus* 5s DNA. This work was supported by the NSF GB 31 119-XI.

References

BICK, M.D., HUANG, H.L., THOMAS, C.A., Jr.: J. molec. Biol. **77**, 75 (1973).

BONNER, J., WU, J.-R.: Proc. nat. Acad. Sci. (Wash.) **70**, 535 (1973).

CALLAN, H.G.: J. Cell Sci. **2**, 1 (1967).

CALLAN, H.G., LLOYD, L.: Phil. Trans. roy. Soc. (Lond.) B **243**, 135 (1960).

DAVIS, R.W., HYMAN, R.W.: J. molec. Biol. **62**, 287 (1971).

LAIRD, C.D., CHOOI, W.Y., COHEN, E.H., DICKSON, E., HUTCHINSON, N., TURNER, S.H.: Cold Spring Harbor Symposium. In press (1973).

LEE, C.S., THOMAS, C.A., Jr.: J. molec. Biol. **77**, 25 (1973).

MERTZ, J.E., DAVIS, R.W.: Proc. nat. Acad. Sci. (Wash.) **69**, 3370 (1972).

PYERITZ, R. E., THOMAS, C. A., Jr.: J. molec. Biol. **77**, 57 (1973).
RICHARDSON, C. C.: Ann. Rev. Biochem. **38**, 795 (1969).
SZYBALSKI, W., IYER. V. N.: Fed. Proc. nat. Acad. Sci. **23**, 946 (1969).
THOMAS, C. A., Jr., DANCIS, B. M.: J. molec. Biol. **77**, 43 (1973).
THOMAS, C. A., Jr., HAMKALO, B. A., MISRA, D. N., LEE, C. S.: J. molec. Biol. **51**, 621 (1970).
THOMAS, C. A., Jr., ZIMM, B. H., DANCIS, B. M.: J. molec. Biol. **77**, 85 (1973).

Discussion

P. KARLSON (Marburg): Thank you very much for your interesting report about "two or three experiments". I think it must have been a lot more. The paper is open for discussion.

R.K. ZAHN (Mainz): You have shown with your statistics of ca. 1000 molecules in electron microscopy that a certain percentage of the molecules are converted to the circular form.

Did you substantiate your conclusions by centrifugation analysis, since we do not know how representative the electron microscopic pictures are?

C.A. THOMAS (Boston): We have been perfecting non-electron microscopic methods of assessing fragment ring frequency, but all of our data to this point are based on EM counts. We know that for phage DNA's EM counts and sedimentation analysis agree quite well, but for the fragment rings, which are heterogeneous in size, we do not have a non EM measurement of ring frequency as yet.

J. PAUL (Glasgow): One of the questions raised is the agreement between the Cot-values and these findings. We made some observations which may help on this. If you look quite carefully at some of the extrapolations that people have done from *E. coli* to eucaryotic DNA, you come out in the unique fraction with a repetiveness of five instead of one. This is the case when the DNA is sheared down to the usual length of about 1000 nucleotides. What we found is that if you continue to shear down to a length of about 300 nucleotides and apply the proper corrections, you do get repetitiveness of about one. This may favour the alternating model over the tandem model.

C.A. THOMAS: We are aware of this fact and we know the corresponding plots you have made. I would summarize our knowledge in saying that we know enough to know that it is a very difficult problem. After all, we have to draw a structural conclusion from the rate of a reaction which is ill defined.

K. SCHERRER (Lausanne): Didn't you say that the *g*-region comprises about 50% of the chromomere? However, you did not give the number of nucleotides which are present in the closing region. How can you compromize the purely tandem model with this?

C.A. THOMAS: I think I should have done this. If we interpret our findings on the basis of the purely tandem model, then this region here should contain about 1200 to 1800 nucleotides. The *g*-region here, as a whole, is for Drosophila about 5 u, or 15000 nucleotides. We believe that about half of the DNA is comprised of unique sequences, and the other half is *g*-regions.

K. Scherrer: Where do you believe that the messenger sequences are, the ones coding for protein?

C.A. Thomas: We do not know where the informative regions are. We are now isolating rings to see whether the messengers would stick to them or not. If I should take a guess, I would guess that they are in the unique regions.

Friedrich (Jülich): There is a paper from the Taylor-Group (1972) dealing with the subunit structure of chromosomal DNA. The authors have shown by differential melting that there are melted loops in a distance of 0.4 μ, MW $\sim$ 300000 to 500000. It would be of interest to show with your method if these low-temperature-melting-regions are repetitious regions and the regions in between non repetitious. The melted regions can be cut by endonuclease.

C.A. Thomas: Partial denaturation maps of chromosomal DNA can be made, and we have done quite a lot of this, and our results agree qualitatively with those of Taylor that you mention. I would prefer to think that the periodicity that appears is a reflection of the tandemly-repeating regions. They have about the right wavelength. However we feel that extreme caution should be exercised in interpreting a supposedly periodic partial denaturation map. The human brain is tempted to see periodicity where none may in fact exist: for example in randomly located denatured regions. For this reason we have been reluctant to draw conclusions from our experiments.

Maas (Hannover): In determining the thermal stability and estimating the length of the rings, how do you achieve the separation of those non-repetitive regions which give base stacking and therefore contribute to the thermal denaturation profile?

C.A. Thomas: Our theoretical extrapolations are calculated on the basis of no stacking, because we believe that stacking is unlikely in these rings. Experimentely, we just count in the electronmicroscope the linear structures of DNA and the rings.

K. Scherrer: Do you think that the closing sequences are all the same over the chromosome, or that there are differences?

C.A. Thomas: They are completely different, because these sections become too long, the frequency of ring drops off. I cannot say that they are all different, but there must be at least five or ten different types, and it might well be that there are 5000 different types.

P. Chambon (Strasbourg): Did you try to correlate your model with denaturation maps of DNA?

C.A. Thomas: Indeed we have. Tom Brown has made a denaturation map of 5 S DNA from *Xenopus*, and we have studied the formation of rings as a function of degree of dissection. We calculated a repeating length of about 700—800 nucleotides. The thermal denaturation map of Brown gives 750 bases for repeat, so we have a good agreement.

Gene Masking in Cell Differentiation: The Role of Non-Histone Chromosomal Proteins

J. Paul, R. S. Gilmour, I. A. R. More, A. J. MacGillivray and D. Rickwood

The Beatson Institute for Cancer Research, Glasgow, Scotland

With 1 Figure

Evidence that the proteins associated with DNA in chromatin might perform a regulatory role in transcription first emerged in some studies in the early 1960's in which the efficiencies of DNA and nucleoprotein as templates for the synthesis of RNA by bacterial RNA polymerase were compared. In all these experiments it was found that DNA was a much less efficient template when it was present as nucleoprotein than when it was free. These early experiments were seriously criticised because of the limited solubility of nucleoprotein in the buffer solutions used in the assays. To obviate this objection, Paul and Gilmour [1] isolated RNAs transcribed from the different templates and compared these by hybridisation back to DNA. These studies showed clear-cut differences between the RNA derived from DNA and RNA derived from nucleoprotein; they indicated that a restricted subset of sequences in DNA was transcribed in nucleoprotein. Moreover, it was demonstrated that the RNAs transcribed from chromatins from different tissues behaved differently and, in competition experiments, tended to resemble the natural RNAs isolated from nuclei from the tissues from which the chromatin preparations were made.

These studies were conducted at a time when the complexity of eukaryotic DNA was not fully appreciated. With a better understanding of this, it became apparent that the information obtained in these experiments could be interpreted only in terms of the repetitive sequences in the genome; this raised doubts about their general validity. Moreover, as more was learned about the differences between bacterial and eukaryotic polymerases, some investigators raised doubts about the usefulness of experiments in which a bacterial polymerase was used to transcribe from mammalian nucleoprotein. In this connection, however, it should not be overlooked that attempts to obtain faithful transcripts from *Xenopus laevis* ribosomal DNA using *Xenopus* RNA polymerase have been unsuccessful [2] whereas faithful copies have been obtained using RNA polymerase from *Micrococcus luteus* [3].

Recently, at the Beatson Institute, we have conducted experiments which seem to put the issue beyond doubt. In these experiments, we sought to determine whether globin genes can be transcribed by *E. coli* RNA polymerase from chromatin from erythropoietic and non-erythropoietic tissues. As a probe for globin RNA sequences, we used complementary DNA synthesised from globin messenger RNA using reverse transcriptase [4]. From theoretical studies of the kinetics of RNA/DNA hybridisation, we were able to define conditions which permitted the titration of globin cDNA against RNA in which globin mRNA sequences might be present in extremely small amounts. When we compared RNA transcribed from mouse brain chromatin with RNA transcribed from mouse foetal liver (erythropoietic) chromatin, we found that RNA transcribed from chromatin from erythropoietic tissue contained globin messenger RNA sequences whereas RNA transcribed from brain chromatin did not (Fig. 1). Errors might arise in interpreting these experiments if the chromatin were contaminated with RNA containing globin messenger sequences which might be liberated during the enzymic reaction. To exclude this, experiments were conducted in which the nascent RNA was labeled with ^{32}P. The hybrid formed with ^{3}H-cDNA was then isolated by exploiting its buoyant density in $CsCl_4$. It was demonstrated that most of the RNA in the hybrid was newly synthesised *in vitro*. These experiments confirmed, therefore, that non-DNA components of chromatin confer a constraint on transcription which results in the copying of some sequences and not others.

Based on observations of this kind, we tried to perform reconstruction experiments in which we took the components of chromatin and recombined them by an annealing procedure which involved dissolving the components in high salt and urea followed by progressive lowering of salt concentration in the presence of urea [1]. We were able to show that, whereas the template activity of DNA was reduced to very low values when combined with histones alone, it was reduced to values similar to those in chromatin when non-histone proteins were also present. These experiments also showed that the addition of histones to chromatin did not reduce the template activity to that of nucleohistone; they there-

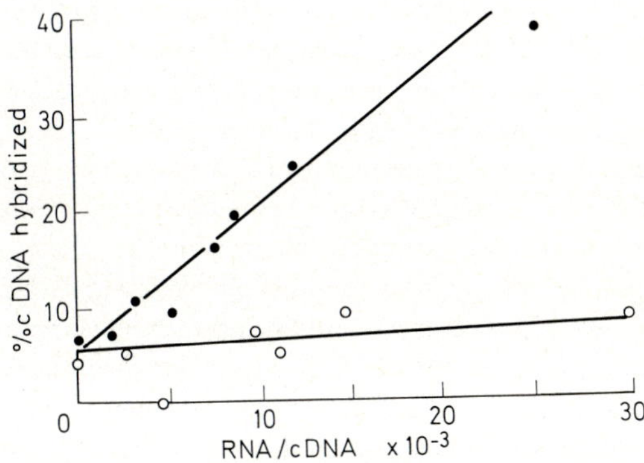

Fig. 1. The titration of RNA, transcribed from chromatin, against cDNA copied from mouse globin mRNA. ●—● RNA transcribed from mouse foetal liver chromatin, ○—○ RNA transcribed from mouse brain chromatin

fore indicated that a non-histone component of the proteins antagonised the inhibitory effects of the histones in a quantitative way [5]. Evidence that this antagonism was not only quantitative but qualitative came from experiments of similar nature in which the quality of the RNA product was tested by hybridisation. These experiments showed that reconstituted chromatin behaved like native chromatin and that both histones and non-histone proteins were essential to produce qualitatively similar products. The key experiments performed some years ago are subject to some of the criticisms mentioned earlier [1]. More critical experiments are currently in progress.

The above experiments strongly suggested that the inhibitory effect of the histones was non-specific but that the antagonistic effect of non-histone proteins was specific, that is to say, the non-histone proteins facilitated transcription from certain sequences. Accordingly, within the past four or five years, a considerable amount of research has been devoted to the non-histone proteins [6]. It turns out that these are a complex mixture and this is not surprising when it is recollected that both DNA and RNA polymerases, many nucleases, phosphatases, proteases and the kinases found in chromatin are all classified as non-histone proteins. There is also evidence for RNA binding proteins (informofers [7]) besides the hypothetical positive and negative regulatory proteins suggested by the above experiments. These proteins, as a class, have proved particularly difficult to study because of their propensity to aggregate and to interact with histones, DNA and RNA. However, reasonably satisfactory techniques have now been evolved for handling them. It has proven necessary to use a two dimensional system utilising isoelectric focusing in one direction and electrophoresis in the other to produce a reasonable separation [8].

It was pointed out some years ago by KLEINSMITH et al. [10] that chromosomal proteins are highly phosphorylated and RICKWOOD, RICHES and MACGILLIVRAY have studied the phosphorylation patterns in fractions of the non-histone proteins [10]. The interesting fact which becomes apparent is that most of the non-histone proteins are quite highly phosphorylated and much more highly phosphorylated than any of the histones. Moreover, although there is a great deal in common in the patterns observed in different tissues, individual tissues do reveal some unique characteristics. Differences in phosphorylation patterns are more striking than differences in the distribution of individual proteins although in a few instances during gene activation, clear-cut differences in the non-histone proteins have been observed in a number of laboratories (*e.g.* [11, 12]). Perhaps the most striking results have been obtained by immunological methods which have shown that when all the histones and many of the readily extractable non-histone proteins are removed from chromatin, the residual complex of DNA with some non-histone proteins has antigenic specificity which is characteristic of its source [13]. Moreover, TENG et al. [14] and KLEINSMITH et al. [15] have conducted experiments to determine whether any non-histone proteins have a particular affinity for special DNA sequences and they have provided evidence that some non-histone proteins exhibit a species specificity, an observation which suggests a specific structural relationship between certain non-histone proteins and the corresponding DNA.

The experiments briefly summarised above all suggest that some components of the non-histone proteins of the chromosome play a regulatory role and that this is probably rather different from the role of the known regulatory classes of proteins in prokaryotes since the isolated DNA-containing complex itself determines the pattern of transcription.

A number of models have been proposed to explain the possible action of non-histone proteins. It seems unlikely that they act like sigma factors since they seem to be quite tightly bound in chromatin. A hypothesis which is currently favoured is that histones prevent transcription from DNA by forming a complex into which polymerase molecules cannot penetrate whereas non-histone proteins interact in a specific way with the nucleohistone complex to alter its configuration so that polymerase can attach. As a result of some of our earlier work, we speculated that the non-histone proteins might loosen the binding of histones over the entire transcriptional unit [16], but Paul is inclined to think that non-histone proteins may exert a positive regulatory effect by binding to sites adjacent to promoter regions and causing these to unwind so that polymerase molecules and other regulatory molecules can approach DNA in that region [17]. Transcription of DNA might then lead to further unwinding of the transcriptional unit by the accumulation of poly-anions (RNA and RNA-binding protein). A structure based on this proposal is compatible with the banded appearance of giant chromosomes and the chromomeres of the lampbrush chromosome. It could also explain quite readily the appearance of puffs in the former and loops in the latter.

References

1. Paul, J., Gilmour, R. S.: J. molec. Biol. **34**, 305 (1968).
2. Roeder, R. G., Reeder, R. H., Brown, D. D.: Cold Sprg. Harb. Symp. quant. Biol. **35**, 727 (1970).
3. Hecht, R. M., Birnstiel, M. L.: Europ. J. Biochem. **29**, 489 (1972).
4. Ross, J., Aviv, H., Scolnick, E., Leder, P.: Proc. nat. Acad. Sci. (Wash.) **69**, 264 (1972).
5. Paul, J., More, I. A. R.: Nature (Lond.) New Biol. **239**, 134 (1972).
6. MacGillivray, A. J., Paul, J., Threlfall, G.: Advan. Cancer Res. **15**, 93 (1972).
7. Samarina, O. P., Lukanidin, E. M., Molnar, J., Georgiev, G. P.: J. molec. Biol. **33**, 251 (1968).

8. MacGillivray, A. J., Rickwood, D.: Biochem. Transactions (in press) (1973).
9. Kleinsmith, L. J., Allfrey, V. G., Mirsky, A. E.: Proc. nat. Acad. Sci. (Wash.) **55**, 1182 (1966).
10. Rickwood, D., Riches, P. G., MacGillivray, A. J.: Biochim. Biophys. Acta **299**, 162 (1972).
11. Hill, R. J., Poccia, D. C., Doty, P. J.: J. molec. Biol. **61**, 445 (1971).
12. Helmsing, P. J., Berendes, H. D.: J. Cell Biol. **50**, 893 (1971).
13. Chytil, F., Spelsberg, T. C.: Nature (Lond.) New Biol. **233**, 215 (1971).
14. Teng, C. T., Teng, C. S., Allfrey, V. G.: Biochem. Biophys. Res. Commun. **51**, 621 (1970).
15. Kleinsmith, L. J., Heidema, J., Carroll, A.: Nature (Lond.) **226**, 1025 (1970).
16. Paul, J., Gilmour, R. S.: Symp. Inter. Soc. Cell Biology. (Differentiation and Immunology. Warren, K. B., Ed.) vol. 7, p. 135. New York-London: Academic Press. 1968b.
17. Paul, J.: Nature (Lond.) **238**, 444 (1972).

Discussion

K. Scherrer (Lausanne): There are two facts that puzzle me with this model. The first one is that you cannot build a human body with only 5000 structural genes.

J. Paul (Glasgow): 5000 per chromosome. You must multiply with 45. This gives a lot more. I think that there is a good deal of evidence that 5000 is enough.

K. Scherrer: I want to make a second point. In our model, a certain gene, *i.e.*, the one coding for globin in haemaglobin, would duplicate and further multiply and then mutate and give rise to non-informative sequences. However, in that case one would expect that the antimessenger would hybridize to a certain extent with these duplicated and mutated genes, but it would not fully hybridize and therefore not be protected against the attack of nucleases. However, the fact is that antimessenger can be completely hybridized to DNA.

J. Paul: I think this argument is somewhat misleading. The decay would occur randomly over the whole length, not only in one part; and when you have about 20% of substitutions, the antimessenger would no longer hybridize to this stretch of DNA.

K. Scherrer: I did not mean that all the DNA is made of cistrons, *i.e.*, that it contains structural information. It is my idea that one transcriptional unit bears 2 or 3 or 5 stretches of information, and that this was later enlarged for reasons of regulation and of structural organization. I shall discuss this in my talk in the afternoon.

P. Karlson (Marburg): May I make a comment here. In the lecture, I missed one point. We all know that genes change by mutation, and have changed during evolution. Now, if you have a gene that mutates rather frequently like the globin gene, it is very hard to visualize that this gene is present in more than one copy; it would be hard to understand that all copies mutate simultaneously in the same way. On the other hand, one often has wondered why the histone molecules show so little changes during evolution; there are but a few amino acid changes between higher plants and higher vertebrates. Could this be due to the fact that these genes are present in the genome, let us say in 100 copies, and that therefore we find very little change in the gene product?

J. Paul: This is a very valid argument. We know that some genes are present only in one copy; the globin gene you mentioned belongs to these. On the other hand the histone genes are often present in multiple copies. That by itself could be a reason for stability. But I believe that another reason, and perhaps the real reason, is selective pressure. A number of theories have been put forwards to explain this, including Callan's master-slave hypothesis. I think we do not need these complicated hypotheses. It could just be ordinary selective pressure having something to do with the function of histones in double-strand interchanges. We cannot go into this argument here. However, I believe that the reason that some genes have mutated during evolution more frequently than others rests in the specificity of the gene product, the protein. In some proteins, very little change in amino acid sequence is allowed without impairment of functions, while in other proteins you may exchange many more sites without loss of function.

P. Karlson: This is very hard to visualize for me, especially in the case of histones. But I think we cannot settle this question.

C.A. Thomas (Boston): Let me comment on this. I do sympathize with the idea of separating selective pressure from multiple representation. But let us just focus on a single fact. We do know that in the nucleolus organizer there are multiple representations of the genes coding for the proteins of the ribosomes. We also know that the DNA coding for the 5 S-RNA is present in several copies. However, there is only one gene product according to the sequence determination of this RNA. It is also known that the several copies of DNA are all replicated in semiconservative fashion; this suggests that they are free to mutate independently. Accordingly to their product, they have not mutated. The question you posed is therefore a very serious one. I expect that this paradox demands that we can expect to discover a new process which I call rectification. After replication, there must be a yet unknown process which I call rectification that makes sure that all of these copies are the same.

J. Paul: Of course, with a function like this one can explain many things. However, there is no evidence for such a process. And I believe that the fact that some genes are present in one copy and others in multiple copies can be explained only on the basis of gene dosage necessary for its function. In case of the sea urchin egg, when in early development large amounts of histones are to be made it seems reasonable that many copies of the histone gene are necessary for this high production. On the other hand, you can easily explain the amount of globin synthesis with just one copy of the globin gene.

FRIEDRICH (Julich): I want to add some experimental remarks on the acidic nuclear phosphoproteins. In our lab Dr. WELSH and Dr. VYSKA are working on the question of DNA-subunits according to the "Bendich-Modell". They find when splitting purified native DNA into subunits of the size of 5—14 S by means of EDTA-treatment that during the splitting acidic phosphopeptides are released from the DNA. The question remains if these phosphoproteins are linked covalently or non covalently to the DNA. It seems that they are linked covalently. Are you able to correlate these observations to your results? In addition it should be mentioned that this fraction of phosphoproteins or -peptides is very small compared to the total nuclear phosphoproteins.

J. PAUL: This is an interesting observation. One cannot say without more evidence what its relevance may be either to our results or to my hypothesis.

Biosynthesis and Enzymatic Modifications of Histones

D. Gallwitz

Physiologisch-Chemisches Institut der Universität Marburg, Lahnberge, Marburg, W. Germany

With 11 Figures

In higher organisms DNA is closely associated with a class of basic proteins, the histones. During the last few years it has become evident that there are only five major histones found in all nucleated cells regardless of their origin. The primary sequences of four histones, F2a1, F2b, F3, and F2a2, have been determined [1–4], and the sequence of one of the subfractions of histone fraction F1 has been worked out partially [5]. A comparison of the sequence of histone F2a1 derived from two species, pea and calf, revealed that only two conservative changes in amino acid composition occurred since the divergence of the two species 1.5 billion years ago [6]. From the nearly identical peptide maps of histone F3 derived from pea and calf, a high degree of sequence conservation has also been inferred for this histone. In contrast, the rates of mutation acceptance of cytochrome C is 50-times faster, that of pancreatic ribonuclease is 500-times faster [7]. This is only to illustrate the exceptional evolutionary stability of the histones suggesting that the entire polypeptide chains are required for the function of these proteins in the chromosomes of higher organisms.

Other interesting features common to all histones are the clustering of specific groups of amino acids and the occurrence of modified amino acid residues in the primary DNA binding sites.

This paper is concerned with two aspects of histone biochemistry: the biosynthesis of histones as studied in *in vitro* systems and the postsynthetic acetylation of amino acid side chains catalyzed by nuclear enzymes.

Biosynthesis of Histones

Since histones are always found in close association with DNA, it was believed for many years that the synthesis of histones occurred inside the cell nucleus. That this assumption was not correct was first shown biochemically by Robbins and coworkers [8, 9] who performed pulse-chase experiments in synchronized HeLa cells showing that histones are synthesized in the cytoplasm and subsequently transported into the nucleus. In addition, these and earlier studies [10, 11] established that histones are synthesized synchronously with DNA replication.

Since nucleated cells contain roughly equal amounts of DNA and histones, one can calculate that a cell synthesizes about 2×10^8 molecules of histones during S-phase. This means that about 4×10^5 histone molecules are synthesized per cell per minute. In HeLa cells approximately 5 % of the total cell protein are histones whose synthesis, as just mentioned, is precisely coupled to DNA replication. From these figures it is immediately apparent that histone biosynthesis represents an ideal system to study the regulation of synthesis of a specific class of well characterized proteins. All studies which will be described in the following have been carried out with HeLa cells synchronized either with thymidine, amethopterin or hydroxyurea.

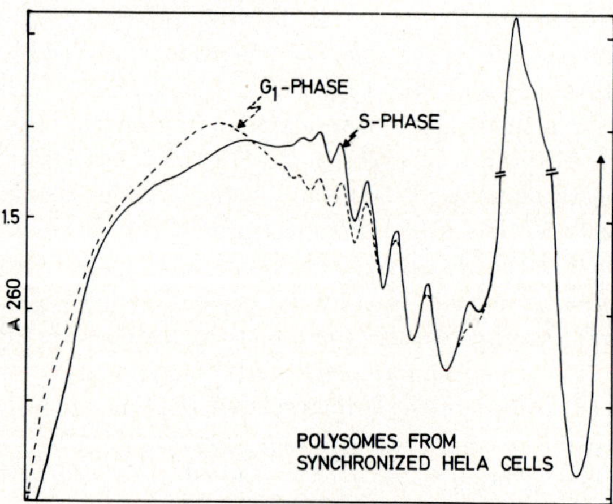

Fig. 1. Sucrose gradient centrifugation of polyribosomes from synchronized HeLa cells. HeLa cells in suspension culture were synchronized with 1×10^{-3} M hydroxyurea. Polyribosomes from unreversed cells (G1-phase) and cells 4 hours after initiation of DNA synthesis (S-phase) were treated with 0.5% sodium deoxycholate. For comparison, the polyribosomal profiles are superimposed

In Fig. 1 the polyribosome profiles of HeLa cells in Gl- and S-phase are compared. As first described by ROBBINS et al. [8], there is a significant enrichment of small polyribosomes during S-phase. Polyribosomes were isolated from cells in S-phase and incubated *in vitro* under conditions of protein synthesis. In such systems, nascent polypeptide chains are finished whereas initiation of new chains is a rare event. Acid-soluble proteins were extracted with 0.25 N HCl and separated on polyacrylamide gels together with marker histones. By these criteria all histones could be shown to be synthesized *in vitro* [12, 13].

In Fig. 2 one can see that when synchronized cells are labeled with uridine and the polyribosome-bound mRNA is separated on polyacrylamide gels, at least three RNA species appear on polyribosomes engaged in histone biosynthesis [14, 15]. The observed molecular weights of these RNA species are 1.55×10^5, 1.8×10^5 and 2.1×10^5 daltons. The calculated molecular weights for the mRNAs coding for the five histones are 1.1×10^5 daltons for histone F2a1, $1.2-1.35 \times 10^5$ daltons for histones F2a2, F3, and F2b and 2.1×10^5 daltons for histones F1. Assuming a linear relationship between the logarithm of the molecular weight of a RNA and the distance moved in polyacrylamide gels, our data suggest that at least some of the histone mRNAs have higher molecular weights then expected. To explain this it could be inferred that, as in many eukaryotic mRNAs, poly A stretches of different length are attached to the 3′-end of histone mRNAs. But neither in HeLa cells [16] nor in L-cells [17] is there a major tract of poly A bound to the histone mRNA. If by other techniques the measured molecular weights of the RNAs can be substantiated, it then would mean that untranslated polynucleotide stretches do exist in histone mRNAs.

That the RNA which appears on polyribosomes only during S-phase represents histone mRNA could be shown by its capacity to direct the synthesis of histones in different *in vitro* protein synthesizing systems. Histone mRNA can be demonstrated as a discrete peak sedimenting in sucrose gradients in the 8–10 S region [15]. This RNA is not present in G1-phase cells and is immediately lost from polyribosomes of cells in S-phase after interruption of DNA synthesis [9, 13–15]. 8–10 S RNA was isolated from polyribosomes of cells which were harvested four hours after initiation of DNA replication. This RNA was added to a rabbit reticulocyte lysate which was chosen for several reasons: a) rabbit reticulocytes are cells without nucleus and probably do not synthesize any histones, which facilitates the interpretation of the experimental

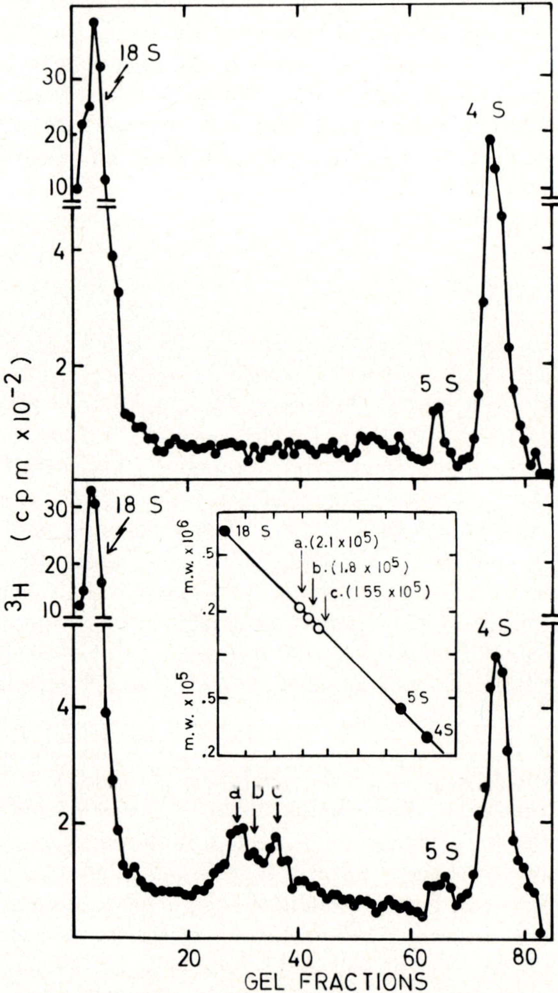

Fig. 2. Molecular weight of histone mRNA. HeLa cell polyribosomal RNA, labeled with uridine, was isolated from thymidine-synchronized cells and separated on 4% polyacrylamide gels. The radioactivity profiles are compared from cells blocked with thymidine (upper part of Fig.) and cells which were 1 hour in S-phase [15]

results; b) about 95 % of the endogenous product is hemoglobin which is easily separated from histones on CM-cellulose; and, c) the reticulocyte lysate has good initiation properties and a number of heterologous mRNAs have been successfully translated in this *in vitro* system. Acid-soluble proteins were extracted from the incubation mixture with 0.25 N HCl and fractionated on CM-cellulose. Histones synthesized under the direction of 8–10 S RNA were then identified by coelectrophoresis in polyacrylamide gels with authentic HeLa cell histones (Fig. 3) and by high voltage paper electrophoresis of tryptic peptides [15, 18]. In our earlier experiments [15, 18] a trace of labeled globin made it difficult to identify histone fraction

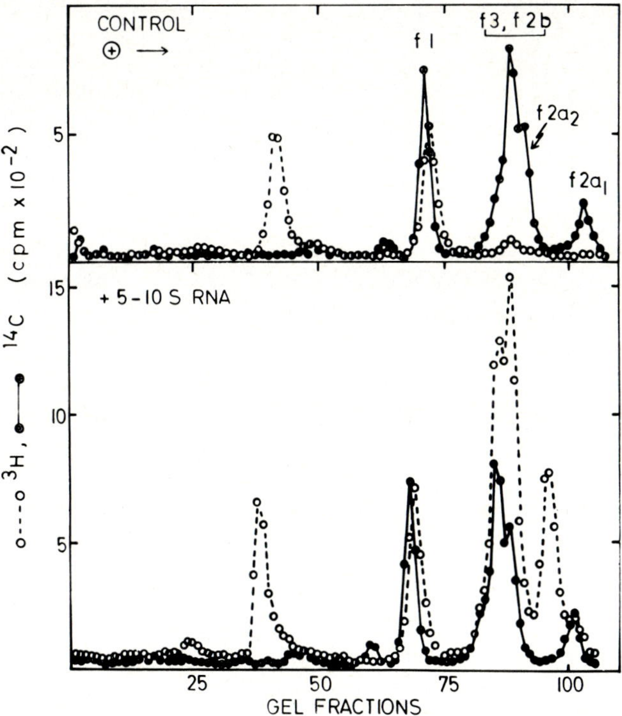

Fig. 3. Polyacrylamide gel electrophoresis of acid-soluble proteins synthesized in a rabbit reticulocyte lysate under the direction of 8—10 S RNA from HeLa cells in S-phase. Before electrophoretic separation the ^{3}H-labeled (○——○) *in vitro* products were mixed with ^{14}C-labeled (●—●) authentic HeLa cell histones [15]

F1 in the *in vitro* product since the two proteins have similar mobilities in acidic polyacrylamide gels. In SDS-gels, which separate these proteins, no labeled material coincided with histone fraction F1 (Fig. 4). From the appearance in the acidic gels of a radioactive peak migrating between histones F2a2 and F2a1, it was concluded that histone F1 was probably prematurely terminated in the reticulocyte lysate [15]. In recent experiments, however, we have consistently oberved the synthesis of all histones in the rabbit reticulocyte lysate.

Our finding that polyribosomal 8–10 S RNA from HeLa cells in S-phase directs the synthesis of histones *in vitro* was subsequently

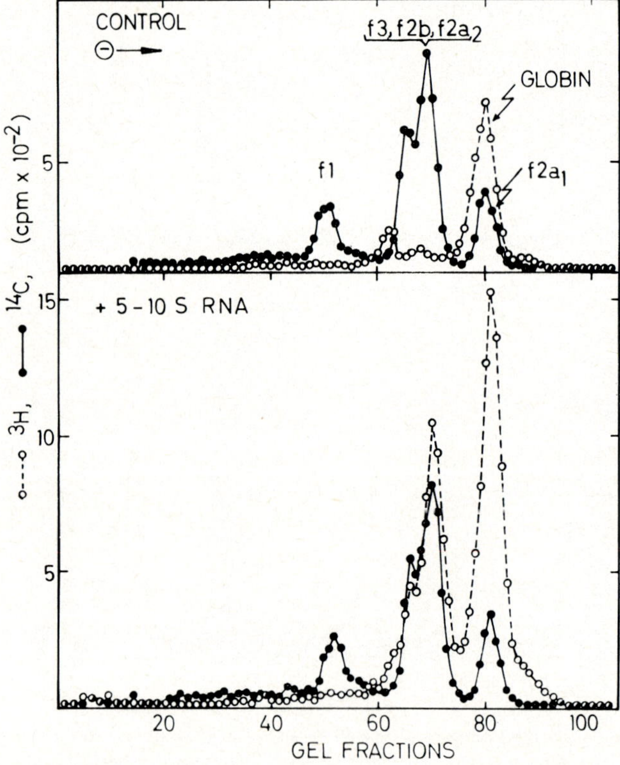

Fig. 4. SDS-gel electrophoresis of acid-soluble *in vitro* products (○——○) formed under the direction of 8—10 S HeLa cell RNA. See also Fig. 3 [15]

confirmed by JACOBS-LORENA et al. [19] who translated histone mRNA in a Krebs ascites tumor cell-free system.

For studying the regulation of histone biosynthesis in which a translational control mechanism seems to be involved [13], a reconstituted cell-free system is required which allows the measurement of the translational efficiency of histone mRNA by substituting different protein factors. Histone mRNA could be efficiently translated in an *in vitro* system prepared according to SCHREIER and STAEHELIN [20]. This system consists of ribosomal subunits and a pH 5-fraction from rat liver and initiation factors either from rabbit reticulocytes or HeLa cells on which the translation of mRNA depends completely. As shown in Fig. 5, all histone fractions are synthesized in the reconstituted system under the direction of 8–10 S polyribosomal RNA from HeLa cells in S-phase. In addition to coelectrophoresis of the *in vitro* products with authentical HeLa cell histones, the *in vitro* synthesized proteins were also identified as histones by column chromatography and high voltage paper electrophoresis of tryptic peptides (BREINDL and GALLWITZ, unpublished).

The experiments described have conclusively shown that the 8–10 S RNAs which appear on polyribosomes during S-phase are indeed histone mRNAs. The possibility to identify histone mRNA biologically presents an important tool for the study of the regulation of histone synthesis during DNA replication.

Since DNA and histone synthesis proceed synchronously, it is expected that the maximum amount of histone mRNA is present on polyribosomes when DNA synthesis is at its maximum. This assumption precludes that the actual number of mRNA molecules rather then the efficiency with which a certain amount of mRNA is translated, determines the rate of histone synthesis. The labeling kinetics of histone mRNA during S-phase and the association of the labeled mRNA with polyribosomes bear, therefore, valuable information about the regulation of histone synthesis.

HeLa cells were synchronized by a double thymidine block. Under our experimental conditions synchronizing up to 10 liters of cells, the peak of DNA synthesis was reached between 5 and 6 hours after release from the block. The kinetics of appearance of the newly synthesized polyribosomal RNA are shown in Fig. 6.

A significant amount of labeled 8–10 S RNA was found on polyribosomes up to the sixth hour after cells were released from thymidine block, *i.e.*, at the time DNA synthesis had reached its

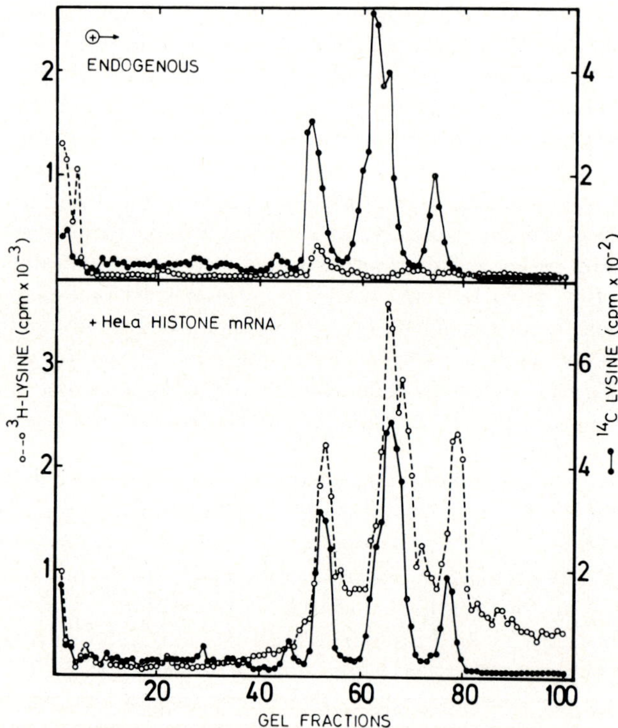

Fig. 5. Translation of HeLa cell histone mRNA in a reconstituted cell-free system. The cell-free system consisted of ribosomal subunits and pH 5-fraction from rat liver and initiation factors from rabbit reticulocytes. The ^{3}H-labeled acid-soluble proteins synthesized *in vitro* were mixed with ^{14}C-labeled HeLa cell histones (●—●) and separated at pH 2.8 in 15% polyacrylamide gels containing 2.5 M urea

maximum. The amount of newly synthesized 8–10 S RNA entering the polyribosomes during a 60 min pulse was calculated by relating the actual radioactivity in the 8–10 S region of the gels to that of the 18 S and 4 S RNA. Using this method, it is assumed that the degree of synthesis of ribosomal and transfer RNA at all

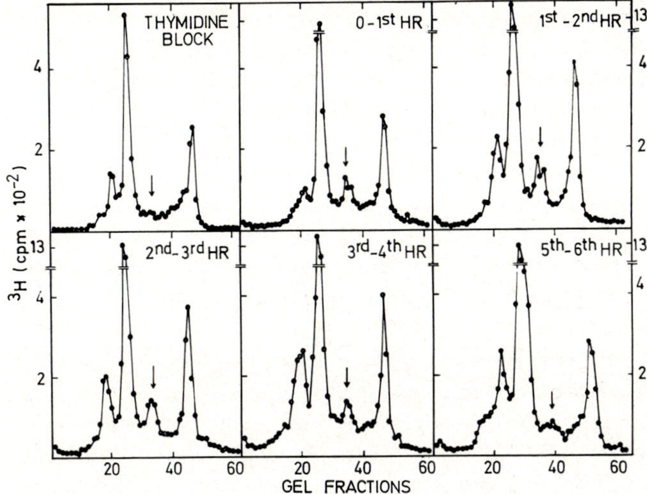

Fig. 6. Appearance of histone mRNA on polyribosomes during S-phase. HeLa cells were synchronized with thymidine and pulse-labeled with ^{3}H—5-uridine for 60 min. immediately before and at different times after thymidine removal. Polyribosomal RNA was separated in 2% polyacrylamide gels containing 0.5% agarose [15]

times of S-phase is not significantly different. Table 1 illustrates that the largest amount of 8–10 S RNA appearing on polyribosomes is synthesized during the first hour following thymidine

Table 1. Appearance of newly synthesized 8—10 S RNA on polyribosomes. Polyribosomal RNA pulse-labeled for 60 min with ^{3}H—5-uridine was separated on polyacrylamide gels. The radioactivity in the 8—10 S region of the gels is compared with that in the 18 S and 4—5 S regions [15]

Time of 60 min pulse	Radioactivity of 8—10 S mRNA	
	cf. 18 S (in %)	cf. 4—5 S (in %)
Thymidine block	10	21
S-phase 1st h	35	75
2nd h	22	62
3rd h	22	69
4th h	19	48
5th h	15	44
6th h	11	26

removal. It then declines, reaching the basal level at about or shortly after the maximum of DNA synthesis.

From the calculation concerning the number of histone molecules synthesized per cell during S-phase, it follows that an enormous number of mRNA molecules has to be synthesized in the first few hours of S-phase. There are in principle two ways for a cell to adapt to such a demand, gene amplification and gene reiteration. Birnstiel and collaborators [21, 22] have presented evidence that the genes for histones are indeed reiterated a few hundred-fold. This is the only example where genes coding for proteins have been found to be reiterated.

There is evidently adaptation and regulation of histone synthesis at the level of transcription; the mechanism by which the transcription of histone genes is controlled is, of course, unknown.

We had shown earlier that after interruption of DNA replication, histone synthesis ceases immediately whereas interruption of RNA synthesis did not affect histone synthesis for a considerable time [13]. Furthermore, resumption of DNA synthesis was accompanied by resumption of histone synthesis only when RNA synthesis was allowed to proceed [13]. These and other studies suggested that histone synthesis is regulated at the translational level as well.

In order to search for factors specifically affecting histone mRNA translation, cell-free protein synthesizing systems (S-30 extracts) were prepared from synchronized HeLa cells in S-phase and from cells in which DNA synthesis was inhibited by hydroxy urea. These systems were used to translate exogenous HeLa cell histone mRNA and, as a control, rabbit globin mRNA. Surprisingly, the electrophoretic analysis of the *in vitro* products (Fig. 7) revealed that histone mRNA and globin mRNA were equally well translated in the two systems. It is, of course, still open to question whether specific initiation factors can indeed modify the translational efficiency of histone mRNA. Preliminary experiments in a reconstituted cell-free system, however, using initiation factors prepared from S-phase cells and from hydroxyurea-treated cells (which *in vivo* do not make histones) did not reveal any differences in the efficiency of translation of histone mRNA (Breindl and Gallwitz, unpublished). The question of the existence of specific initiation factors involved in the translation of certain mRNAs has recently become of interest since the translation of EMC virus mRNA seems to require such a factor [23]. In addition, SCHREIER and STAEHELIN [24] have described quantitative dependency of an initiation factor necessary for translation of duck but not of rabbit globin mRNA in a reconstituted cell-free system.

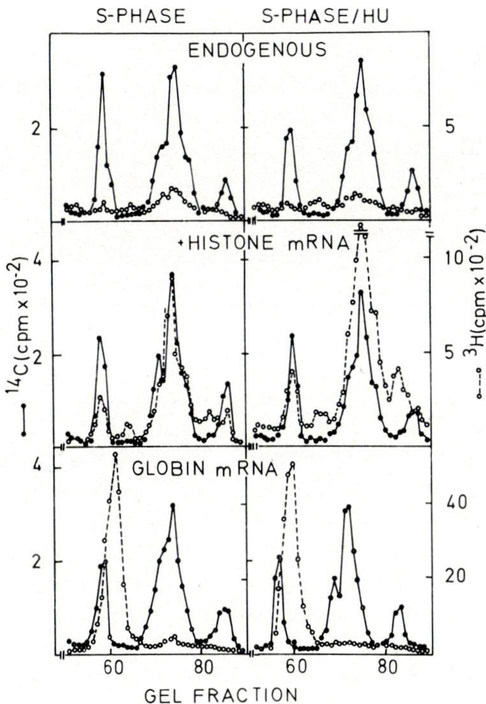

Fig. 7. Translation of HeLa cell histone mRNA and rabbit globin mRNA in cell-free systems (S-30 extracts) prepared from HeLa cells in S-phase and cells in which DNA replication and histone synthesis was interrupted with 1×10^{-3} M hydroxyurea. ^{3}H-labeled *in vitro* products (○---○) and ^{14}C-labeled HeLa cells histones (●—●) were mixed and separated electrophoretically in 15% polyacrylamide gels

Besides a specific initiation factor requirement for histone synthesis, which seems not very likely, the control of translation of histone mRNA could somehow involve newly synthesized histones themselves. It has been suggested by BUTLER and MUELLER [25] that, by blocking DNA replication, newly synthesized histones accumulate in the cytoplasm and prohibit reinitiation of histone synthesis followed by degradation of histone mRNA. This process could then be explained by a complex formation of the basic histones with histone mRNA having a highly ordered secondary structure.

The translational control of histone synthesis could also be the consequence of the action of other factors of low molecular weight which, in our experiments described above, were lost during the preparation of HeLa cell S-30 extracts in which histone mRNA was translated. The exact translational control mechanism remains to be elucidated.

Enzymatic Modifications of Histones

After histones are synthesized in the cytoplasm, they are transported into the nucleus and bind to DNA. The binding of histones to DNA is mainly achieved by ionic forces between the phosphate groups of the DNA backbone and the basic amino acid residues of the polypeptide chains.

According to sequence studies, the basic, acidic and hydrophobic residues in histones are not evenly distributed but clustered in certain regions of the polypeptide chains. Two examples, calf thymus histones F2a1 and F3, are shown in Table 2. It is suggestive from these data that there are primary DNA binding sites in histone molecules whereas other regions have the potential to form secondary structure and may loop away from the DNA helix, being free for contact with other cationic and anionic molecules in the cell nucleus. Histones are probably involved primarily in the structural organization of chromosomes, *i.e.*, in packaging of DNA in the cell nucleus. But it should be stressed that such a role has to be seen as a dynamic one since the physical properties of DNA-protein complexes almost certainly change during different phases of the cell cycle and very likely during transcriptional processes as well.

It is therefore of interest to note that histone polypeptide chains can be enzymatically modified by acetylation, methylation and phosphorylation. These modifications occur in the primary

DNA binding sites and might therefore be of importance for the tightness of binding of these basic proteins to DNA. The arginine-rich histone F2a1 of calf thymus is acetylated in the ε-N-position of the lysine residue 16 and in pea seedlings this histone, in addition to residue 16 [1], is partially acetylated at a second lysine residue at position 5 or 8 [6]. Sung and Dixon [26] have demonstrated multiple sites of acetylation (lysines 5, 8, 12 and 16) occurring in the same histone F2a1 from trout testis. Calf thymus histone F3 is partially acetylated at lysine residues 14 and 23 [3]. From model building it has been concluded that the acetylation of multiple lysine residues in the NH_2-terminal region of histone F2a1 could extremely modify the binding of the histone to DNA [26, 27], a process which is expected to have functional consequences.

Clearly, multiple sites of acetylation at specific positions within histone molecules very likely require a specific enzymatic machinery located in the cell nucleus where these modification reactions take place. We have studied the enzymes involved in histone acetyl-

Table 2. Distribution of amino acid residues in calf thymus histones F2a1 and F3 [1, 3]

Residue	N-terminal region	middle region	C-terminal region
F2a1			
Total	20	39	43
Basic	10	10	8
Acidic	—	3	4
Net positive charge	+10	+7	+4
Hydrophobic	1	13	16
F3			
Total	42	41	52
Basic	15	10	8
Acidic	—	5	6
Net positive charge	+15	+5	+2
Hydrophobic	3	14	17

ation. From a variety of mammalian cell nuclei, two enzymatic activities, histone acetyltransferases A and B, could be isolated and separated on DEAE-cellulose [28–30]. In Fig. 8 and 9, the

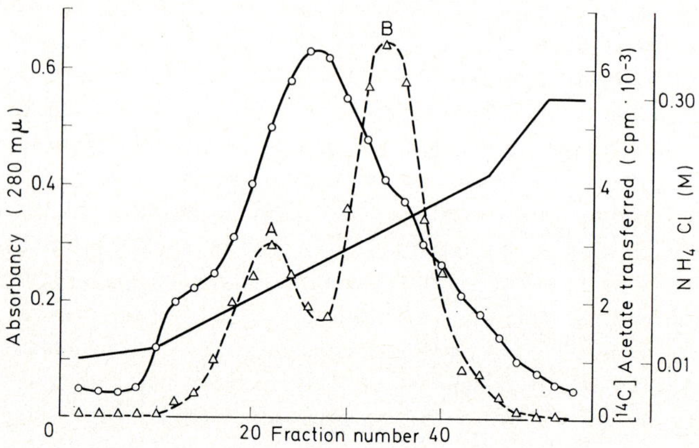

Fig. 8. DEAE-cellulose chromatography of histone acetyltransferases *A* and *B* from rat thymus nuclei [29]

chromatographic behavior of the enzymes derived from rat liver and thymus are shown. At pH 7.9 rat thymus histone acetyltransferases A and B are clearly separated (Fig. 8) whereas the liver enzymes are separated only by rechromatography with a shallower salt gradient (Fig. 9, upper part). The differing chromatographic patterns of histone acetyltransferases from different organs indicate a certain organ specificity of these enzymes [30]. Recently we have purified histone acetyltransferases A and B from calf thymus (Sures and Gallwitz, unpublished) in order to explore which histone and which site in a given histone molecule is recognized by the two enzymes.

Although, according to sequence studies, only histones F2a1 and F3 from calf thymus are acetylated to a significant degree, the two purified enzymes transfer acetate from acetyl-CoA to all histones. Acetyltransferase A, however, prefers histone fraction F1 as substrate, whereas histone F2a1 is the best substrate for acetyltransferase B. We then compared tryptic peptides of histones F2a1 and F3 which were acetylated *in vivo*, *i.e.*, in calf thymus

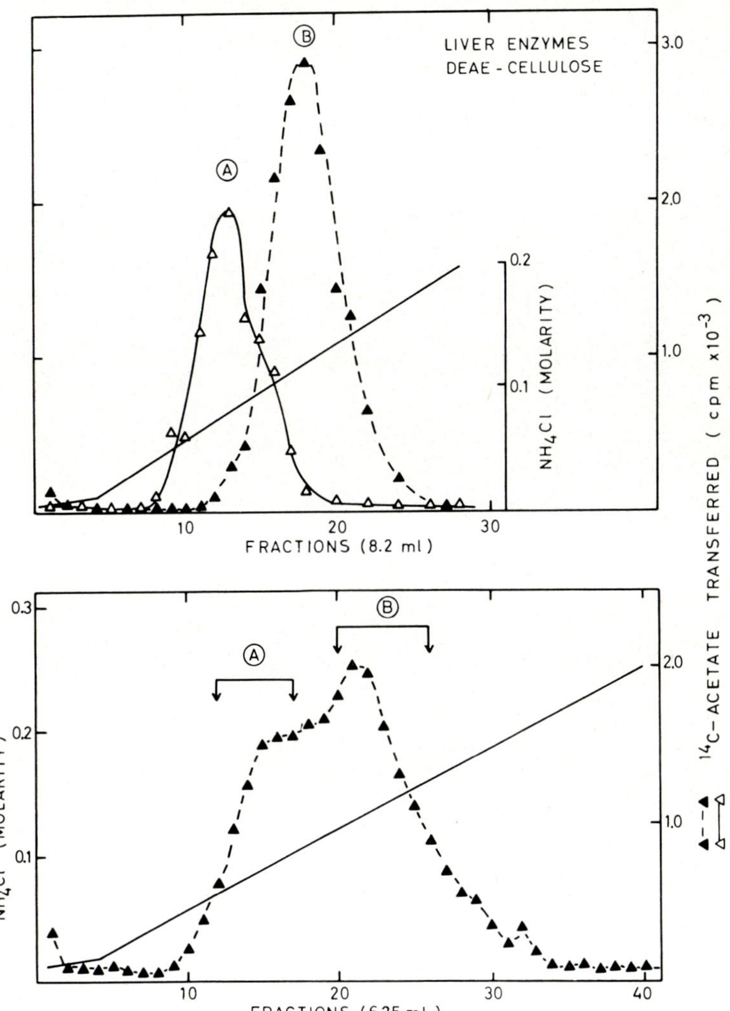

Fig. 9. DEAE-cellulose chromatography of histone acetyltransferases *A* and *B* from rat liver nuclei. After chromatography under the same experimental conditions as shown for thymus enzymes (Fig. 8), enzymes *A* and *B* were pooled as indicated and rechromatographed with a shallower NH_4Cl-gradient (upper part of Fig.)

cells in suspension, and *in vitro* with the purified enzymes (Fig. 10 and 11). A surprising finding was that three labeled peptides resulted from *in vivo* acetylation of histone F 2a 1 suggesting that, similar to trout testis histone F 2a 1, more than one site can be acetylated. *In vivo* acetylation of calf thymus histone F 3 resulted in four labeled tryptic peptides which could be separated by high voltage paper electrophoresis at pH 4.7 (not shown). This again leads to the conclusion that, contrary to the expectation from the sequence

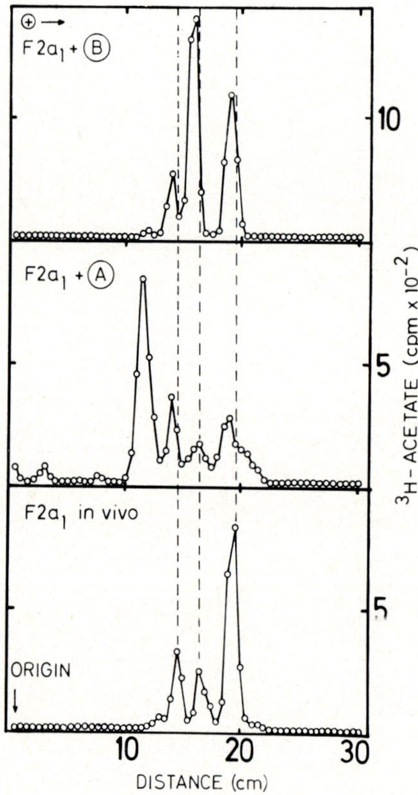

Fig. 10. High voltage paper electrophoresis at pH 1.9 of ^{3}H-acetate-labeled tryptic peptides from calf thymus histone F 2a 1. The peptide pattern resulting from *in vivo* acetylation is compared with the patterns resulting from *in vitro* acetylation of histone F 2a 1 with calf thymus histone acetyltransferases *A* and *B*

studies, not only two but possibly four lysine residues of calf thymus histone F3 can be acetylated *in vivo*.

As seen from Fig. 10 and 11, the *in vitro* acetylation of calf thymus histones F2a1 and F3, using purified calf thymus enzymes A and B, resulted in a quite different peptide pattern although a few labeled peptides correspond to some obtained by *in vivo* acetylation. This has been confirmed by autoradiography of high voltage paper electrophoretograms done at pH 1.9 and pH 4.7 (Sures and Gallwitz, unpublished).

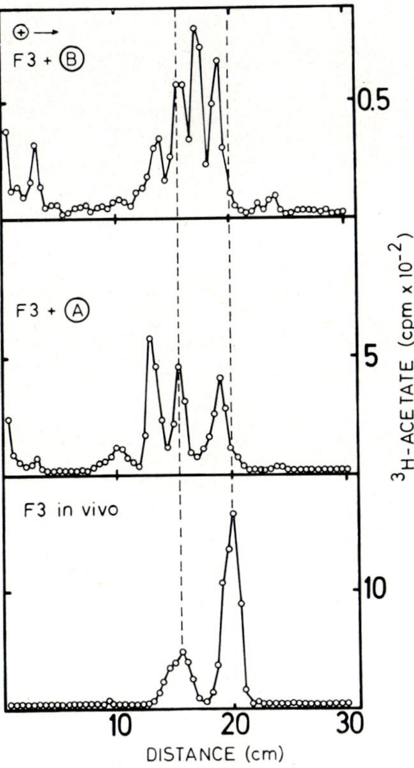

Fig. 11. High voltage paper electrophoresis at pH 1.9 of tryptic peptides from calf thymus histone F3 labeled with ^{3}H-acetate *in vivo* and *in vitro* (see Fig. 10). The two radioactive peaks resulting from *in vivo* acetylation were separated into four by electrophoresis at pH 4.7 (not shown)

The obvious differences between *in vivo* and *in vitro* acetylation make it likely that for the specificity of the histone acetyltransferases a highly specific conformation of histones is required which probably results from their binding to DNA.

Acknowledgements

The experimental work described is the result of the joint effort with M. BREINDL and I. SURES to whom I am very grateful. The work was supported partially by the Deutsche Forschungsgemeinschaft.

References

1. DELANGE, R. J., FAMBROUGH, D. M., SMITH, E. L., BONNER, J.: J. biol. Chem. **244**, 319 (1969).
2. IWAI, K., ISHIKAWA, K., HAYASHI, H.: Nature (Lond.) **226**, 1056 (1970).
3. DELANGE, R. J., SMITH, E. L.: Proc. nat. Acad. Sci. (Wash.) **69**, 882 (1972).
4. TEOMAN, L. C., OLSON, M. O. J., SUGANO, N., JORDAN, J. J., TAYLOR, C. W., STARBUCK, W. C., BUSH, H.: J. biol. Chem. **247**, 6018 (1973).
5. RALL, S. C., COLE, R. D.: J. biol. Chem. **246**, 7175 (1971).
6. DELANGE, R. J., FAMBROUGH, D. M., SMITH, E. L., BONNER, J.: J. biol. Chem. **244**, 5669 (1969).
7. DAYHOFF, M. O.: Atlas of Protein Sequence and Structure, Vol. 5. National Biomedical Research Foundation (1972).
8. ROBBINS, E., BORUN, T. W.: Proc. nat. Acad. Sci (Wash.) **57**, 409 (1967).
9. BORUN, T. W., SCHARFF, M. D., ROBBINS, E.: Proc. nat. Acad. Sci (Wash.) **58**, 1977 (1967).
10. PRESCOTT, D. M.: J. Cell Biol. **31**, 1—9 (1966).
11. SPALDING, J., KAJIWARA, K., MUELLER, G. C.: Proc. nat. Acad. Sci. (Wash.) **56**, 1535 (1966).
12. GALLWITZ, D., MUELLER, G. C.: Science **163**, 1351 (1969).
13. GALLWITZ, D., MUELLER, G. C.: J. biol. Chem. **244**, 5947 (1969).
14. GALLWITZ, D., MUELLER, G. C.: FEBS Letters **6**, 83 (1970).
15. BREINDL, M., GALLWITZ, D.. Europ. J. Biochem. **32**, 381 (1973).
16. ADESNIK, M., DARNELL, J. E.: J. molec. Biol. **67**, 397 (1972).
17. PERRY R. P., LATORRE, J., KELLEY, D. E., GREENBERG, J. R.: Biochim. biophys. Acta **262**, 220 (1972).
18. GALLWITZ, D., BREINDL, M.: Biochem. Biophys. Res. Commun. **47**, 1106 (1972).
19. JACOBS-LORENA, M., BAGLIONI, C., BORUN, T. W.: Proc. nat. Acad. Sci. (Wash.) **69**, 2095 (1972).
20. SCHREIER, M. H., STAEHELIN, TH.: J. molec. Biol. **73**, 329 (1973).
21. KEDES, L. H., BIRNSTIEL, M.: Nature (Lond.) New Biol. **230**, 165 (1971).
22. WEINBERG, E. S., BIRNSTIEL, M., PURDOM, I. F., WILLIAMSON, R.: Nature (Lond.) New Biol. **240**, 225 (1972).
23. WIGLE, D. T., SMITH, A. E.: Nature (Lond.) New Biol. **242**, 136 (1973).

24. Schreier, M. H., Staehelin, Th., Steward, A., Gander, E., Scherrer, K.: Europ. J. Biochem. **34**, 213 (1973).
25. Butler, W. B., Mueller, G. C.: Biochim. biophys. Acta **294**, 481 (1973).
26. Sung, M. T., Dixon, G. H.: Proc. nat. Acad. Sci. (Wash.) **67**, 1616 (1970).
27. Pardon, J. F., Richards, B. M.: Chromosomes Today **3**, 38 (1972).
28. Gallwitz, D.: Biochem. Biophys. Res. Commun. **40**, 236 (1970).
29. Gallwitz, D., Sures, I.: Biochim. biophys. Acta **263**, 315 (1972).
30. Gallwitz, D.: FEBS Letters **13**, 306 (1971).

Discussion

C. A. Thomas (Boston): What evidence do we have that histones or acidic proteins bind to specific sequences of the DNA? This is often tacitly assumed, but where is the evidence?

D. Gallwitz (Marburg): We have not done any work in this respect and I can only refer to the literature. Some authors say that histones indeed could bind to specific sequences of DNA [Clark, R. J. and Felsenfeld, G.: Nature New Biology **240**, 226 (1972)], and even model building has shown that some base sequences would be preferred by certain histones [Smythies, J. R. et al.: J. theor. Biol. **37**, 151 (1972)]. But it is not really known how histones bind to DNA, and I would not like to speculate on this.

Pretranslational Control in Embryonic Differentiation

H. Tiedemann

Institut für Molekularbiologie und Biochemie, Freie Universität Berlin, Berlin, W. 6 Germany

With 9 Figures

The topic of this symposium is the regulation of transcription and translation in eukaryotes. This topic is closely related to cell differentiation which may be regulated either at the level of transcription or translation or intermediate levels. Because of their high rate of RNA and protein synthesis, embryonic tissues and rapidly regenerating tissues like the hemopoietic cell lines are most suitable for such investigations.

At first let me describe some features of the synthesis of ribosomal RNA and of heterogeneous RNA in embryos and then discuss whether the translation of different messenger RNA's is restricted in distinct tissues. In this case, tissue-specific factors should be involved in the translation of a distinct messenger RNA, the globin messenger for example, which should therefore not be translated in heterologous systems. On the other hand a differential transcription of RNA could take place in different embryonic stages and in different tissues. The supply of distinct messenger RNA's would control then to some extent tissue differentiation.

The different tissues of an organism are derived from "anlagen" whose differentiation potencies are less limited. The erythropoietic cells for instance arise from stem cells for the erythropoietic and the leucopoietic cell lines, and these cells again have their origin in the mesoderm anlage which probably arises from a common mesoderm-endoderm anlage. Special morphogenetic factors are involved in the determination of these anlagen. Their chemical nature and possible interference with transcriptional or posttranscriptional processes will be considered in the second part of this lecture.

The *ribosomal RNA's* constitute about 90% of the total cell RNA and can be analyzed easily. It has been shown by Brown and

Dawid [1], and by Wallace and Birnstiel [2], that in *Xenopus* oocytes ribosomal DNA is amplified many times and distributed into a large number of small nucleoli which actively synthesize ribosomal RNA. During the first meiotic division the synthesis of ribosomal RNA stops or proceeds at a very low level and the extranucleoli disappear. A rapid synthesis of ribosomal RNA is not resumed until the embryo reaches the gastrula stage [3]. Crippa has reported that an acidic protein which could be isolated from mature oocytes inhibits the synthesis of ribosomal RNA by combination with ribosomal DNA when injected into rRNA-synthesizing oocytes [4]. The synthesis of ribosomal RNA during oogenesis and early embryogenesis provides an example for gene amplification and gene regulation at the transcriptional level.

Fischberg et al. [5] have discovered a nucleolus-less mutant of *Xenopus* which in the homozygous state cannot synthesize ribosomal RNA. This mutant develops normally until the tailbud stage with the store of rRNA (and ribosomes) which were synthesized during oogenesis [6]. Obviously the synthesis of new ribosomes is not needed for early embryogenesis. The situation in the very large (and yolk-rich) amphibian oocytes, however, can not be generalized. In mammalians, where the eggs are much smaller and the embryos supplied by the maternal circulation, rRNA synthesis is resumed much earlier [7].

The *heterogeneous RNA*, which is in part a precursor of messenger RNA, seems to be more important for cell specialization. In the nuclei of amphibian and chicken embryos, as in the nuclei of other cells – *e.g.* HeLa cells and erythroblasts [8–10] – pulse-labeled non-ribosomal RNA is found in part as high molecular weight RNA with sedimentation constants up to 100 S. However, when 11 day chicken embryos were labeled for 15′ by injection of ^{3}H-uridine into the chorioallantoic vein, only a small percentage of the pulse-labeled RNA was high molecular weight RNA. By far the main part of the rapidly labeled RNA from nuclei or whole embryos sediments between 28 S and 10 S. This is also true when care was taken to exclude RNAase action during the RNA extraction as far as possible. Whether the high molecular weight RNA consists in part of noncovalently linked aggregates must be further investigated. The lower weight RNA's could be primary transcription products or they could originate by rapid segmentation of nascent RNA's which could perhaps take place when the RNA is still bound to chromatin.

When the intracellular distribution of pulse-labeled RNA was measured, about one third was found in the nuclei. When the nuclei were treated with deoxycholate to remove the nuclear as well contaminating endoplasmic membranes, a considerable part of the nuclear RNA was found in the chromatin sediment. Part of the RNA in the sediment resists extraction with 0.7 M NaCl. This RNA consists of ribosomal precursor RNA [11], but may also contain some heterogeneous RNA. When the chromatin was incubated with pancreatic deoxyribonuclease (0.125 M KCl, 0.002 M $MgCl_2$), about one third of the DNA was DNAase resistant [12]. A similar

Table 1. Percentage of DNA and of pulse-labeled RNA in chromatin and DNAase-treated chromatin of chicken embryos. 11 days chicken embryos were homogenized with 0.25 M sucrose, 0.02 M Tris, 0.125 M KCl, 0.025 M NaCl, 0.002 M $CaCl_2$, 0.001 M $MgCl_2$, 0.5% Triton-X-100, and the nuclei isolated by several centrifugations through 0.8 M and 2.05 M sucrose. The concentration of deoxycholate (DOC) was 0.5%. The amount of DNA resp. RNA in the chromatin was set 100% (arrows)

	DNA (24^h ^{14}C-thymidine)		RNA ($15'$ ^{3}H-uridine)	
	Total Cpm incorp. (corr.) (in %)	chromatin Cpm (in %)	Total Cpm incorp. (corr.) (in %)	chromatin Cpm (in %)
Nuclei (2.05 M sucrose)	95		34	
Chromatin (DOC treated)	72 →	100	24 →	100
Chromatin (DOC-0.7 M NaCl treated)	16	22	13	54
Chromatin (DOC treated) sediment:				
incub. 45′ with 20 μg DNAase/ml; 25°	22	30	22	92
incub. 90′ with 20 μg DNAase/ml; 25°	19	26	21	88
Chromatin (DOC-0.7 M NaCl treated) sediment:				
incub. 3′ with 20 μg DNAase/ml; 25°	3	4	14	58

part of thymus chromatin-DNA was found to be resistant to DNAase [13]. Most of the pulse-labeled RNA is found in the DNAase-resistant sediment (Table 1). This suggests that this RNA is in part nascent RNA which is bound to chromatin in a DNAase-resistant complex. But other explanations are not excluded. The DNAase-resistant chromatin fraction contains also histones. A part of them may be released from the degraded chromatin and then adsorbed to the DNAase-resistant chromatin. When chromatin was extracted with 0.7 M NaCl and then incubated with DNAase for a short time, 96% of the DNA is degraded.

The part of the heterogeneous chicken RNA which is transferred from the nucleus to the cytoplasm appears, as KNÖCHEL has shown in our laboratory, in the cytoplasm as ribonucleoprotein particles, called informosomes by SPIRIN. They are transported into the cytoplasm independently of ribosomal subunits [14]. Only a small portion of these particles is bound to the small ribosomal subunit. The incorporation of the RNP-particles into polysomes is a slow process as compared to transcription. When in chicken embryos the transcription of RNA was completely stopped by actinomycin D, the incorporation of RNP-particles into polysomes proceeds for about 60 min as KNÖCHEL [15] has found. A similar observation was made with HeLa cells by PENMAN et al. [11].

The existence of informosomes has been questioned. It was argued that the ribonucleoprotein particles are formed artificially from protein and RNA during homogenization. However there is no argument against the assumption that these proteins bind to RNA already in the cell under physiological conditions. Further proof for the physiological role of the informosomes has been provided by SPIRIN [16].

Whether an RNA is actually a messenger can only be shown by its function in a protein synthesizing system and the analysis of the protein product. A messenger which can now easily be obtained is the messenger for globin chains, which was isolated for the first time by CHANTRENNE, MARBAIX, and collaborators from rabbit erythroblasts [17, 18]. This messenger has been added to heterologous protein synthesizing systems to show whether cell-specific factors are needed for its translation. When the messenger is injected into oocytes or unfertilized eggs of *Xenopus*, rabbit globin is synthesized [19]. The messenger is also translated in a cell free extract from Ehrlich ascites tumor cells [20–22]. It could be argued that in oocytes or tumor cells translation is less restricted than in normal somatic cells. In our laboratory D. HENDRICK has therefore compared the translation of globin messenger RNA in cell free systems from Ehrlich ascites tumor cells, chicken embryo liver and chicken embryo brain. The extracts, prepared from homogenates by centrifugation at 30000 rpm, were preincubated to reduce the endogenous protein synthesis and then incubated with rabbit globin messenger. The result is shown in Table 2. The stimulation by the added globin messenger is the largest in the ascites tumor cell extract, intermediate in the chick embryo brain extract, whereas the smallest stimulation is obtained in chick embryo liver extract. This must however not indicate the degree of tissue specificity. The degree of stimulation by globin messenger is correlated to the reduction of endogenous protein synthesis after preincubation of the tissue extracts (Table 2). In embryo liver, where the lowest percentage of stimlation by globin messenger is observed, the endogenous protein synthesis is the least reduced by preincubation. The radioactive protein which was synthesized was analyzed after addition of unlabeled rabbit globin chains by gel filtration on sephadex G 100 followed by C-M cellulose chromatography (mod-

Table 2. Translation of rabbit globin messenger in heterologous systems. The extracts were prepared by centrifugation at 30000 rpm and filtration through sephadex G 25. 50 μl assays contained 0.4 A_{260} units of extract, 80 mM K^+, 3 mM Mg^{++} and 20 mM Tris pH 7.5

Extracts of	Cpm		% Stimulation	Endog. incorp. Leucine not preincub. / Endog. incorp. Leucine preincub.
	Endogenous incorporation	+ 20 μg Globin mRNA/ml		
Ehrlich ascites cells	173	3475	1900	15
Chick embryo liver	3340	4900	47	2,7
Chick embryo brain	3220	13100	305	4

ified after DINTZIS [23] as shown in Fig. 1). When globin messenger was added, the ^{3}H-leucine label follows the absorbance of the α- and β-chains, whereas in the control without added globin messenger no label is found in the chains. To prove definitely that the label is

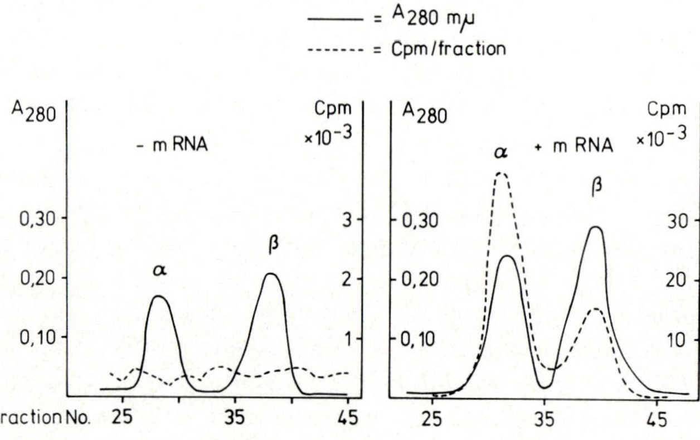

Fig. 1. Synthesis of ^{3}H-leucine labeled α- and β-globin chains in extracts from 11 day chicken embryo brain after addition of rabbit globin messenger. Unlabeled rabbit globin chains were added before chromatography on carboxymethyl-cellulose

actually incorporated in newly synthesized globin chains, the isolated chains were degraded by trypsin and the peptides separated chromatographically. The ^{3}H-leucine label follows the elution pattern of the leucine-containing globin peptides.

The heterologous translation of rabbit globin messenger in chicken embryo brain is especially interesting because the anlagen for erythroblasts, which are of mesodermal origin, and for brain, which is of ectodermal origin, are already separated at the first steps of determination in very early stages of embryogenesis. The results indicate that the factors which are involved in translation have a limited specificity[1]. Furthermore it was shown that within the limits of the sensitivity of the mRNA test, free globin mRNA-

[1] The results are compatible with experiments of ANDERSON, OCHOA, STAEHELIN and their collaborators which have shown that isolated initiation factors have no tissue specificity [25—27]. A higher specificity has been observed with some viral RNA [24].

protein complexes are not found in other tissues except the erythropoietic ones [23a].

The synthesis of globin chains depends primarily on the availability of the globin messenger. This may suggest that in different tissues different populations of messenger RNA's are synthesized by selective transcription or the transcribed RNA's may be selectively stabilized. It would therefore be of interest to analyze whether in different developmental stages, or in different organs, different populations of RNA are found. Such investigations have been carried out by several investigators by competitive DNA-RNA hybridization. Differences in the non-ribosomal RNA population of different organs have been observed by PAUL [27a]. RNA preparations from different stages of loach embryos differ in their composition [28]. In *Xenopus* tailbud stages, RNA species are present which are not yet transcribed in the gastrula stage [29]. In similar experiments with RNA transcribed from chromatin of different developmental stages of sea urchin embryos, even larger differences in hybridization have been observed [30].

The interpretation of these hybridization experiments is hampered by the fact that only more or less reiterated DNA sequences, or DNA sequences to which a large number of copies of complementary RNA exist in the RNA mixture, are hybridized under the conditions employed [31]. Therefore a differential transcription of repeated DNA sequences is preferentially observed in these experiments.

Several different RNA polymerases [32] have been extracted from *Xenopus* oocytes and embryos and their activities measured. It was found that these enzymes are present in large excess and that they do not limit transcription [33].

The examples I have given above suggest that the synthesis of tissue-specific proteins is primarily controlled at the level of transcription or of RNA processing. This does of course not exclude that translation is also regulated. For instance, the ratio of α- to β-globin chains which are synthesized varies in different cell-free systems. An impressive example for translational control has been observed in sea urchin eggs. After fertilization, protein synthesis increases rapidly. This may depend on a modification of the ribosomes which more easily dissociate after fertilization [34] so that pre-existing messengers may now be translated more efficiently.

When the differentiation is primarily controlled at the level of transcription or of the further processing of RNA in the nucleus, the question can be asked whether the nuclei specialize early in development. Nuclear transplantation experiments which have been carried out in amphibians by Briggs and King, by Fischberg et al., as well by Gurdon [35–37] and in insects *(Drosophila melanogaster)* by Illmensee [38], have shown that in early developmental stages nuclei taken from cells which are already determined to a special pathway in development are still omnipotent. Obviously cytoplasmic constituents play an important role in tissue determination. On the other hand the reported experiments suggest a differential RNA transcription. This could be explained if cytoplasmic factors, which are distributed very early in development in a distinct pattern, are in part (reversibly) incorporated into the nuclei which come to lie in particular regions during cleavage. Recently some evidence has been obtained that morphogenetic factors, the so-called inducing factors, are involved in nuclear-cytoplasmic interactions.

Fifty years ago the discovery was made by Hans Spemann and Hilde Mangold [39] that the mesodermal upper blastoporal lip of amphibians induces muscle and notochord as well as neural tube and brain when transplanted into the ventral ectoderm of an early gastrula[2]. This shows that the blastoporal lip has an inducing capacity and that the ectoderm of early gastrula stages can still be channeled into different pathways of differentiation.

Anlage plan and developmental potentials in the early amphibian gastrula are outlined in Fig. 2 [40]. The anlagen for mesodermal and endodermal tissues are located in the vegetal half, the anlagen for ectodermal tissues – the epidermis and the neural system – in the animal half (Fig. 2a). In Fig. 2b the developmental fate of different regions is shown when they were isolated and cultured in salt solution. The anlagen for mesoderm and endoderm are determined already in the morula and blastula stage [41–43]. The presumptive anlagen for epidermis and the neural system, on the other hand, form loose complexes of cuboidal cells when isolated and cultured in salt solution. Recently Grunz has shown by electronmicroscopy [44] that this so called undifferentiated ectoderm

[2] In normal embryogenesis, the brain and neural tube develop in the dorsal ectoderm.

forms ciliated epithelial cells with a characteristic vacuolous zone. The cells cannot be distinguished from the cells of epidermis except for the alignment of the cells which depends on the mesenchyme which underlies the epidermis [45].

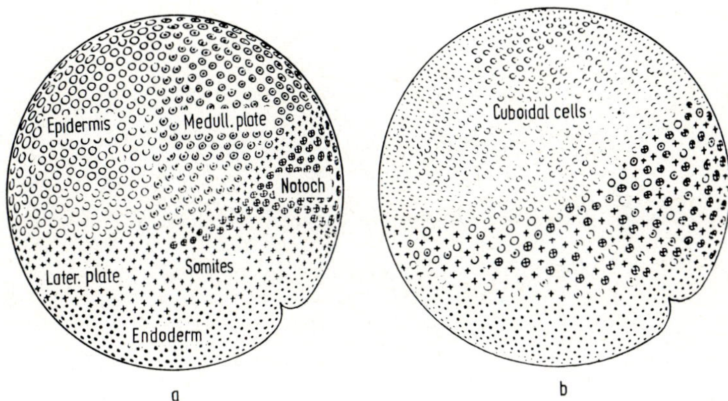

Fig. 2. Anlage plan and developmental potentials in the early amphibian gastrula; side view of a gastrula. a) Differentiation of the prospective regions in normal development; b) differentiation obtained when small pieces of the areas were isolated and cultured in salt solution. From [40]

Obviously the ectoderm of early amphibian gastrulae is already determined to form epidermal cells, but the developmental program of these cells can still be changed to form the neural anlage or tissues derived from mesoderm or endoderm after addition of appropriate inducing factors. In this respect the change of determination of the amphibian ectoderm is related to transdetermination in insects which was discovered some years ago by Hadorn [46].

Because the differentiation of amphibian ectoderm can still be channeled into different pathways, it can be used as a test system for morphogenetic factors. These factors can either be tested on isolated ectoderm or by the implantation method on whole gastrulae [47, 48]. A pellet which contains a certain amount of inducing factor is implanted into the blastocoel of early gastrula stages and by the gastrulation movements comes into contact with the ventral ectoderm where it can exert its inducing action.

The so-called "vegetalizing factor" which induces mesodermal as well endodermal tissues has been obtained from guinea pig bone marrow by YAMADA [49] and in a highly purified form from 11 day old chicken embryos in our laboratory [50–52]. The purification procedure is briefly outlined in Fig. 3. The final purification was

Homogenate of chick embryo (11 day) trunks
centrifuged at 19000 g
Ammonium sulphate precipitation (0·40 S)

Extraction of proteins from supernatant
with phenol

Chromatography of proteins on carboxymethylcellulose (6 M urea)
|
Electrophoresis on Sephadex G 100 (6 M urea)

Electrofocusing (pH 3—10)
|
Electrofocusing (pH 7—9)

Fig. 3. Purification of the vegetalizing inducing factor. From [72]

achieved by electrofocusing. Most of the protein is located in the pH 5 region, whereas most of the inducing activity is found at pH 7.5–8.5 in a region which has no absorbance at 280 mμ. When iodinated protein was subjected to electrofocusing to increase the sensitivity of protein analysis, a number of peaks is seen in this region (Fig. 4). The ^{125}I activity profile is of course not directly proportional to protein concentration because the degree of iodination and the tyrosine content differ in the protein fractions. A large amount of chromatographically and electrophoretically purified factor was then repeatedly focused in the pH 7–9 region. An absorbance peak is now observed at about pH 8 where the factor has its isoelectric point. Approximately 40 μg of inducing protein are obtained from 1 kg of chicken trunks. The factor is purified about 10^6-fold. About 1×10^{-4} μg per gastrula of the most highly purified factor cause inductions in about 50% of the cases when tested by the implantation method. The molecular weight in 6 M urea is about 35000 [52][3].

[3] The neuralizing factor which is very probably also protein in nature has not yet been obtained in a highly purified state.

The vegetalizing factor induces, when tested on whole gastrulae by the implantation method, organs which are derived from the mesoderm (muscle, notochord, renal tubules, blood cells). In addition, in isolated ectoderm the factor also induces tissues which are derived from endoderm (endodermal epithelium, intestine, liver, etc. [53, 54]). All these tissues arise by secondary interactions from a joint anlage.

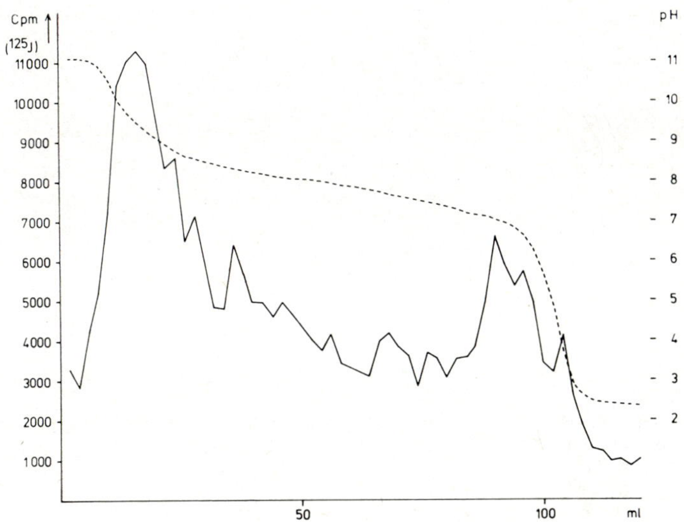

Fig. 4. Electrofocusing of an iodinated protein fraction containing the vegetalizing inducing factor

If a similar factor acts in normal development, it must be present in amphibian embryos. Faulhaber [55] in our laboratory has extracted vegetalizing factor together with neuralizing factor from cleavage as well from gastrula stages of *Xenopus*. Because of the small amount of material which is available, the purification is very difficult and only crude fractions have been obtained. The vegetalizing factor in these fractions is protein in nature and bands in electrofocusing between pH 6–8 [56].

In chicken embryos most of the vegetalizing factor is bound to cellular structures, especially endoplasmic membranes. The vegetalizing factor is also present in the 105000 $\times$ g supernatant, but in an inactive form. It can be activated either by electrophoresis or by extraction with phenol (Fig. 5). The active factor can be isolated from the phenol layer together with other proteins, whereas an

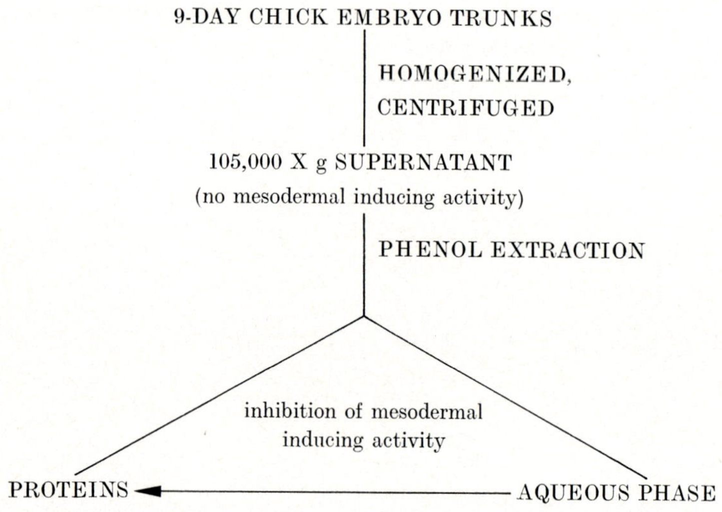

Fig. 5. Activation of the vegetalizing factor in the supernatant of chicken embryo homogenates by extraction with phenol and reinhibition by recombination with the aqueous phase. After [48]

inhibitor has been found in the aqueous layer. When the aqueous layer is combined with the inducing factor in presence of 6 M urea and the mixture dialyzed, the inducing factor is inhibited. With the combination test the inhibitor has been partially purified (Fig. 6). The partially purified inhibitory fraction consists of glycoproteins and mucopolysaccharides. About 90% of the fraction is carbohydrates and about 10% proteins. The acidic amino acids as well as serine and threonine prevail in the protein moiety [57]. The inhibitor is inactivated after treatment with preincubated pronase (but not by trypsin) [58], as well endo-N-hexosaminidase. This suggests that the protein as well the carbohydrate portion is needed for inhibitory activity.

The inhibitor is not found in the nuclei and has no affinity to DNA. It is therefore unlikely that the inhibitor acts as a gene repressor. It may rather regulate the level of free inducing factor by forming an inducer-inhibitor complex.

In gastrula ectoderm of amphibians, some vegetalizing and neuralizing factor are present in a masked form [59]. It is however not known whether amphibian gastrulae contain an inhibitor with properties similar to the chicken inhibitor. An inhibitor has not been found in the 105000 × g supernatant from the homogenate of *Xenopus gastrula*, but the amount may be too small to be detected.

Homogenate of chick embryo (11 day) trunks
centrifuged at 105000 g (2 h)

|

Extraction of supernatant with phenol
constituents of aqueous layer precipitated with 2 vol. ethanol
and dissolved in 0.1 M NaCl

|

centrifugation at 116000 g

|

chromatography of supernatant on hydroxyapatite

|

incubation of not-adsorbed-peak with pancreatic ribonuclease,
gel filtration on sephadex G 50 (6 M urea),
(inhibitor eluted in exclusion peak)

Fig. 6. Purification of an inhibitor for the vegetalizing factor. After [73]

For a further analysis of the mechanism of embryonic determination, experiments with metabolic inhibitors have been undertaken [60]. When actinomycin D, which stops DNA-dependent RNA synthesis, was applied to embryonic tissues, a differential sensitivity was observed. In explants from gastrula stages, newly determined tissues like the neural anlage are more sensitive than the mesodermal tissues which are determined much earlier in development (Fig. 7) [61, 62]. However, when mesodermal tissues are newly induced in gastrula ectoderm by addition of vegetalizing factor, this induction is also very sensitive to actinomycin D as Toivonen, Vainio, and Saxén have shown [63]. In sea urchin embryos the process of skeleton formation is extremely sensitive to actinomycin D treatment [64]. Bromodeoxyuridine (BUdR) substitutes thymidine in DNA. Before gastrulation amphibian embryos are very sensitive to treatment with BUdR when this drug is injected into cleaving eggs as was shown by TENCER and BRACHET [65]. It was suggested by these authors that BUdR causes a change in the chromosome structure.

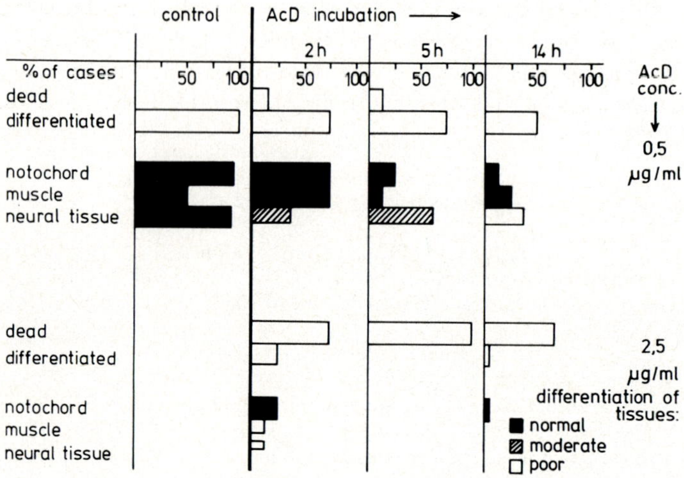

Fig. 7. Differentiation of blastoporal lip and ectoderm of early gastrula stages (Amphibia) after incubation with 0.5 µg/ml and 2.5 µg/ml actinomycin D for 2 to 14 hours]

The experiments with actinomycin D do not allow a conclusion that the inducing factors act *directly* at the level of transcription or RNA processing. However, an observation which favors such a hypothesis was made when different polyanions were tested for their ability to inhibit the vegetalizing factor [66]. It was found that chicken embryo DNA partially inhibits this factor (Table 3). Equivalent amounts of the crude fraction as well the highly purified factor were inhibited to the same extent. The small amount of neuralizing factor which is still present in crude preparations is not inhibited. Single-stranded DNA and sonicated DNA have the same inhibitory activity as high molecular weight double-stranded DNA. It is however not excluded that a partial renaturation occurs under the conditions of the test. DNA from T4 coliphages also inhibits the vegetalizing factor. This seems to indicate that the specificity of the DNA is not very high, but *in vivo* the specificity of interaction with chromatin could perhaps be enhanced by other proteins.

The inhibition could result from binding of the factor to DNA. To test this hypothesis, ^{3}H-acetyl-labeled factor was combined with sonicated DNA and centrifuged in a sucrose gradient [66].

Table 3. Inhibition of the vegetalizing factor by different polyanions.[a] Statistical significance (calculated by χ^2 method): $p \quad 0.03 - < 0.001$. (After [66])

Amount combined with 2.5 mg inducing protein fraction	400 μg			700 μg		
	No. of cases	Inductions, %		No. of cases	Inductions, %	
		Neural	Trunk-tail		Neural	Trunk-tail
Inhibitory fraction	29	7	**14**	57	0	**5**
tRNA (yeast)	—	—	—	22	23	**77**
DNA (chick)	83	32	**53**[a]	61	39	**25**[a]
Melted DNA (chick)	46	13	**50**[a]	18	17	**22**[a]
Sonicated DNA (chick)	—	—	—	37	27	**24**[a]
DNAase-treated DNA (chick)	—	—	—	30	0	**77**
Poly d(A—T)	40	15	**88**	19	0	**68**
Poly I · poly C	25	16	**84**	23	35	**61**
Control:	No. of cases			Neural	Ind. %	Trunk-tail
Inducing protein fraction	53			15		72

A part of the factor migrates with the DNA at low ionic strength ($I = 0.02$). In the control experiment without DNA, the inducer sediments as a broad slower sedimenting peak which consists of polymers of different size (Fig. 8). At higher ionic strength (as used in the inhibition experiments) it is difficult to determine by centrifugation which portion of the factor is bound because DNA and factor form aggregates which in part sediment already after short centrifugation. At low ionic strength the binding of basic or slightly basic proteins to DNA is thought to be a more or less nonspecific ionic interaction. But even under these conditions, basic proteins are not indiscriminately bound. As shown in Fig. 9, ^{3}H-acetyl-α-chymotrypsin, a slightly basic protein of about the same size as the vegetalizing factor, is not bound. It is, however, difficult to draw definite conclusions from binding experiments only, because proteins which are unrelated to DNA may also bind to DNA (for instance pro-collagen has been reported to be bound to denatured DNA [67]).

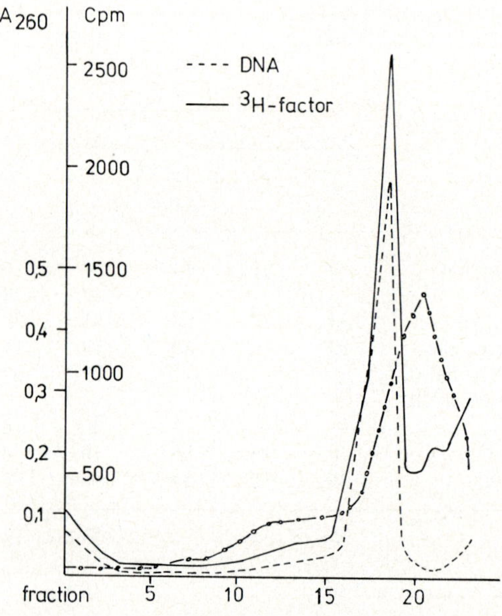

Fig. 8. Sedimentation of labeled vegetalizing factor (————) with sonicated DNA (– – – – –). In the control run without DNA the (aggregated) factor (—•—•—•—) sediments more slowly. About 1000 μg chick DNA and less than 1 μg factor were combined in 6 M urea, dialyzed, layered on a 5—20% sucrose, 0.01 M NaCl, 0.01 M Tris pH 7.5 gradient and centrifuged 5.5 h at 40000 RPM (Spinco SW 41 rotor). Sedimentation is from right to left

Furthermore, it should be considered that heterogeneous RNA may act as the physiological receptor in the target cells. That is, if the factor has an affinity to single-stranded DNA, it may also have an affinity to RNA which is transcribed from the complementary strand of DNA[4]. In this way the vegetalizing factor could be involved in the selection of certain parts of the transcribed RNA for translation.

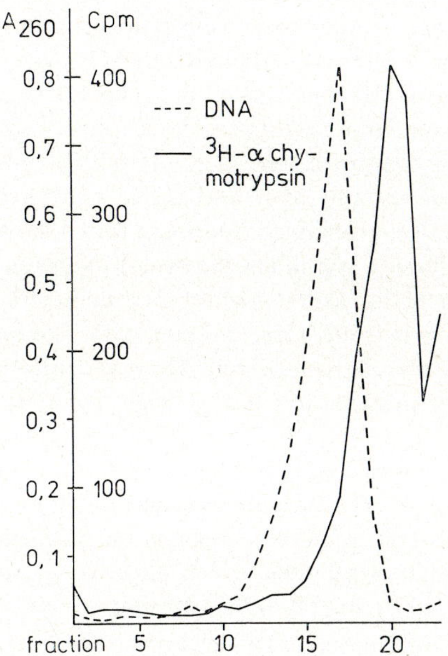

Fig. 9. Sedimentation of labeled α-chymotrypsin (————) and sonicated DNA (– – – – – –). 1000 μg chick DNA and 1 μg α-chymotrypsin were combined and centrifuged as described in Fig. 8

Besides its action at the level of transcription or processing of informational RNA, quite other mechanisms are conceivable and must be considered. The inducer could change the ion permeability of cytoplasmic membranes. However, no change of K^+ and Cl^- and only a slight increase, if any, in the concentration of Na^+ was observed 5 hours after induction with the vegetalizing factor [68]. An increase in the exchange of intracellular sodium and calcium with extracellular ions occurs in the gastrula and neurula stage,

[4] *Note added in proof:* Further experiments have shown that a crude RNA fraction obtained by phenol extraction (60°, without SDS) from nuclei of *Xenopus* gastrulae inhibits the vegetalizing factor. Ribosomal RNA has no inhibitory activity.

respectively, which is much later than the determination of the mesoderm-endoderm anlage [69]. Differentiation certainly depends on the proper ionic environment, but it is difficult to see how ions could directly control differentiation.

Binding of the factors to cytoplasmic membranes could lead to an activation of membrane-bound adenylcyclase and the release of 3′5′-cyclic adenylate as a second messenger. Cyclic AMP as well monobutyryl- and dibutyryl-cyclic-AMP were tested on gastrula ectoderm in the concentration range of $10^{-2}/10^{-4}$ M [70]. In some explants a small amount of neural tissue was formed. The main part of the ectoderm remained undifferentiated, however. The possibility that different low molecular weight substances, which are synthesized on endoplasmic membranes under the direction of certain morphogenetic factors, interact with different sites of the genome should also be considered. However, attempts to isolate small molecular weight substances with inducing activity have failed so far.

After induction by the vegetalizing factor, the mutual affinities of the cells change. The induced ectoderm acquires an affinity to endoderm [71]. This probably depends on biochemical processes in the cytoplasmic membranes which seem not to be the cause but rather a consequence of cell differentiation.

In conclusion, experiments which approach the molecular mechanism of determination under different aspects support the hypothesis that the regulation of RNA transcription or the processing of informational RNA is most important. It is, however, not yet possible to correlate single steps in differentiation with certain biochemical events, *i.e.*, a selection of certain genes for transcription, a modification of chromatin structure in a more general way, a selection and stabilization of certain parts of the transcribed RNA, or perhaps a modification of DNA or chromosomal proteins.

These questions cannot all be solved on the same object. But by a combination of the appropriate methods it should be possible to answer these questions.

Acknowledgement

The investigations included in the summary, which were carried out in our laboratory, were supported by the Deutsche Forschungsgemeinschaft (Sonderforschungsbereich Embryonale Entwicklung und Differenzierung).

References

1. Brown, D. D., Dawid, I. B.: Science **160**, 272 (1968).
2. Wallace, H., Birnstiel, M. L.: Biochim. biophys. Acta **114**, 296 (1966).
3. Brown, D. D., Littna, E.: J. molec. Biol. **8**, 688 (1964).
4. Crippa, M.: Nature (Lond.) **227**, 1138 (1970).
5. Elsdale, T. R., Fischberg, M., Smith, S.: Exp. Cell Res. **14**, 642 (1958).
6. Gurdon, J. B., Brown, D. D.: J. molec. Biol. **12**, 27 (1965).
7. Woodland, H. R., Graham, C. F.: Nature (Lond.) **221**, 327 (1969).
8. Scherrer, K., Latham, H., Darnell, J. E.: Proc. nat. Acad. Sci. (Wash.) **49**, 240 (1963).
9. Scherrer, K., Marcaud, J.: J. Cell. Comp. Physiol. **72**, *Suppl.* **1**, 181 (1968).
10. Georgiev, G. P., Samarina, O. P., Lerman, M. I., Smirnov, M. N., Severtzov, A. N.: Nature (Lond.) **200**, 1291 (1963).
11. Penman, Sh., Vesco, C., Penman, M.: J. molec. Biol. **34**, 49 (1968).
12. Knöchel, W., Born, J., Tiedemann, H.: unpublished experiments (1972).
13. Mirsky, A. E.: Proc. nat. Acad. Sci. (Wash.) **68**, 2945 (1971). — Clark, R. J., Felsenfeld, G.: Nature (Lond.) New Biol. **229**, 101 (1971).
14. Knöchel, W., Tiedemann, H.: Biochim. biophys. Acta **269**, 104 (1972).
15. Knöchel, W.: Dissertation, Freie Universität Berlin 1972.
16. Spirin, A. S.: Europ. J. Biochem. **10**, 20 (1969).
17. Chantrenne, H., Marbaix, G., Burny, A.: Prog. Nucleic Acid Res. Molec. Biol. **7**, 173 (1967).
18. Huez, G., Burny, A., Marbaix, G., Lebleu, B.: Biochim. biophys. Acta **145**, 629 (1967).
19. Gurdon, J. B., Lane, C. D., Woodland, H. R., Marbaix, G.: Nature (Lond.) **233**, 177 (1971).
20. Mathews, M. B., Korner, A.: Europ. J. Biochem. **17**, 328 (1970).
21. Mathews, M. B., Osborn, M., Lingrel, J. B.: Nature (Lond.) New Biol. **233**, 206 (1971).
22. Metafora, S., Terada, M., Dow, L. W., Marks, P. A., Bank, A.: Proc. nat. Acad. Sci. (Wash.) **69**, 1299 (1972).
23. Dintzis, H. M.: Proc. nat. Acad. Sci (Wash.) **47**, 247 (1961).

23a. Hendrich, D., Knöchel, W.: unpublished experiments (1973).

24. Lebleu, B., Nudel, U., Falcoff, E., Prives, G., Revel, M.: Febs Letters **25**, 97 (1972).
25. Crystal, R. G., Nienkuis, A. W., Prichard, P. M., Picciano, D., Elson, N. A., Merrick, W. C., Graf, H., Shafritz, D. A., Laycock, D. G., Last, J. A., Anderson, W. F.: Febs Letters **24**, 310 (1972).
26. Zasloff, M., Ochoa, S.: Proc. nat. Acad. Sci. (Wash.) **69**, 1796 (1972).
27. Schreier, M. H., Staehelin, T.: Proc. nat. Acad. Sci. (Wash.) **70**, 462 (1973).

27a. Paul, J.: Curr. Topics Develop. Biol. **5**, 317 (1970).

28. Kafiani, C.: Advan. Morphogenesis **8**, 209 (1970).
29. Jantzen, H., Gasc, J.-M., Tiedemann, H.: Hoppe-Seylers Z. physiol. Chem. **351**, 579 (1970).

30. CHETSANGA, C. J., POCEIA, D. L., HILL, R. J., Doty, P.: Cold Spr. Harb. Symp. quant. Biol. **35**, 629 (1970).
31. BRITTEN, R. J., KOHNE, D. E.: Science **161**, 529 (1968).
32. ROEDER, R. G., RUTTER, W. J.: Proc. nat. Acad. Sci. (Wash.) **65**, 675 (1970).
33. ROEDER, R. G.: Carnegie Inst. Year Book **71**, 17 (1972).
34. MAGGIO, R., VITTORELLI, M. L., CAFFARELLI-MORMINO, I., MONROY, A.: J. molec. Biol. **31**, 621 (1968).
35. BRIGGS, R., KING, T. J.: J. Morph. **100**, 269 (1957).
36. FISCHBERG, M., GURDON, J. B., ELSDALE, T. R.: Exp. Cell Res. Suppl. **6**, 161 (1958).
37. GURDON, J. B.: J. Embryol. exp. Morph. **10**, 622 (1962).
38. ILLMENSEE, K.: Wilhelm Roux Arch. Entwickl.-Mech. Org. **171**, 331 (1973).
39. SPEMANN, H.: Experimentelle Beiträge zu einer Theorie der Entwicklung. Berlin: Springer 1936.
40. HOLTFRETER, J., HAMBURGER, V.: In: WILLIER, B. H., WEISS, P. A., HAMBURGER, V., (Eds.): Analysis of Development. Philadelphia: W. B. Saunders 1955.
41. TIEDEMANN, HILDEGARD: unpublished experiments (1970).
42. NIEUWKOOP, P. D.: Wilhelm Roux' Arch. Entwickl.-Mech. Org. **166**, 105 (1970).
43. NAKAMURA, D.: Ann. Embryol. Morphogenese Suppl. **1**, 261 (1969).
44. GRUNZ, H.: Wilhelm Roux' Arch. Entwickl.-Mech. Org., in press.
45. HOLTFRETER, J.: Arch. exp. Zellforsch. **23**, 169 (1939).
46. HADORN, E.: Brookhaven Symp. Biol. **18**, 148 (1965).
47. MANGOLD, O.: Wilhelm Roux' Arch. Entwickl.-Mech. Org. **100**, 198 (1924).
48. TIEDEMANN, H.: J. Cell. Comp. Physiol. **72**, Suppl. 1, 129 (1968). (Schematic representation of the test methods).
49. YAMADA, T.: J. Cell. Comp. Physiol. **60**, Suppl. 1, 49 (1962).
50. GEITHE, H. P., TIEDEMANN, H., TIEDEMANN, H.: Biochim. biophys. Acta **208**, 157 (1970).
51. BORN, J., GEITHE, H. P., TIEDEMANN, H., TIEDEMANN, H.: Hoppe-Seylers Z. physiol. Chem. **353**, 1075 (1972).
52. TIEDEMANN, H.: In: WEBER, R. (Ed.): The Biochemistry of Animal Development III. New York: Academic Press, in press.
53. KOCHER-BECKER, U., TIEDEMANN, H.: Nature (Lond.) **233**, 65 (1971).
54. GRUNZ, H.: unpublished experiments (1971).
55. FAULHABER, I.: Wilhelm Roux' Arch. Entwickl.-Mech. Org. **171**, 87 (1972).
56. FAULHABER, I., GEITHE, H. P.: Rev. Suisse Zool. **79**, 103 (1972).
57. BORN, J., TIEDEMANN, H., TIEDEMANN, H.: Biochim. biophys. Acta **279**, 175 (1972).
58. BORN, J., TIEDEMANN, H., TIEDEMANN, H.: J. Embryol. exp. Morph. **28**, 77 (1972).
59. TIEDEMANN, H., BECKER, U., TIEDEMANN, H.: Embryologia (Nagoya) **6**, 204 (1961).

60. Denis, H.: Develop. Biol. **9**, 473 (1964).
61. Tiedemann, H., Born, J., Tiedemann, H.: Z. Naturforsch. **22** b, 649 (1967).
62. Duprat, A. M., Zalta, J. P., Beetschen, J. C.: Exp. Cell Res. **43**, 358 (1966).
63. Toivonen, S., Vainio, T., Saxén, L.: Rev. Suisse Zool. **71**, 139 (1964).
64. Giudice, G., Mutoto, V., Donatuti, G.: Wilhelm Roux' Arch. Entwickl.-Mech. Org. **161**, 118 (1968).
65. Tencer, R., Brachet, J.: Differentiation **1**, 51 (1973).
66. Tiedemann, H., Born, J., Tiedemann, H.: Wilhelm Roux' Arch. Entwickl.-Mech. Org. **171**, 160 (1972).
67. Tsai, R. L., Green, H.: Nature (Lond.) New Biol. **237**, 171 (1972).
68. Grunz, H.: unpublished experiments (1972).
69. Barth, L. G., Barth, L. C.: Develop. Biol. **28**, 18 (1972).
70. Grunz, H., Tiedemann, H.: unpublished experiments (1972).
71. Kocher-Becker, U., Tiedemann, H., Tiedemann, H.: Science **147**, 167 (1965).
72. Tiedemann, H.: Symp. Soc. exp. Biol., Cambridge, at the university press, **25**, 223 (1971).
73. Tiedemann, H.: In: Advan. Biosciences **6**, 477 (1971).

Discussion

C. A. Thomas (Boston): Thank you very much for this interesting paper. I'm sure that there will be many questions.

Zachau: In the binding of your factor to DNA, you have used homologous DNA. What about binding and inhibition with heterologous DNA?

H. Tiedemann (Berlin): With heterologous DNA — we have used T 4 phage DNA — there is also some inhibition.

Unidentified speaker: In bacteria, there are two protein factors bound to the membrane that regulate cell division. Their molecular weight is around 35000 daltons. Is there any relationship between these factors and the ones you mentioned?

H. Tiedemann: I do not believe that there is any relation between these mitosis regulating factors and the factors we have studied. They behave differently against actinomycin D.

C. A. Thomas: I have a question regarding the bioassay. It would be important to know if only those cells very close to the transplant respond to the inducing factor or if also very remote cells would undergo differentiation. In that case one would have to explain whether these cells carry their own inducer, or whether they get a small amount of it in the distance.

H. Tiedemann: It is our impression that the factor has really to penetrate into the cell. We have done some experiments in which the factor was bound to bromoacetyl-cellulose; this completely abolishes the activity.

Chr. Scholtissek (Gießen): What is the evidence that the factor has really to enter the cell? Could it act from the outside, by absorption to the cell surface?

H. Tiedemann: I think the experiment just mentioned is quite conclusive. We have bound the factor to bromoacetyl-cellulose according to the procedure of Sela. This preparation shows no activity, although it can come in contact with the cell surface. Secondly, in preliminary experiments Dr. Grunz has prepared labeled factor, and radioautographs show that the factor really enters the cell.

Transcription I

Formation and Regulation of Messenger RNA in Animal Cells

Klaus Scherrer

Molecular Biology Department, Swiss Institute for Experimental Cancer Research, Lausanne, Switzerland

With 4 Figures

Introduction

For many years the problem of messenger RNA (mRNA) formation in animal cells has resisted the efforts of investigators to comprehend this basic mechanism of cellular life. Although the discovery of messenger RNA in bacterial cells [1] touched off almost immediately the search for its analogue in animal cells, the basic pattern of its biogenesis in eukaryotic cells was only recently clarified. However, at a relatively early stage of this investigation, a correct interpretation of this fundamental mechanism of gene expression in higher organisms was possible. The basic observations concerning RNA synthesis in animal cells [2–8], cytogenetic data and theoretical considerations led us to propose at that time a pattern of mRNA synthesis which differed fundamentally from that observed in bacteria [5, 6, 9–12]. Nevertheless, much more precise knowledge had to be acquired to permit a comprehensive account of the life history of a specific messenger RNA from transcription to translation [13].

Three recent review articles deal with some basic aspects of messenger RNA formation in eukaryotic cells. The "life history" of duck globin messenger RNA was described in the Karolinska Symposium 1973 [13]. The regulative aspects of intracellular information transfer were dealt with in a lecture given at the Oholo Symposium 1973 [14] in which the "cascade regulation hypothesis" [10, 12] was reviewed. Synthesis and turnover as well as the structure of nuclear pre-mRNA were discussed in a conference at the Fogarty Symposium 1973 [15]. Thus, we shall restrict ourselves in this lecture to review the basic pattern of messenger RNA formation and regulation, referring to the corresponding original papers as well as to the three articles cited above.

The Precursor-Product Relationship of Giant Nuclear RNA and Polyribosomal Messenger RNA

That eukaryotic giant nuclear RNA with sedimentation values of more than 50 S represents in fact a precursor to the messenger RNA translated in polyribosomes could be concluded on the basis of the following early evidence:

1. Even the earliest (technically still deficient) analysis of RNA synthesis demonstrated that more than 90% of the transcription products of animal cells consists of RNA molecules with molecular weights of $1-20 \times 10^6$, considerably larger than the polyribosomal messenger RNA ($1-10 \times 10^5$ daltons). In avian erythroblasts which synthesize haemoglobin to the extent of 80–90% of total protein synthesis, the corresponding 9 S globin mRNA could not be identified among the nuclear RNA molecules, but was found only in polyribosomes [5, 6, 9, 10, 16].

2. In duck erythroblasts about 10% of the genome, corresponding to about 10^5 times more DNA than that of the three globin genes, was found to be transcribed into giant RNA [10], but more than 30% of the polyribosomal mRNA consists of the 9 S globin messenger [6, 9, 10, 17]. Thus, if the bulk of RNA synthesized in the nucleus had any relation to the messenger RNA accumulating in the cytoplasm, a dramatic reduction in size and in mass of the transcription products had to be postulated.

3. Early cytogenetic evidence showed that the genetically and morphologically characterized bands in polytene chromosomes of diptera correspond to units of transcriptional regulation. The amount of DNA in these bands (dividing the haploid DNA content by the number of bands) permits the calculation of the amount of DNA in the transcriptional unit: it corresponds by order of magnitude to the size of the giant RNA found in all animal cells [5].

4. Early observations in Edström's laboratory [18] indicated that giant chromosomes of *Chironomus* with single puffs synthesize giant RNA similar in size to that found in erythroblasts. In this case the transcript of genetically mapped genes was involved, *i.e.*, a type of nuclear RNA that necessarily carries the information of

mRNA. Thus, the logical conclusions were that this giant chromosomal RNA contained the messenger RNA translated in polyribosomes, and that this giant nuclear RNA corresponded to the obligatory transcription products of regulational units in chromosomes [5, 6, 10].

5. The pattern of ribosomal RNA synthesis already known in its basic form at that time [2–4, 19] showed that in the case of an identifiable gene product, the cell operates according to a particular mechanism: a giant RNA molecule is produced first; later the piece of functional RNA needed to form ribosomes is excised from the precursor by a complicated multi-step process.

There was no logical reason to believe that the above pattern is restricted to ribosomal RNA. This processing procedure, *i.e.*, to produce excess and then to select and regulate with precision in a second phase, is equally advantageous for both types of RNA, mRNA and rRNA. Thus a general principle of RNA synthesis particular to the animal cell and fundamentally different from the pattern of mRNA and rRNA synthesis in prokaryotes emerged. Moreover, it was reasonable to assert that this particular organisation of chromosomes in transcriptional units – which may be the consequence of genetic recombination and evolutionary selection – was the same for both types of eukaryotic RNA. Therefore it was legitimate to consider seriously the model case of the pre-rRNA – rRNA relationship for messenger RNA as well.

Recent experiments [20] justify this assertion: globin messenger RNA which directs *in vitro* and *in vivo* the synthesis of all three duck globin chains [21–23] can be transcribed with the RNA-dependent DNA polymerase from avian myeloblastosis virus into antimessenger DNA (amDNA). This amDNA represents by size about 75% of the 9 S mRNA length. The copy is of highest fidelity: the amDNA/globin mRNA hybrid melts with a Tm of 86 °C, expected in the case of an RNA composed of 52% guanosine and cytidine. The crucial observation is that a hybrid of almost equal thermal stability (Tm of 83 °C) can be obtained by hybridizing the globin amDNA to giant nuclear RNA from duck erythroblasts.

This experiment demonstrates that nuclear RNA molecules of high sedimentation constants, corresponding to molecular weights of $1-20 \times 10^6$, contain the globin messenger sequence. To prove that the globin messenger sequence is indeed part of the homogeneous covalent structure of giant molecules, this RNA can be sedimented through 99% DMSO gradients (which disrupts all non-covalent bonds in nucleic acids). RNA molecules of more than 2×10^6 MW isolated from DMSO gradients contain about 0.03% globin messenger sequence. Although most of the labeled (*i.e.*, newly synthesized) pre-mRNA molecules sediment in DMSO according to this MW, 90% of thc unlabeled ("aged" ?) giant RNA (measured by absorbancy) breaks down in DMSO to molecules of smaller MW; these hybridize with globin amDNA to the same extent as the "giant" RNA isolated originally from the sucrose gradients.

In formamide (under conditions which melt the amDNA/mRNA hybrid) giant RNA molecules migrate electrophoretically on polyacrylamide gels to the zone of a 45 S pre-rRNA marker (4.9×10^6 MW); they still hybridize with globin amDNA. However, higher concentrations of globin messenger sequence are contained in the zones of smaller RNA [24].

We draw the following conclusions from these experiments:

1. Globin messenger sequences are transcribed as an integral part of giant pre-mRNA molecules of up to 2×10^7 MW. These molecules are represented by the pulse-labeled RNA. 2. The seemingly small amount of globin mRNA sequence in these molecules is reasonable considering a) the size relation of giant pre-mRNA to 9S mRNA, b) the rapid decay of pre-mRNA which limits the amount of accumulating nascent molecules (c.f. following paragraph), and c) the theoretical representation of globin mRNA sequences (about 6×10^{-6}) in the transcribed fraction of the DNA. 3. An artifactual aggregation of (labeled) gobin mRNA with rapidly sedimenting nucleic acid could not be observed (using low salt gradients!). Thus, the fraction of hybridizable globin mRNA sequence sedimenting rapidly in sucrose gradients but decaying in DMSO to smaller molecules may be contained in partially processed pre-mRNA molecules which contain "hidden breaks". Since the giant RNA molecules are usually extracted by hot (65 °C) phenol, the regions of base complementarity which maintain these nicked (aged) pre-mRNA molecules must have a considerable thermal stability. Double-stranded sequences of high G+C content and more than 30 base-pairs indeed occur in loops of nuclear pre-mRNA [25, 43].

Before concluding this chapter we must point out that, although the experiments reported show unequivocally that giant nascent pre-mRNA is an *informative* precursor of mRNA, no stringent proof can be given that it represents also the *physical* precursor.

Kinetics of Synthesis and Turnover of Pre-mRNA

Employing sucrose gradients and exponential polyacrylamide gel electrophoresis [26], we studied the synthesis and decay of nuclear pre-mRNA in HeLa cells and in duck erythroblasts [13, 27, 28]. The following conclusions are drawn from this extensive investigation:

There exist three metabolic fractions of nuclear pre-mRNA: 1. giant *nascent* pre-mRNA molecules with $5-20 \times 10^6$ MW; these decay with a half-time of less than 30 min to smaller molecules and to acid soluble material (about 50%), 2. *intermediate-size* pre-mRNA molecules with apparent MWs of $1-5 \times 10^6$ and half-decay times of 3–7 h, and 3. *small or messenger-size* pre-mRNA molecules with $1-10 \times 10^5$ MW and half-decay times of more than 15 h [13, 15, 27, 28].

The analysis of nuclear RNA from erythroblasts indicates clearly that no 9 S globin mRNA in its final size accumulates in the nucleus. Thus, the globin messenger sequence must leave the nucleus very rapidly when processed from its giant precursor [28].

In HeLa cells where the nuclear pre-mRNA and cytoplasmic mRNA can be labeled into steady state and rRNA synthesis is suppressed by low doses of actinomycin D, as much as 80% of the total cellular RNA exists in the form of molecules with sedimentation and electrophoretic migration properties indicating molecular weights greater than those of functional polyribosomal mRNA [15, 28]. The qualitative comparison by high C_0t hybridization/competition experiments of nuclear pre-mRNA to cytoplasmic mRNA from duck erythroblasts [10] and HeLa cells [27] indicate that the cytoplasmic mRNA contains less than 10% of the sequences occurring in the nuclear transcription products.

Considering the total RNA (and not a selected specific mRNA), we conclude that quantitatively as well as qualitatively much more pre-mRNA is in the nucleus than there is mRNA in the cytoplasm and that the bulk of nuclear RNA consists of giant molecules.

Therefore simple pulse-chase experiments are not adequate to demonstrate that the pre-mRNA, an *informational* precursor to mRNA, is also the direct *physical* precursor molecule in which the globin messenger sequence is simply stabilized, excised and transported to the cytoplasm. The size and mass relations of pre-mRNA and mRNA, the rapid decay of nuclear RNA, the large size spectrum of pre-mRNA types produced and the relatively small amount of mRNA transferred to cytoplasm and polyribosomes practically exclude such conclusions. The same basic difficulty is encountered when pulse-chasing specific sequences such as poly (A) (c.f. next chapter).

Characteristic Structural Features in Pre-mRNA

Nuclear pre-mRNA contains oligo (A) sequences [6, 13, 15, 29, 30, 31] as well as poly (A) sequences (c.f. review by Weinberg, [32]). Sequences rich in deoxyadenosine (dA) and thymidine (T) occur in the nuclear DNA [6, 29, 33, 34] which is transcribed into nascent pre-mRNA. Probably after the first cleavage of nascent pre-mRNA, intermediate-size pre-mRNA molecules are adenylated post-transcriptionally. The resulting poly (A) sequences are eventually up to 200 nucleotides long [35–37]; small (processed) pre-mRNA contains the same size poly (A) sequence.

Adenosine labeling and pulse-chase experiments seem to indicate a precursor-product relationship between nuclear and cytoplasmic poly (A) sequences [35]. However, in view of the delayed synthesis [38] and the turnover of poly (A) in the nucleus [39] and considering the possibility of its synthesis in the cytoplasm (Brawermann, in [69]) as well as the pools of nuclear and cytoplasmic poly (A), it seems to me impossible to draw a stringent conclusion concerning such a precursor-product relationship.

The observed oligo (A) sequences in pre-mRNA could play an important role related to biochemical mechanisms of regulative signals involved in transcription and the first post-transcriptional processing step. The post-transcriptional poly-adenylation must certainly be related to post-transcriptional stabilization of pre-mRNA, to transport mechanisms or to the regulation of processing.

In addition to oligo (A) and poly (A), nuclear pre-mRNA contains oligo (U) sequences [40, 41] which occur once in every giant pre-mRNA molecule of the nascent or the intermediate-size classes described above. The function of this oligo (U) fraction in pre-mRNA is not understood.

Furthermore, the giant pre-mRNA contains double-stranded sequences [27, 42–43]. This double-stranded RNA falls into two size classes, one of more than 200000 MW (excluded on Sephadex G-200 [6]) and the other of a 20000–50000 MW range. These sequences seem to occur about ten times per nascent pre-mRNA molecule, and one of them may be present in every intermediate-size pre-mRNA molecule. The role of these fairly large hairpin loops are not understood at present. In view of the occurrence of such loops in f 2 phage RNA and considering their role in translational regulation [44], they may represent signals in the programming sequence [13, 14] of pre-mRNA and/or of mRNA. They could also represent attachment sites for stabilizing proteins, or expose specific RNA fractions to cleaving enzymes. This latter possibility is particularly attractive in view of the previously mentioned "hidden breaks" in pre-mRNA.

The Overall Structure of Pre-mRNA

On the basis of the pattern of synthesis and turnover of pre-mRNA and in view of its structural features mentioned above, we propose a model (Fig. 1) of the general structure of pre-mRNA. According to this model giant nascent pre-mRNA molecules, the direct transcript of the full unit of transcriptional regulation, would be subdivided into several sub-units corresponding to intermediate-size pre-mRNA molecules. The nascent pre-mRNA would contain at its 5′ end a transcript of promotor and operator regions related to transcriptional regulation; furthermore, this molecule may contain internal signal sequences which direct the first processing steps. The intermediate-size pre-mRNA units would contain, in addition to one or several messenger sequences, "programming" sequences [13–15] including such structural features as terminal oligo (A), oligo (U) sequences and double-stranded regions representing signals for further processing and for association with specific proteins.

Thus we propose to distinguish in pre-mRNA between *coding* (messenger) sequences and *programming* sequences; the latter relate to the mechanisms of post-transcriptional processing and modification and may contain signals for transcriptional and post-transcriptional regulation, for transport and translation.

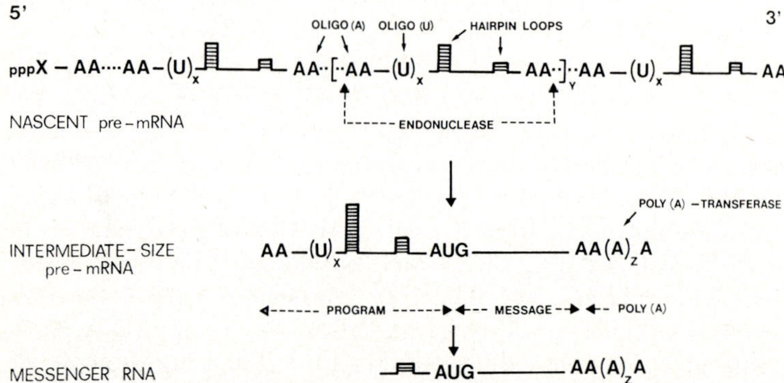

Fig. 1. *Tentative scheme of pre-mRNA structure and processing.* According to this model the *nascent* pre-mRNA contains one or several segments of *intermediate-size* pre-mRNA which are excised by the processing enzyme(s). Each intermediate-size pre-mRNA molecule may contain one (or several) mRNA sequences which are excised (or further processed by re-transcription) and transported to the cytoplasm, or are stored in the nucleus as small pre-mRNA. The pre-mRNA is composed of messenger sequences, true no-sense segments and *programming sequences.* The latter contain regulative signals directing, through association with proteins, the stabilization of pre-mRNA, its processing and transport to the cytoplasm as well as its translation. Thus, the extra, non-coding nucleotide sequence observed in mRNA may represent the remaining part of the programming sequence(s). The regulative signals in pre-mRNA and mRNA may reside in particular RNA sequences as oligo (U), oligo (A), and poly (A) or in regions of particular secondary structure leading eventually to the formation of hairpin loops. These regions are recognized by specific proteins which constitute the communicating system with the intracellular environment. No-sense and programming sequences of no further use would be solubilized and could account in part for the decay of pre-mRNA. In addition, entire units composed of programming and messenger sequence may be destroyed in such cases where an mRNA, transcribed as part of a unit of transcription, is not needed in a given cell at a given time

The Ribonucleoprotein Nature of Nuclear Pre-mRNA

SAMARINA and GEORGIEV reported several years ago that pre-mRNA is associated with a specific protein called *informofer* [46]. This observation was corroborated in many other systems as well as in duck erythroblasts [47, 48]. As we will relate later, these nuclear pre-mRNA-bound proteins could not be observed associated with cytoplasmic mRNA.

Each informofer (800000 MW) is composed of up to 20 subunits of 40000 MW (called *informaton* by SAMARINA and GEORGIEV), and is associated with a fraction of the pre-mRNA molecule of about 200000 MW. Thus, the pre-mRNA molecule contains up to 100 informofer proteins which can specifically be dissociated from and reassociated to the RNA backbone. The apparent universality and stability of the informofer proteins make it likely that their function is related to the liberation of nascent pre-mRNA from chromatin, in stabilization and eventually processing of pre-mRNA. In view of their generality and abundance, the informofer proteins cannot relate to the regulation of individual mRNA sequences. However we cannot exclude that the pre-mRNP structure still hides specific messenger-sequence related proteins that, obscured by the abundantly occurring informofer, are below the level of detection. As an illustration we may point out that one of the larger messenger-specific proteins (described in the next chapter) associated with one mRNA sequence in an intermediate-size pre-mRNA molecule would represent at most 1% of the informofer protein.

The Cytoplasmic Messenger RNA

Although the cytoplasmic existence of giant RNA similar to nuclear pre-mRNA has been postulated in HeLa cells and duck erythroblasts, we never observe such RNA in significant amounts if carefully controlled lysis conditions are maintained. Thus, we tend to interpret giant cytoplasmic RNA as an artifact of preparation. However, very small amounts of truly cytoplasmic pre-mRNA may exist below the level of detection. (If, for example, the last processing steps of pre-mRNA occur at the nuclear membrane, a small fraction would be expected in the cytoplasm.) At present we conclude that all cytoplasmic mRNA falls into the size spectrum of translated messenger molecules ($1-10 \times 10^5$ MW) as they are found in polyribosomes.

Thus, upon entering the cytoplasm, the messenger has already acquired its functional state. However, processing of mRNA continues right through the translation process. As we related elsewhere [13], erythroblast polyribosomes contain a heterogeneous population of globin mRNA; on polyacrylamide gels, the newly synthesized molecules migrate slower than the bulk of the translatable mRNA.

Cytoplasmic mRNA exists in two compartments: free in the cytoplasm and associated with ribosomes in the polyribosomal translation complex. In both compartments the mRNA molecule is associated with three times its molecular weight of protein; these mRNA-protein complexes are termed messenger-ribonucleoprotein particles (mRNP).

Free Cytoplasmic Messenger Ribonucleoprotein Particles

The early observations of Schapira et al. [49] showed that globin mRNA in reticulocytes could exist free in the cytoplasm outside the translation complex. Spirin proposed in his *informosome* scheme that the *untranslated* messenger RNA of oocytes exists in the form of ribonucleoprotein particles free in the cytoplasm [50]. The inherent contradiction in these two observations – Spirin speaks of "masked" (*i.e.*, untranslated) forms of messenger RNA whereas Schapira et al. refer to messenger RNA actually being translated in the cell – seems to be resolved by the accumulated evidence suggesting that in fact the pool of free mRNA in most cells contains *both* temporarily masked and actually translated messenger RNA.

The protein association *per se* is not the reason for the masking of mRNA. As we will relate in the following, a more subtle interplay of protein "factors" with the mRNA molecule may be the reason for their maintenance in a free, untranslated state or their association with ribosomes.

Spohr et al. [51] and Jacobs-Lorena and Baglioni [52] recently isolated free ribonucleoprotein particles containing a specific globin mRNA actually translated in the respective reticulocyte. In both cases the deproteinized globin mRNA could be translated in an *in vitro* ribosome system. These observations demonstrate unambigously that the free mRNP represent not primarily masked forms of mRNA but constitute a free pool of the message actually translated. Kinetic labeling experiments of Spohr et al. [17, 53] suggest that the messenger RNA entering the cytoplasm passes through this transient pool before associating with ribosomes. In the case of duck erythroblasts, the pool of free mRNP may incorporate as much as 25% of the total cytoplasmic globin messenger RNA [54].

Furthermore, the comparison of the messenger population in polyribosomes to that contained in free mRNP particles suggests that there are not only *quantitative* differences in respect to the relative representation of specific types of messenger RNA in both cytoplasmic compartments but *qualitative* ones as well. The spectrum of free mRNA in duck erythroblasts suggests that some types of free cytoplasmic mRNA are not represented in polyribosomes. These could represent truly masked messenger RNA. Thus, the observations of masked maternal mRNA in oocytes represents only a particular case – specific by the amount of message involved and the prolonged time of its inactivation – of a quite general mechanism of regulation characteristic for the animal cell [55].

The Polyribosomal Messenger Ribonucleoprotein

The association of free messenger RNA with proteins is not surprising considering the necessity of messenger stabilization and regulation. However the observations of PERRY [56], HENSHAW [57], CARTOUZOU [58] and BURNY [59] indicating that the translated message in polyribosomes is also associated with proteins was quite surprising. In fact it is difficult to understand how translated mRNA could be permanently associated with proteins. However, the accumulated evidence leaves no doubts: specific types of mRNA are associated with specific proteins [47, 48, 60–62].

In the case of the duck erythroblasts (which synthesize haemoglobin to the extent of 80–90% of their total protein synthesis), the treatment of polyribosomes with EDTA releases a 15 S ribonucleoprotein particle [47] which contains the translatable globin messenger RNA [21, 22]. The three mRNA molecules coding for the three duck globin chains are associated with two major proteins of 49000 and 73000 MW and with 6 minor components in the 60000 and 100000 MW range. Some of these proteins are phosphorylated (c.f. Fig. 3. and Ref. [47, 48]). The mRNA-protein complex is highly stable: it resists gel electrophoresis and salt concentrations which fully dissociate artificial RNA-protein associations. The electron micrographs of M. HERZBERG and J. DUBOCHET [62] show that the proteins are distributed over the full length of the mRNA molecules, as indicated in the model of Fig. 2.

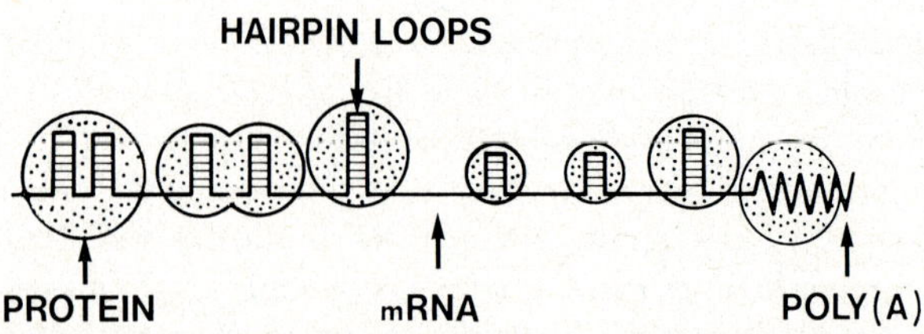

Fig. 2. *Model of a messenger-ribonucleoprotein particle.* This model gives a schematic representation of a globin mRNP particle. The structure proposed derives essentially from electron microscope observations of DUBOCHET and HERZBERG [62] and from the first results of a structural analysis (A. FAVRE, unpublished results). According to this model, the messenger RNA is composed over its full length of specific regions of high secondary structure that are recognized by the specific proteins found to be associated with the mRNA (cf. Fig. 3)

On the basis of these and further observations, we conclude that the association in polyribosomes of the translated mRNA with specific proteins corresponds to a physiological reality.

The Comparison of Free Cytoplasmic and Polyribosome-Bound mRNP and the Possible Significance of the Messenger-Associated Proteins

The free 20 S globin-mRNP particle isolated from duck erythroblasts by GANDER et al. [63] contains a discrete population of proteins. Most of them fall into a MW-range of 15000–24000; a 21000 MW protein is phosphorylated. With the possible exception of one minor component of 51000 MW, all of these proteins differ from those associated with the translated messenger in the polyribosome (c.f. Fig. 3 and Ref. [63]). None of the proteins associated with free cytoplasmic or polyribosome-bound globin mRNA could be identified among the informofer proteins associated with nuclear pre-mRNA.

The obvious conclusion is that during the last processing steps of pre-mRNA the informofer proteins are lost and that upon entry of mRNA into the cytoplasm the mRNA associates with specific cytoplasmic proteins. Prior to, or upon association with the ribosomal subunits, the mRNA exchanges this protein population for that found with the translated mRNA.

Theoretically such a protein exchange could be associated with the requirement for "initiation" factors reported for prokaryotic (c.f. review by M. GRUNBERG-MANAGO in Ref. [69]) and eukaryo-

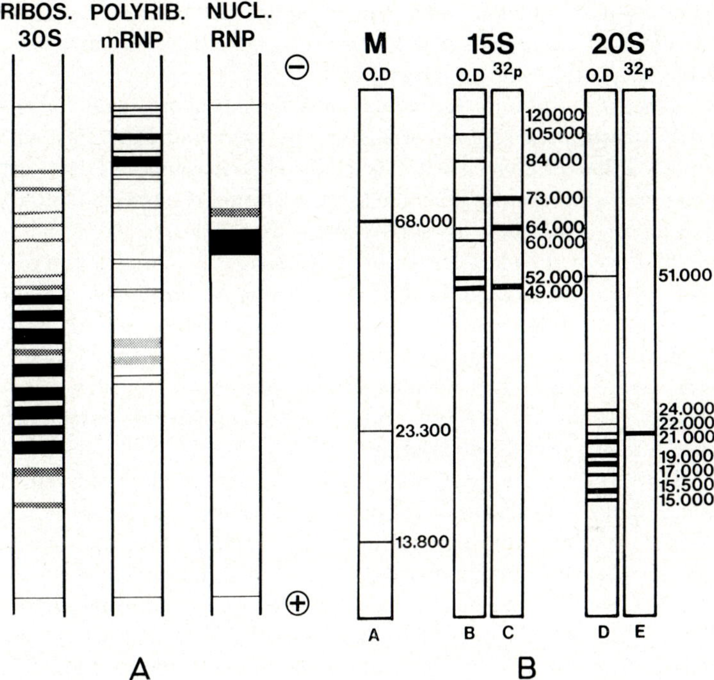

Fig. 3. *Polyacrylamide gel electrophoresis of proteins associated with nuclear pre-mRNA, free cytoplasmic and polyribosomal mRNA. Panel (A)*: Polyacrylamide-gel electrophoresis in urea (pH 4.5) of proteins from the 30 S ribosomal subunit, the polyribosomal mRNP and the nuclear pre-mRNP (informofer acc. to SAMARINA et al. [46]). Schematic drawings on the basis of original photographs by MOREL et al. [47,48]. *Panel (B)*: Comparison by polyacrylamide gel electrophoresis in 0.5% SDS (pH 7) of proteins from ^{32}P-labeled 15 S polyribosomal globin mRNP (*B, C*), and the 20 S free cytoplasmic globin mRNP (*D, E*) with marker proteins (*A*). Markers were: bovine serum albumin, trypsin, pancreatic RNase. *A, B, D* represent gels stained with Coomassie blue; C, E give the results of ^{32}P autoradiography. Schematic drawings after original photographs by GANDER et al. [63]

tic (c.f. reports of Revel [64] and those of Lebleu, of Staehelin, and of Anderson in Ref. [69]) protein synthesis. Testing for a possible factor role of the mRNA-bound proteins, we found that the polyribosomal mRNP isolated from functional erythroblast polyribosomes can be translated in a heterologous *in vitro* ribosome system as efficiently as – but no better than – the deproteinized globin mRNA [65]. To our knowledge, no analogous experiments have been conducted with free cytoplasmic mRNP particles; therefore we do not know if the masking effect is maintained in an acellular protein synthesizing system.

Thus far, the only indication of a biological function for the mRNA-associated proteins was given by Ilan and Ilan [66] who reported that proteins isolated from the polyribosomal mRNP may be an essential requirement for the translation of messenger RNA in a partially purified ribosome system from insects.

This is not the place for further speculation; in addition, the present state of knowledge is such that it may be impossible to correlate the proteins associated with messenger RNA and the initiation or translation factors isolated from several eukaryotic systems. The essential fact is that a specific messenger, the duck globin mRNA, is found in two different functional states associated with two different protein populations.

However, we shall relate another observation which may shed a new light on the discussion about specific or non-specific initiation factors. Fundamental differences in translation factor requirement can be observed in comparing duck globin messenger RNA to globin mRNA of rabbit [21]. The purified duck globin messenger RNA added to a rabbit reticulocyte lysate is translated to less than 5 %; surprisingly the converse experiment shows that purified rabbit globin mRNA added to a duck erythroblast lysate suppresses almost completely the synthesis of the endogenous duck globin chains. Tested in a purified mouse ribosome system, adding both mRNA's at the same time, the rabbit globin mRNA is again translated much better than the duck messenger [22]. However, if added separately to the same system (optimized for the duck mRNA), both messengers are translated with equal efficiency. The crucial observation is that the *duck* globin message requires a higher concentration of a particular *rabbit* supernatant factor than is needed for optimal translation of the rabbit globin mRNA.

Mammalian and avian globins are far enough separated on an evolutionary scale to be considered as products of different types of specific mRNA. The fact that in a mouse ribosome system a higher concentration of a rabbit factor is necessary to translate the duck globin message as efficiently as the rabbit globin mRNA indicates that no absolute messenger specificity *per se* exists for initiation or translation factors.

The solution to this enigma may be found in the proposition that all messenger RNA's include a programming sequence which can be recognized by translational factors. The efficiency of these translational factors may depend on the *dissociation constants* of the corresponding mRNA-protein complex; consequently the translation of a given mRNA will require a certain *concentration* of a given factor. If we assume a certain set of programming sequences, each sequence of which may be contained in more than one message (but in another set), it is possible to envisage how several mRNA's may be stimulated by the same factors. The programming sequence is also subject to evolutionary screening; thus the rabbit message may have optimized a programming sequence which is also present in duck message, but working with lower efficiency in the latter molecule.

Intracellular Information Transfer

Summarizing the facts pertinent to messenger RNA formation and translation in animal cells, we may propose the model of intracellular information transfer shown in Fig. 4. According to this model, the mRNA is transcribed as a small fraction of a unit of transcriptional regulation fixed in the structure of the chromosome. The full transcript of this unit, the nascent pre-mRNA, is associated with informofer proteins and immediately metabolized to intermediate-size pre-mRNA. This first step of processing is accompanied by the loss of more than 50 % of the synthesized RNA. The intermediate size pre-mRNA is of considerable stability and may be processed into small or messenger-size pre-mRNA which accumulates in the nucleus. However, the messenger molecules actually translated in a given cell cannot be observed in the nucleus; they are found exclusively in the cytoplasm.

This fact makes it difficult to conclude that the intermediate-size pre-mRNA is directly cleaved into messenger RNA. Alternative models for the processing or the information in pre-mRNA by RNA-RNA replication or RNA-DNA-RNA re-transcription (involving a RNA-dependent DNA polymerase) cannot be excluded at present. (To recall this fact in our model, we put this phase of intracellular information transfer in a "black box".) Although unlikely, it is also not excluded that the DNA actually transcribed is exported to the cytoplasm. Cytoplasmic DNA has been found in many systems and a cytoplasmic DNA-dependent RNA polymerase has been described [68].

In relating these still confusing experimental observations, I purposely play *advocatus diaboli*. However, I wish to impress upon the reader that the straightforward model of direct utilization of the mRNA sequence cleaved from pre-mRNA has not yet been proven.

The cytoplasmic mRNA, isolated as free mRNP particles in masked or in actually translated form, is translatable after deproteinization. We may conclude that the mRNA molecule *per se* is in its final translatable form. A more complicated structure, the messenger ribonucleoprotein complex, allows the cell further and subtle ways to regulate gene expression. Thus, the cell may discriminate between those mRNA types that will be translated and phenotypically expressed and those which must remain (or will be) masked at a given moment of cellular life or differentiation.

Fig. 4. *Flow scheme of messenger RNA formation and intra-cellular information transfer in eukaryotic cells from the DNA to the polypeptide.* The transcriptional unit in the DNA corresponding to the chromomere is transcribed into giant pre-mRNA molecules including one or several messenger sequences. This *nascent pre-mRNA* is rapidly cleaved into *intermediate size pre-mRNA* (cf. Fig. 1). During this process about 50% of the nascent pre-mRNA decays. The further processing until the appearance of the final mRNA in the cytoplasm is hypothetical and therefore placed in a "*black box*". However it is certain that in addition to the pool of intermediate size pre-mRNA which decays with a half-times of about 3—7 h, a pool of very stable *small pre-mRNA* exists. The mRNA appears in the cytoplasm first in a transient pool from which the mRNA is either transferred into polyribosomes and translated or stored in the "informosome" pool. In order to draw attention to the fact that the "black box" may still hide alternative ways of information processing, we depict as an example the possibility of the transfer of DNA to the cytoplasm, following the proposition of Bell [67]. (R represent possible regulative steps)

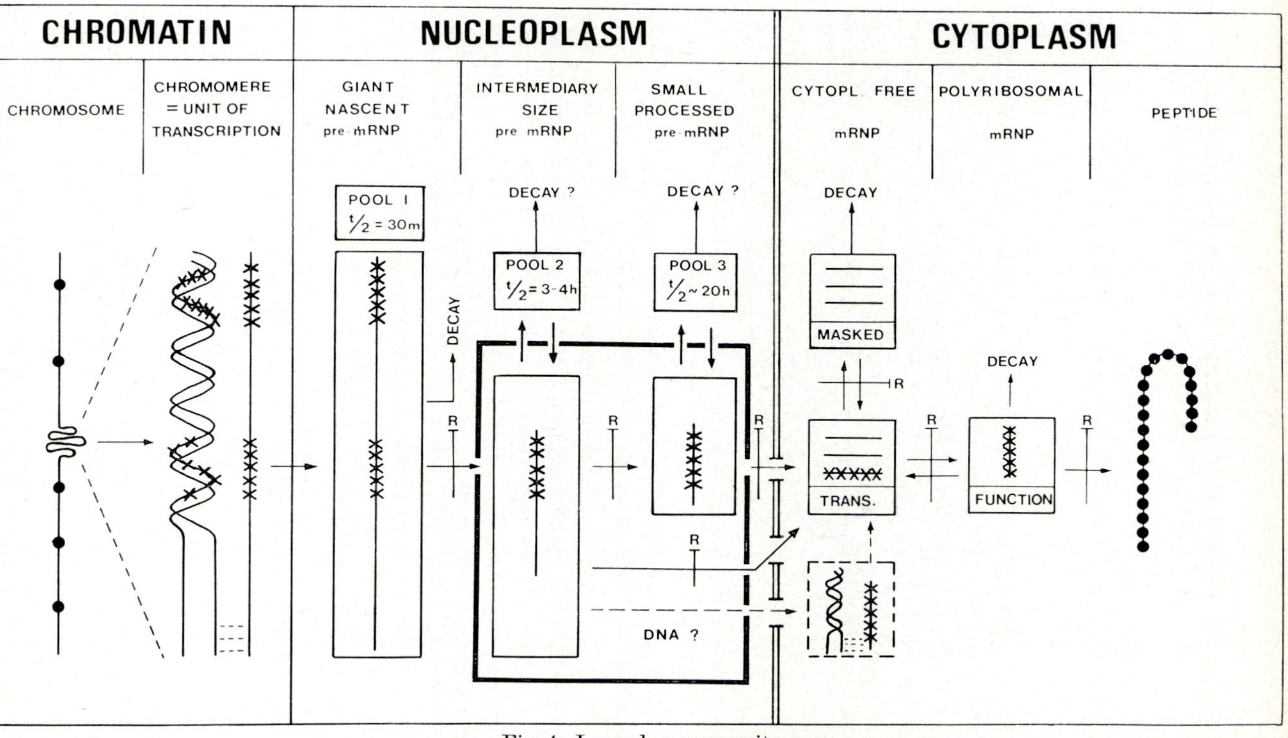

Fig. 4. Legend, see opposite page

The Control of Intracellular Information Transfer and the "Cascade Regulation" Hypothesis

Summarizing the "life history" of mRNA with the model proposed in Fig. 4, we shall speculate about the significance of this long and complex process. Indeed, on first thought such a complicated system of mRNA formation by which most of what the genome transcribes from the DNA is lost seems illogical and wasteful. However, the reason for such an apparently illogical process may reside in the formal logics and the mechanics ("soft ware" and "hard ware" in computer language) of an information processing system adapted to the eukaryotic cell. We will discuss elsewhere [14] the theoretical implications of such a regulational system. Here we will just outline some essential considerations.

We postulated previously [10, 12] the importance of a *multi-key system* of regulation for cell *economy*. However the formal logics of a simple multi-key system *per se* (as proposed by some other regulation models, which imply almost necessarily an all transcriptional system of control [70–73]) alone do not suffice to fulfill the regulational requirements of an animal cell.

Two of these further requirements (in addition to economy) are those of *efficiency* and *versatility*. In the case of the diauxic growth of *E. coli* [74], the bacterial regulation system has to retrieve new information from the central genetic memory in order to adapt to new growth requirements. It takes a bacterial population about 20 min to adapt its enzyme systems to a new substrate. Such a system is inadequate for the requirements of a multi-cell and multi-organ organism which must adapt rapidly and locally its physiological or functional state. In the formal consideration of optimal function there is an advantage in storing genetic information in *peripheral memories* rather than having always to call the central memory. In fact, the passage to and the storage of bits of the genetic information in peripherical memories may be one of the essential features of the process of differentiation. Is it a coincidence that the information needed for the first phases of embryonic development is stored in the oocyte peripherally in the form of cytoplasmic mRNP-particles which are activated upon fertilization [75, 76]? It may be actually a formal requirement of biological information transfer and storage that the program for retrieving information from the central memory (the DNA genome) must be stored peripherally. We will discuss this consideration elsewhere in more detail [14]. Here we may just conclude that it seems reasonable for cell function to store mRNA close to translation.

Another requirement for a regulation system in eukaryotic cells in addition to *economy* and *efficiency* is that such a system must be able to operate in temporal sequence, *i.e.*, in *the dimension of time*. In fact, the spacial dimension in the transport of the genetic information from a site in the chromatin to the polyribosome at the periphery of the animal cell already introduces a time factor in the expression of genetic information. Furthermore, we know that all multi-cell organisms are subject to a precise regulation in time. Thus a multi-key system in which all the keys are turned at the same time is theoretically not sufficient. Rather the cell operates according to a modification of the multi-key system that allows the turning of one key after the other at precise moments in time (and in space!). Considering also the above mentioned fact that in animal cells information may be stored peripherally in form of mRNP-particles, we must accept as a consequence that the mechanism of the multi-key system is linked to and transported with the message.

A last requirement of such a regulational system resides in the necessity that somehow *the center has to be informed of the situation at the periphery* in order to be able to react and to adapt to metabolic or physiological conditions. Feed-back mechanisms are well known in prokaryotic cells. There, due to the close coupling of transcription and translation in the DNA-mRNA-ribosome complex, any interference with peptide synthesis can have a direct biochemical effect on the movement of the RNA polymerase. The biochemical formulation of such feed-back mechanisms is, however, much more difficult in the case of the animal cell; again, this difficulty resides in the separation in space and time of the sites of the phenotypic expression from the central genetic memory. Therefore we postulate that at any time of this transfer through the animal cell, the mRNA must be able to receive signals directing its further progress. Somehow a feed-back loop has to engulf not only phenotypic expression and genetic information (direct feed-back of product or substrata levels to the chromatin) but also must be able to influence the peripheral memories, the mRNA during its transfer.

We therefore propose that the transcription product of the animal cell genome represents not only the structural information for proteins (as, possibly, in the case of the bacterial message) but a whole *program* that permits, as a sort of "mailing tag", guidance of the mRNA in space and time from the moment of its liberation from the DNA to the moment of its translation in the polyribosome. In fact the program does not leave the genetic information there: in cases as pro-insulin, trypsinogen, etc., the polypeptides synthesized still carry that part of the "program" which allows the control of the time of activation of the enzyme (cf. Ref. [14]).

In attempting to explain the size of the transcriptional unit in eukaryotes, we may postulate that, due to the particularities of the regulational system in animal cells, the *programming sequence* became larger than the message and accounts for a part of the giant pre-mRNA. True "no-sense" sequences which are derived from the architectural arrangement of individual genes within the chromatin (as *e.g.* the localization of the ribosomal 28S and 18S rRNA within its precursors) may also be present. In addition, due to evolution and selection and the mechanisms of genetic recombination, related or apparently unrelated units of genetic information may have been packed into one transcriptional unit. As a corollary we may recall that related sequences as α and β haemoglobin chains were separated and stored in different chromosomes, whereas β and δ chains seem to be linked [77]. That several units of genetic information (cistrons) might be contained in one transcriptional unit is suggested by the fact that only about 2000 genetically and morphologically identifiable units of transcription exist in a dipteran species as *Chironomus* (PELLING, pers. comm.). It seems impossible to me to envisage that only 2000 polypeptides could build such an organism.

In summary, I see in the fact that globin mRNA is first transcribed as a part of a giant precursor, and in its long and complicated life history, the consequences of a complicated system of gene regulation whose intrinsic logics may still escape us. According to this view, the pre-mRNA which combines programming, no-sense and informational sequences represents the biochemical consequence of a multi-key regulational system that must be able to operate in time and space; in addition, the pre-mRNA manifests the physical characteristics of the organization or the chromomere.

Thus, the functional message became the appendage of a *programming sequence* written like the genetic code in nucleic acids. This second *programming code* could be expressed by the secondary structure of the RNA. The direct readers of this program are the proteins associated with the RNA; they represent the communication system between mRNA and the controlling environment. In an attempt to formulate these ideas years ago, we tried to rationalize the observed pattern of mRNA synthesis together with the biochemical and genetic requirements of the eukaryotic cell in the system of "cascade regulation" [10, 12, 14] which embraced the four requirements postulated above. According to this hypothesis, the genetic information potential of the eukaryotic genome is reduced gradually in the most organized and well controlled steps to what has to be expressed phenotypically at a given time in a given cell.

Discussion, see p. 121

References

1. Gros. F., Gilbert, W., Hiatt, H. H., Attardi, G., Spahr, P. F., Watson, J. D.: Cold Spr. Harb. Symp. quant. Biol. **16**, 111 (1961).
2. Scherrer, K., Darnell, J. E.: Biochem. biophys. Res. Commun. **7**, 486 (1962).
3. Perry, R. P.: Proc. nat. Acad. Sci. (Wash.) **48**, 2179—2186 (1962).
4. Scherrer, K., Latham, H., Darnell, J. E.: Proc. nat. Acad. Sci. (Wash.) **49**, 240 (1963).
5. Scherrer, K., Marcaud, L., Zajdela, F., London, I. M., Gros, F.: Proc. nat. Acad. Sci. (Wash.) **56**, 1571 (1966a).
6. Scherrer, K., Marcaud, L., Zajdela, F., Breckenridge, B., Gros, F.: Bull. Soc. Chim. biol. (Paris) **48**, 1037 (1966b).
7. Georgiev, G. P., Samarina, O. P., Lerman, M. L., Smirnov, M. N.: Nature (Lond.) **200**, 1291 (1963).
8. Hiatt, H.: J. molec. Biol. **5**, 217 (1962).
9. Scherrer, K., Marcaud, L.: Bull. Soc. Chim. biol. (Paris) **47**, 1967 (1965).
10. Scherrer, K., Marcaud, L.: J. Cell. Physiol. **72**, 181 (1968).
11. Scherrer, K.: Exp. Biol. Med. **1**, 244 (1967a).
12. Scherrer, K.: Int. Symp. Reinhardsbrunn on "Biochemistry of Ribosomes and mRNA". Abhandl. dtsch. Akad. Wiss. Berlin (Medizin) **1968**, 259 (1967b).
13. Scherrer, K.: In: Diczfalusy, E. (Ed.): Protein Synthesis in Reproductive Tissue. Sixth Karolinska Symposium on Research Methods in Reproductive Endocrinology. Copenhagen: Bogtrykkeriet Forum, 1973a.

14. Scherrer, K.: In: Kohn, A. (Ed.): Strategies for the Control of Gene Expression, Oholo Symposium. London: Plenum Press 1973b.
15. Scherrer, K.: In: Regulation of Gene Expression in Cultured Cells. Fogarty Conference (1973c).
16. Spohr, G., Imaizumi, T., Mirault, M-E., Scherrer, K.: Manuscript in preparation.
17. Spohr, G., Kaybanda, B., Scherrer, K.: Europ. J. Biochem. **31**, 194 (1972).
18. Edström, J.-E.: Proceedings of the Biophysics Congress, Vienna (1966).
19. Weinberg, R., Loening, U., Willems, M., Penmann, S.: Proc. nat. Acad. Sci. (Wash.) **58**, 1088 (1967).
20. Imaizumi, I., Diggelmann, H., Scherrer, K.: Proc. nat. Acad. Sci. (Wash.) **70**, 1122 (1973).
21. Stewart, A., Gander, E., Morel, C., Luppis, B., Scherrer, K.: Europ. J. Biochem. **34**, 205 (1973).
22. Schreier, M. H., Stähelin, T., Stewart, A., Gander, E., Scherrer, K.: Europ. J. Biochem. **34**, 213 (1973).
23. Lane, C. D., Greogory, C. M., Morel, C.: Europ. J. Biochem. **34**, 219 (1973).
24. Imaizumi, T., Spohr, G., Scherrer, K.: unpublished results.
25. Georgiev, G. P.: Personal communication (1972).
26. Mirault, M.-E., Scherrer, K.: Europ. J. Biochem. **23**, 372 (1971).
27. Scherrer, K., Spohr, G., Granboulan, N., Morel, C., Grosclaude, J., Chezzi, C.: Cold Spr. Har. Symp. quant. Biol. **35**, 539. (1970).
28. Spohr, G., Imaizumi, T., Mirault, M.-E., Scherrer, K.: manuscript in preparation.
29. Scherrer, K.: FEBS Letters **17**, 68 (1971).
30. Georgiev, G. P.: Personal communication (1972).
31. Adesnik, M., Salditt, M., Thomas, W., Darnell, J. E.: J. molec. Biol. **71**, 21 (1972).
32. Weinberg, R. A.: Ann. Rev. Biochem. **42**, 329 (1973).
33. Shenkin, A., Burdon, R. H.: FEBS Letters **22**, 157 (1972).
34. Philipson, L., Lindberg, U., Persson, T., Vennström, B.: In: Regulation of Transcription and Translation in Eukaryotes. Proc. 24th Mosbach Colloquium. Berlin-Heidelberg-New York: Springer 1973.
35. Jelinek, W., Adesnik, M., Salditt, M., Sheiness, D., Wall, R., Molloy, G., Philipson, L., Darnell, J. E.: J. molec. Biol. **75**, 515 (1973).
36. Lee, S. Y., Mendecki, J., Brawermann, G.: Proc. nat. Acad. Sci. (Wash.) **68**, 1331 (1971).
37. Edmonds, M., Vaughan, M. H. Jr., Nakazato, H.: Proc. nat. Acad. Sci. (Wash.) **68**, 2806 (1968).
38. Philipson, L., Wall, R., Glickman, G., Darnell, J. E.: Proc. nat. Acad. Sci. (Wash.) **68**, 2806 (1971).
39. Perry, R. P.: J. Cell. Physiol. (Gatlinburg Symposium Supplement) (1973).
40. Molloy, G. R., Thomas, W. L., Darnell, J. E.: Proc. nat. Acad. Sci. (Wash.) **69**, 3684 (1972).

41. Burdon, R. H., Shenkin, A.: FEBS Letters **24**, 11 (1972).
42. Montagnier, L.: C. R. Acad. Sci. D **267**, 1417 (1968).
43. Jelinek, W., Darnell, J. E.: Proc. nat. Acad. Sci. (Wash.) **69**, 2537 (1972).
44. Lodish, H. F.: J. molec. Biol. **50**, 689 (1970).
45. Ryskov, A. P., Saunders, G., Faraskyan, V. R., Georgiev, G. P.: Biochem. biophys. Acta (Amst.) **312**, 152 (1973).
46. Samarina, O. P., Lukanidin, E. M., Molnar, J., Georgiev, G. P.: J. molec. Biol. **33**, 251 (1968).
47. Morel, C., Kayibanda, B., Scherrer, K.: FEBS Letters **18**, 84 (1971).
48. Morel, C., Gander, E., Herzberg, M., Dubochet, J., Scherrer, K.: Europ. J. Biochem. **36**, 455—464 (1973).
49. Schapira, G., Padieu, P., Maleknia, N., Kruh, J., Dreyfus, J. C.: J. molec. Biol. **20**, 427 (1966).
50. Spirin, A. S.: Europ. J. Biochem. **10**, 20 (1969).
51. Spohr, G., Imaizumi, T., Stewart, A., Scherrer, K.: FEBS Letters **28**, 165 (1972).
52. Jacobs-Lorena, M., Baglioni, V.: Proc. nat. Acad. Sci. (Wash.) **69**, 1425 (1972).
53. Spohr, G., Granboulan, N., Morel, C., Scherrer, K.: Europ. J. Biochem. **17**, 296 (1970).
54. Spohr, G., Scherrer, K.: Unpublished observations.
55. Gross, P. R., Gross, K. W., Skoultchi, A. I., Ruderman, J. V.: In: Diczfalusy, E. (Ed.): Protein Synthesis in Reproductive Tissue. Sixth Karolinska Symposium on Research Methods in Reproductive Endocrinology. Copenhagen: Bogtrykkeriet Forum 1973.
56. Perry, R. P., Kelley, D. E.: J. molec. Biol. **35**, 37 (1968).
57. Henshaw, E. C.: J. molec. Biol. **36**, 401 (1968).
58. Cartouzou, G., Poirée, J. C., Liszitzki, S.: Europ. J. Biochem. **8**, 357 (1969).
59. Burny, A., Huez, G., Marbaix, G., Chantrenne, H.: Biochim. biophys. Acta (Amst.) **190**, 228 (1969).
60. Burny, A., Marbaix, G.: Biochim. biophys. Acta (Amst.) **103**, 409 (1965).
61. Blobel, G.: Biochem. biophys. Res. Commun. **47**, 88 (1972).
62. Dubochet, J., Morel, C., Lebleu, B., Herzberg, M.: Europ. J. Biochem. **36**, 465—472 (1973).
63. Gander, E., Stewart, A., Morel, C., Scherrer, K.: Europ. J. Biochem. **38**, 443 (1973).
64. Revel, M., Groner, Y., Pollack, Y., Cnaami, D., Zeller, H., Nudel, N.: In: Diczfalusy, E. (Ed.): Protein Synthesis in Peproductive Tissue. Sixth Karolinska Symposium on Research Methods in Reproductive Endocrinology. Copenhagen: Bogtrykkeriet Forum 1973.
65. Stewart, A., Scherrer, K.: Unpublished results.
66. Ilan, J., Ilan, J.: Nature (Lond.) New Biol. **241**, 176 (1973).
67. Bell, E.: Science **174**, 603 (1972).
68. Bell, E., Brown, J. M.: Biochim. biophys. Acta (Amst.) **269**, 237 (1972).

69. Schapira (Ed.): International Colloquium on "Normal and Pathological Proteins in Higher Organisms". Comptes rendus des Colloques. Paris: Inst. Nat. de la Santé et de la Recherche Médicale 1973.
70. Britten, R. J., Davidson, E. H.: Science **165**, 349 (1969).
71. Georgiev, G. P.: J. theoret. Biol. **25**, 473 (1969).
72. Crick, F.: Nature (Lond.) **234**, 25 (1972).
73. Paul, J.: Nature (Lond.) **238**, 444 (1972).
74. Monod, J., Cohn, M.: Advanc. Enzymol. **13**, 67 (1952).
75. Slater, I., Gillespie, D., Slater, D. W.: Proc. nat. Acad. Sci. (Wash.) **70**, 406 (1973).
76. Gross, K. W., Jacobs-Lorena, M., Baglioni, C., Gross, P. R.: Proc. nat. Acad. Sci. (Wash.) (Manuscript in press).
77. Ingram, V. M.: In: The Hemoglobins in Genetics and Evolution. New York: Columbia University Press 1963.

Transcription and Processing of Adenovirus RNA in Productive Infection

Lennart Philipson, Uno Lindberg, Torgny Persson and Björn Vennström

Department of Microbiology, The Wallenberg Laboratory, Uppsala University, Uppsala, Sweden

With 10 Figures

Introduction

Productive infection of human heteroploid cells with human adenoviruses leads within the first hour after infection to uncoating and passage of parental DNA into the cell nucleus [17]. Transcription of the viral DNA starts immediately and continues throughout the infectious cycle which lasts for around 25 hours [9, 18, 34]. Two functionally different periods of transcription can be recognized: 1. an early phase preceeding viral DNA synthesis, involving transcription of mRNA from 10–20% of the DNA sequences [5, 6] (this mRNA probably codes for the early antigens, "T and P" antigens [7, 30]) and 2. a late phase beginning at the onset of viral DNA synthesis, 5–6 h after infection, where specific adenovirus mRNA is transcribed in increased amounts and from a large fraction of the virus genome (80–90%) [9, 18]. Late in infection the protein synthesizing machinery of the host cell is almost entirely engaged in synthesis of virus-coded proteins necessary for the formation of mature virus [31].

The present communication will summarize our evidence that viral messenger RNA (mRNA) is transcribed and processed by mechanisms similar to those utilized by the host cell in manufacturing mRNA: that is, formation of large nuclear precursors, terminal addition of polyadenylic acid (poly (A)), and cleavage into mature mRNA molecules and finally transport of the mRNA as mRNP to the cytoplasm [2, 9, 11, 12, 26, 32–34]. The viral system with its defined mRNA species synthesized in large amounts [16,

22] may serve as a useful model system for the study of transcription and processing of mRNA in mammalian cells as well as of the control mechanisms probably involved in these steps.

Size of Adenovirus RNA in the Nucleus

Several lines of evidence suggest that polysomal mRNA in animal cells is generated by cleavage of high molecular weight heterogenous nuclear RNA (HnRNA) [14, 24]. Recently this view was strengthened through the demonstration that a large fraction of the HnRNA of all sizes as well as mRNA on polysomes contain poly (A) [2–4, 11–13, 15], and that the addition of poly (A) to nuclear RNA [1, 3] is a prerequisite for the appearance of mRNA on polysomes [3, 25]. To study the analogy between host cell and viral transcription during lytic infection, the size of viral nuclear RNA made at various times after infection was first determined by labeling infected HeLa or KB-cells with ^{3}H-uridine from 1–2, 3–6 and 18–21 h after infection. Figure 1 demonstrates the dis-

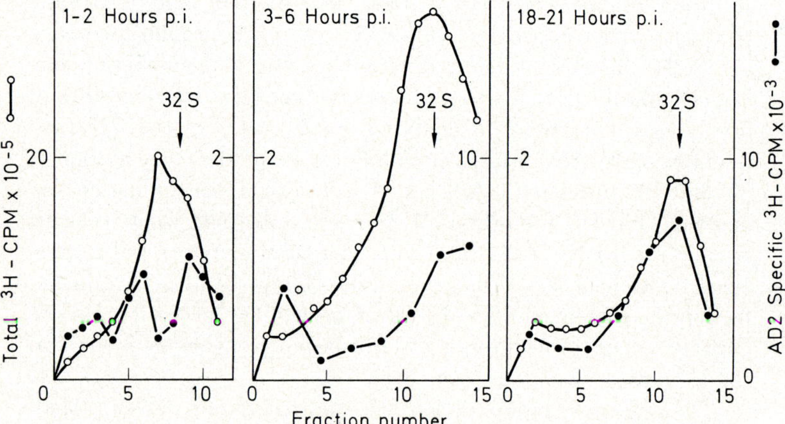

Fig. 1. Zonal sedimentation pattern of nuclear RNA from adenovirus-infected HeLa cells [34]. Nuclear RNA from adenovirus-infected cells extracted with phenol at different times after infection was dissolved in 100% DMSO with 0.01 M LiCl, diluted and sedimented through 15—30% sucrose gradient at 40000 rpm 25 °C for 2—3 h. Total virus-specific RNA in fractions from the gradient was determined by hybridization [34]. Adenovirus RNA binding was scored after treatment with pancreatic and T_1 RNase. Background binding to blank filters not containing viral DNA was subtracted

tribution of adenovirus-specific RNA in sucrose gradients following melting in dimethyl sulfoxide [34]. In all cases the viral RNA was found to be broadly distributed in the gradients. A significant fraction (>50%) of the virus-specific RNA sedimented faster than the 32 S ribosomal precursor RNA (molecular weight 2.5×10^6), which means that the average size of the adenovirus nuclear RNA both early and late after infection appears to be larger than the average size of viral polysomal mRNA at the corresponding times ([19, 22] and see below). A complete transcript of the adenovirus genome would be an RNA molecule approximately 12×10^6 daltons in size sedimenting at around 80 S. Both early and late after infection the most rapidly sedimenting viral nuclear RNA molecules appear to be as large as 12×10^6 daltons, but it remains to be established if they only contain viral or both viral and cellular sequences.

Viral Nuclear RNA and Polyadenylic Acid

To examine the presence of poly (A) segments in adenovirus nuclear RNA, viral RNA labeled with ^{3}H-adenosine was selected by hybridization to adenovirus DNA without RNase digestion [34]. Table 1 demonstrates the specificity of the hybrid selection experiments. The nuclear RNA recovered from the initial hybridization in all cases rehybridized to viral DNA at high efficiency (75–80%)

Table 1. Selection of adenovirus-specific nuclear RNA [a]

RNA	Input cpm	DNA	cpm bound and eluted	Per cent hybridized
Early Ad2	1.6×10^6	Ad2	32000	2.0
2—6 post infection		Blank	430	0.03
Late Ad2	4.05×10^6	Ad2	344000	8.5
16—18 post infection		Blank	9550	0.24
Uninfected	5.19×10^6	Ad2	11200	0.22
		Blank	6300	0.12

[a] Adenovirus-specific RNA was selected on filters containing 40 µg of DNA in 30% formamide 2× SSC buffered at pH 7.5 with TES for 18 hours at 45 °C. Hybrids were not digested with RNase prior to elution in 1 mM EDTA pH 7.5 at 100 °C for 10 min. This table is reprinted from [34].

indicating that this step resulted in a 10–40-fold purification of viral RNA. The adenovirus-specific RNA purified in this manner from either nuclear or polysomal RNA contained poly (A) segments 180–200 nucleotides long [26, 34], which is similar to the poly (A) material isolated from HnRNA and mRNA of uninfected cells [3]. Both early and late after infection 1–2 % of the ^{3}H-adenosine label was in poly (A) when large nuclear RNA (>40 S) was examined as compared to 4–10 % in the smaller RNA molecules (<40 S). This latter value approached that obtained for adenovirus polysomal RNA prepared in the same way (8–10 %) [26, 34].

Polysomes present in cells late in the infectious cycle contain almost exclusively virus-specific mRNA [16]. When the poly (A) content of mRNA purified from such polysomes by affinity chromatography on poly (U)-Sepharose was analyzed, we arrived at an average poly (A) content of 15 % for this fraction of viral RNA [16].

This value agrees well with that expected from the average molecular weight of the RNA assuming that all molecules contain only one poly (A) segment of some 180 nucleotides [1, 15, 16]. We also know that there is probably no virus-specific RNA that lacks poly (A) [16]. Therefore, we are inclined to conclude that the discrepancy observed [16, 26, 34] regarding the poly (A) content of late virus-specific polysomal mRNA purified by the two methods is due to losses of some of the poly (A) during hybridization selection.

Viral RNA Sequences in the Nucleus and on the Polysomes

The possibility that high molecular weight nuclear virus-specific RNA gives rise to mRNA through posttranscriptional cleavage led us to investigate whether all or only a fraction of the viral sequences transcribed in the nucleus could be detected on the polysomes. This was investigated with competition hybridization techniques [6, 8, 18].

DNA filters were incubated with increasing concentrations of unlabeled early or late RNA from nuclei or polysomes and the presaturated filters were treated with RNase prior to the second hybridization where the filters were charged with labeled RNA [18].

Figure 2 shows that unlabeled early nuclear RNA blocks hybridization of all of the RNA sequences present in both nuclei and polysomes at 3–6 h after infection. On the other hand, unlabeled polysomal RNA failed to compete with 25–30% of both early and late nuclear RNA sequences. These results and those of others suggest

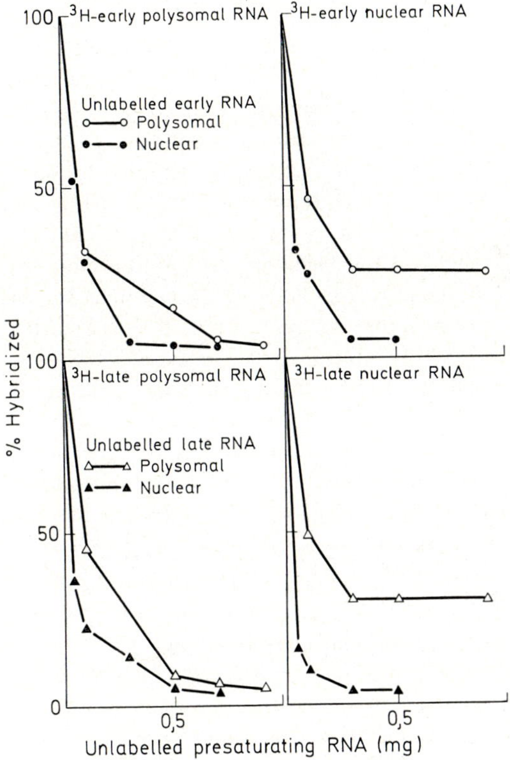

Fig. 2. Competition hybridization between early and late adenovirus polysomal RNA and HnRNA [34]. Purified unlabeled adenovirus polysomal or HnRNA (> 32 S) was annealed to 0.05 μg adenovirus DNA filters for 18 to 20 h at 65 °C. After RNase treatment and washing, the filters were subjected to a second round of hybridization for 18—20 h at 65 °C with saturating quantities of labeled adenovirus polysomal RNA or HnRNA (> 32 S). Saturating levels of hybridization to adenovirus DNA filters not exposed to 1500 cpm for early adenovirus HnRNA, 1700 cpm for early adenovirus polysomal RNA, 1650 cpm for late adenovirus HnRNA and 800 cpm for late adenovirus polysomal RNA. The extent of annealing of labeled RNA was scored after washing and nuclease digestion of filters as described by LUCAS and GINSBERG [18]. Hybridization to adenovirus DNA filters was corrected for binding to blank filters not containing viral DNA

that about a fourth of the adenovirus-specific sequences transcribed early or late never reaches the polysomes. [19, 34].

Resolution of Viral mRNA Species

The majority of messenger RNA both in uninfected and adenovirus-infected cells contain poly (A) sequences and these sequences are readily accessible to hybridization to poly (U) bound to an insoluble matrix [33]. On this basis we developed conditions for isolating mRNA from adenovirus-infected cells by binding to poly (U)-Sepharose [15]. The elution profiles of ^{3}H-uridine-labeled early and late polysomal RNA are shown in Fig. 3. That the poly (U)-Sepharose selects on the basis of poly (A) content was suggested by the fact that more than 95% of the poly (A) was recovered in the fraction that bound to the column. It is shown in Fig. 3 that approximately 20% of the early polysomal RNA was retained compared to 65% of late RNA. This large difference is due to suppression of ribosomal RNA synthesis late in adenovirus infection [10, 29]. Of the material which was retained by the column, that is the bulk of the mRNA, 19% early and as much as 92% late represented adenovirus-specific RNA as measured by exhaustive hybridization [16].

Electrophoresis on composite polyacrylamide gels [20, 23] was used to resolve the species of early and late polysomal mRNA purified by affinity chromatography on poly (U)-Sepharose [15, 16]. Hybridization to viral DNA was used to identify viral sequences. Figure 4 shows that the material *not* retained by poly (U)-Sepharose almost exclusively contained 28 S and 18 S ribosomal RNA, and no species of viral RNA could be detected in these fractions. When the material retained by the poly (U)-Sepharose columns was analyzed, the total radioactivity pattern of the early RNA showed the same degree of heterogeneity as the equivalent RNA from uninfected cells but reproducibly 3 main size classes (size range $0.3-3.5 \times 10^6$ daltons) of virus-specific RNA was revealed by hybridization [16, 21].

Late in infection the pattern of adenovirus mRNA was changed completely [16]. The overall radioactivity pattern coincided with the adenovirus-specific RNA when the material retained by the poly (U) columns was analyzed. Two dominating peaks, one at 26 S and the other at 22 S, was identified. Table 2 summarizes the relative S values and the approximate molecular weights for four of the species, one early and three late, which could easily be identified on polyacrylamide-agarose electrophoresis of material from different fractions of the cytoplasm (see below). It is noteworthy that while as much as 92% of the mRNA in the polysome

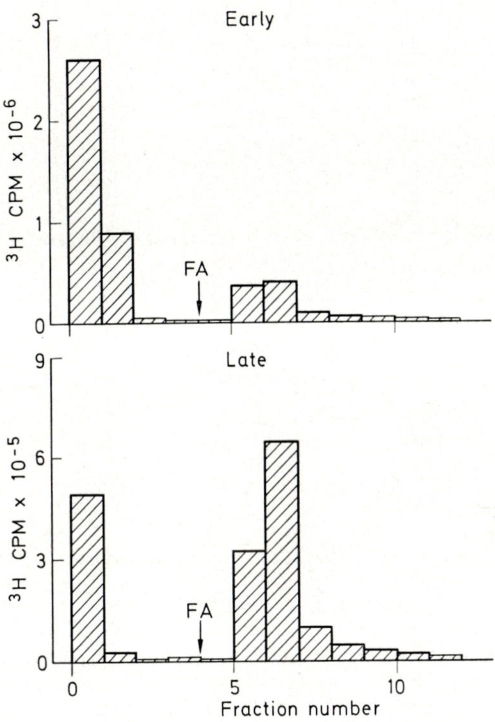

Fig. 3. Fractionation of polysomal RNA from adenovirus-infected cells by affinity chromatography on poly (U)-Sepharose. Cells labeled with 20 μCi/ml of ^{3}H-uridine during 3—5 and 14—16 h after infection with adenovirus were used for the preparation of early and late polysomal RNA, respectively. Polysomes were prepared, dissociated and fractionated on poly (U)-Sepharose as previously described [16]. Total radioactivity in each fraction is shown in the histograms. Twenty per cent of total radioactivity was recovered in the formamide-eluted fraction for early viral RNA and 65% for late viral RNA. Recovery of labeled RNA from the columns was around 95—97%. A detailed technique of the procedure has been published [15]

fraction late in infection is virus-specific [16], only 10–20% of the HnRNA at this time constitutes viral transcription products [27]. This discrepancy in the proportion of virus-specific RNA between the polysomal and the nuclear fractions might indicate that infected

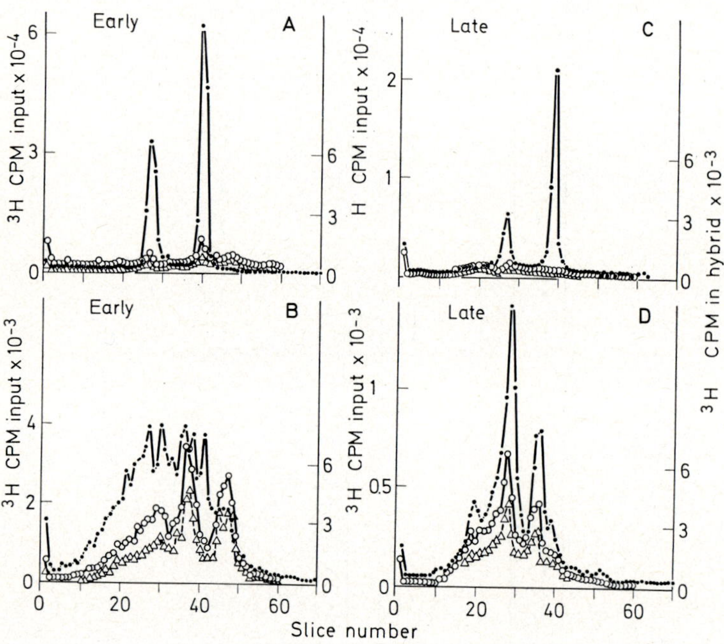

Fig. 4. Size distribution of RNA from the two fractions of poly (U)-Sepharose fractionation. The RNA from the fraction not retained by the column (PUS I) and the fraction retained by the column (PUS II) was precipitated with ethanol at −20 °C, redissolved in 0.1 × SSC containing 0.5% sarcosyl, and analyzed by gel electrophoresis [15]. Gels were analyzed for total radioactivity and virus specific RNA. Panel *A* and *C*: early and late PUS I RNA, respectively; Panel *B* and *D*: early and late PUS II RNA, respectively. Total ^{3}H-cpm ●—●; ^{3}H-cpm in viral mRNA ○—○

cells have a block in processing and transport of host messenger RNA or, alternatively, that the sequences of HnRNA representing host messenger RNA are not transcribed late in infection.

Table 2. Characteristics of some of the viral mRNA species produced during productive infection

		Relative S-value[a]	Approximate molecular weight $\times 10^{-6}$
Early	E 3	20 S	0.8
Late	L 4	22 S	1.0
	L 5	26 S	1.6
	L 7	33 S	2.4

[a] Estimated from the migration rate in polyacrylamide gels relative to ribosomal RNA.

A complete table of the different discernible viral mRNA species has been reported [16].

Kinetics of Appearance of mRNA on Polysomes During Productive Infection

The appearance of early and late mRNA on polysomes during productive infection was studied by isolating polysomes from cells labeled with ^{3}H-uridine for 2 h periods from 0–12 h after infection. Polysomal RNA was selected on poly (U)-Sepharose and material containing poly (A) was separated on polyacrylamide gels [15]. Adenovirus-specific RNA was identified by hybridization [8]. Fig. 5

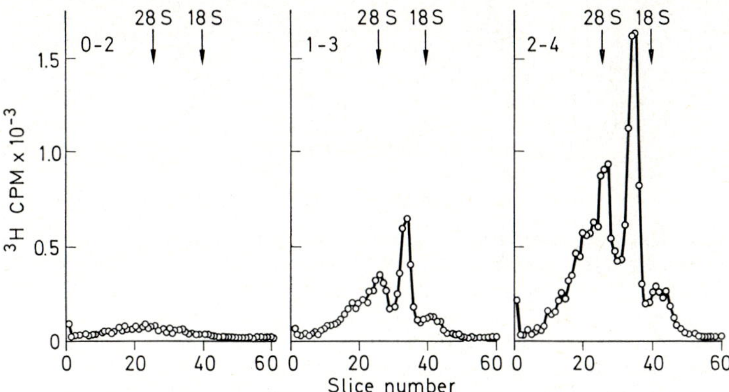

Fig. 5. Kinetics of synthesis of viral polysomal mRNA during the early phase of adenovirus infection. Poly (U)-Sepharose-selected polysomal RNA was analyzed by polyacrylamide gel electrophoresis [20, 23]. After electrophoresis the gels were sliced and analyzed for total acid insoluble radioactivity and for RNA hybridizable to adenovirus DNA [16]. Only radioactivity in virus-specific RNA is shown in the figure. Panel *A*, *B* and *C*: label was introduced from 0—2 h, 1—3 h and 2—4 h after infection, respectively

shows that early viral mRNA is accumulated on polysomes already 1—2 h after infection and the E 3 species is synthesized in large amounts already at 4 h after infection. During a highly synchron-

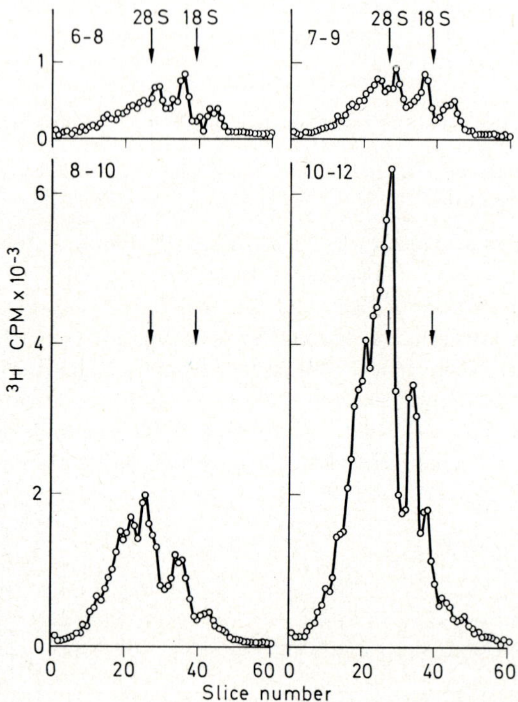

Fig. 6. Kinetics of synthesis of viral polysomal mRNA during the late phase of adenovirus infection. Polysomal RNA was selected on poly (U) Sepharose and analyzed by polyacrylamide gel electrophoresis. After electrophoresis the gels were sliced and analyzed for total acid insoluble radioactivity and for RNA hybridizable to adenovirus DNA [16]. Radioactivity corresponding to virus-specific RNA is given in the figure. Panel *A*, *B*, *C* and *D*: label was introduced from 6—8 h, 7—9 h, 8—10 h and 10—12 h after infection, respectively

ized infection there is a shift from an early to a late pattern at around 8—10 h after infection as shown in Fig. 6. The L 4 and L 5 species are detectable at 10—12 h after infection. The kinetic data of Fig. 5 and 6 have been summarized in Fig. 7, which

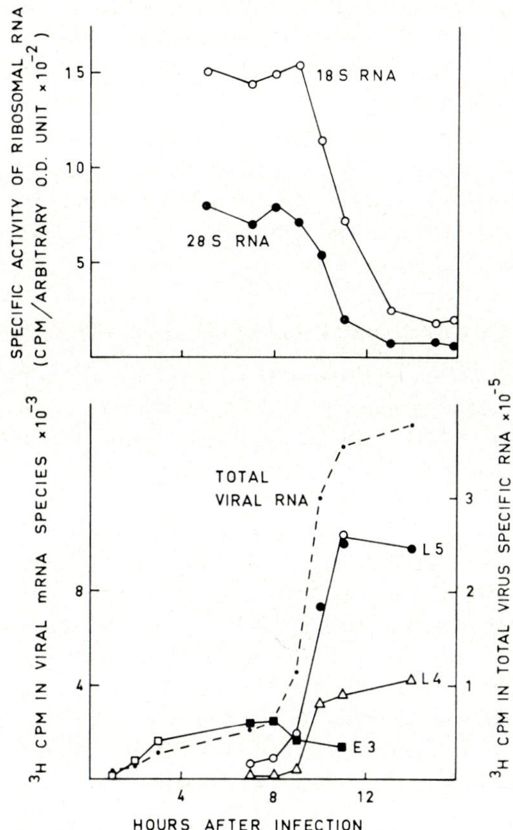

Fig. 7. Inhibition of ribosomal RNA synthesis in temporal relation to synthesis of viral RNA during adenovirus infection. Panel *A* shows the inhibition of 28 S and 18 S ribosomal RNA late in the infectious cycle. Polysomal RNA from cells labeled at different times after infection was subjected to affinity chromatography on poly (U)-Sepharose and the material passing unadsorbed through the columns was analyzed by sucrose density gradient centrifugation. The specific activity of the 28 S and 18 S peaks expressed in cpm/optical density unit was then determined for each gradient. Panel *B* shows the total radioactivity under the peaks of E 3, L 4, and L 5 at different times after infection as defined in Table 2. These resolved species of RNA are seen in Figs. 5 and 6

demonstrates a reduction in the E 3 species concurrent with an exponential increase in synthesis of the L 4 and the L 5 species. The decrease in host cell ribosome synthesis [10, 29] which occurs concurrently with the shift from the early to the late RNA patterns is also shown in Fig. 7.

In conclusion, the kinetics reveal that the pattern of mRNA species is different at early and late periods after infection and some species of early RNA does not appear to be synthesized late although it is difficult to discern small amounts of early mRNA present late in infection.

A Cytoplasmic Pathway for Viral mRNA?

In confirmation of the results of RASKAS and OKUBO [28] we have found that more than 50% of the adenovirus-specific RNA labeled during a 2 h pulse late in infection is present in the postpolysomal fractions. This lead us to examine in greater detail the distribution of viral mRNA between different fractions of the cytoplasm using affinity chromatography on poly (U)-Sepharose [16] to isolate mRNA and electrophoresis on polyacrylamide agarose gels to resolve viral mRNA species. Figure 8 demonstrates that distinct species of viral mRNA can be demonstrated not only in polysomes but also in the postpolysomal region of the sucrose gradient. Fig. 9 shows a plot of the relative amount of three viral mRNA species at different positions in such sucrose gradients.

It has been suggested that mRNA might exist in the cytoplasm of mammalian cells in the form of ribonucleoprotein particles [32].

Fig. 8. Localization of late viral mRNA in different regions of a polysomal sucrose gradient. Cells were labeled with 1 mCi of ^{3}H-uridine from 14—16 h after adenovirus infection and the cytoplasm-prepared as described earlier [15], was subjected to centrifugation in a 7—47% sucrose gradient in isotonic buffer [15]. The top panel shows the optical density and the radioactivity patterns for the sucrose gradient. ^{3}H-cpm ○—○. From each of the six indicated regions RNA was dissociated from protein with sarcosyl and EDTA and the poly (A) containing RNA isolated by poly (U)-Sepharose chromatography [15] (see Fig. 3). After precipitation with ethanol the redissociated RNA was analyzed by polyacrylamide electrophoresis as shown in the lower frames. The poly (A)-containing RNA from regions 1—5 was found to hybridize to adenovirus DNA

Reproducibly 30% of the total cytoplasmic poly (A)-containing RNA in uninfected cells appears to be confined to the post polysomal region of sucrose gradients [11, 15] but it has been difficult

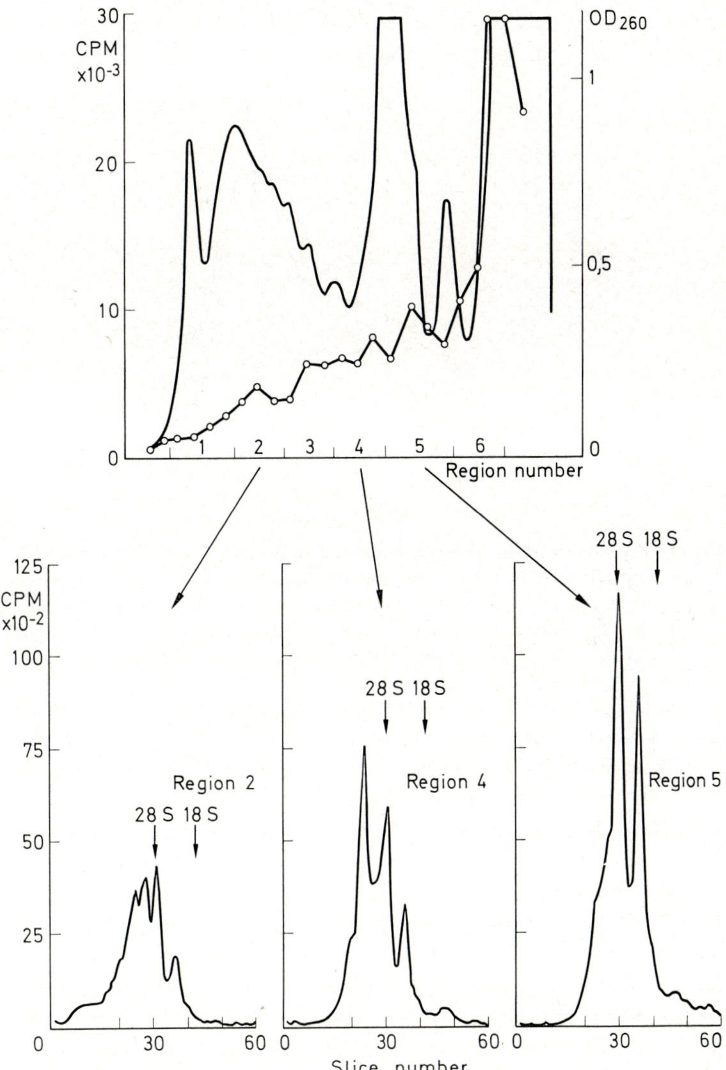

Fig. 8. Legend, see opposite page

to establish whether this cytoplasmic pool of mRNP is part of the normal transport pathway for mRNA going from the nucleus to the cytoplasmic polysomes [11, 32].

Since viral mRNA could be detected in all parts of the polysomal gradient, it might be rewarding to reinvestigate whether mRNA passes through a nonpolysomal mRNA pool in the cyto-

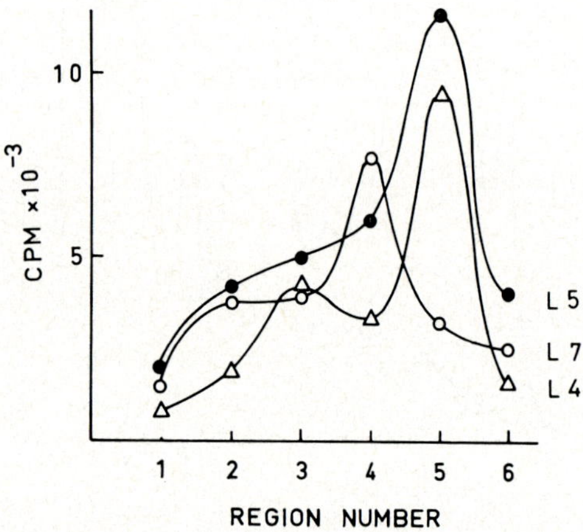

Fig. 9. Relative distribution of three adenovirus-specific mRNA species across the polysomal sucrose gradient of Fig. 8. The species L 4, L 5, and L 7 could be identified on polyacrylamide gels as shown in the lower frames of Fig. 8. The relative amounts present in each species in the six regions have been plotted

plasm before it is engaged in protein synthesis on polysomes in the viral system. The drug cordycepin, 3-deoxyadenosine, strongly interferes with the polyadenylation of nuclear RNA in uninfected and adenovirus-infected cells [3, 26] and at the same time causes a block in the transport of mRNA from the nucleus to cytoplasmic polysomes [25]. This drug was used in experiments designed to examine the kinetics of appearance of viral mRNA in cytoplasmic fractions obtained by sucrose gradient centrifugation. Cells were pulsed for 10 min with ^{3}H-uridine and cordycepin was added at 50 μg/ml after this period. Samples were then taken at different times after the addition of the drug and poly (U)-selected RNA from the polysomal, the pre 80 S and the post polysomal regions of sucrose gradients as in Fig. 5 were analyzed for viral mRNA by exhaustive hybridization to adenovirus DNA. The results (Fig. 10) indicate that indeed viral mRNA appeared first in the postpolysomal region of the gradient and only later in the polysomes. The analysis of this system therefore suggests that mRNA was exported from the nucleus in the form of mRNP, possibly not in combination with ribosomes or subunits, to a pool in the cytoplasm from which it is then drawn to initiate protein synthesis.

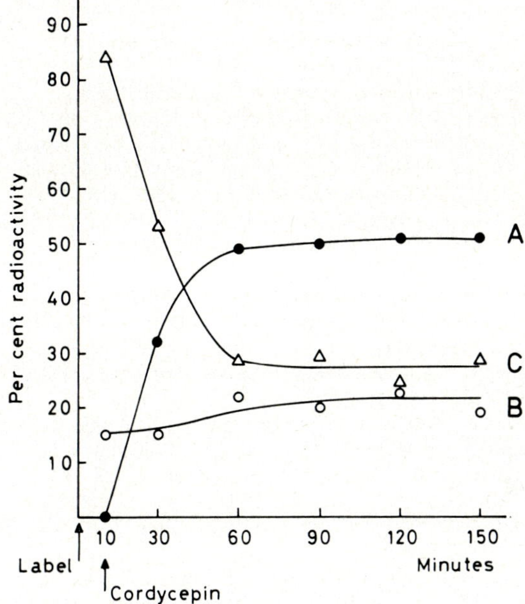

Fig. 10. Relative distribution of adenovirus-specific RNA between three different compartments of the cytoplasm after a pulse-chase label late in adenovirus infection ($1{,}5 \times 10^8$). Cells were labeled with 4 mC of ^{3}H-uridine for 10 min at 14 h after infection whereupon cordycepin was added at 50 μg/ml. Samples of cells were taken at different times, the cytoplasm was prepared and separated on 7—47% sucrose gradients as previously described [15] and as shown in Fig. 8. The polysomal (*A*), the initiation (*B*) and the postpolysomal regions (*C*), approximately corresponding to regions 1—2, 3—4 and 5—6 in Fig. 8, were separately dissociated and fractionated on poly (U)-Sepharose. The fractions retained by the columns were exhaustively hybridized [15] to filters containing 20 μg of adenovirus DNA and the amount of RNase-resistant virus-specific mRNA found in each compartment at each time was used to calculate the relative distribution

References

1. ADESNIK, M., SALDITT, M., THOMAS, W., DARNELL, J. E.: J. molec. Biol. **71**, 21—30 (1972).
2. DARNELL, J. E., WALL, R., TUSHINSKI, R.: Proc. nat. Acad. Sci (Wash.) **68**, 1321—1325 (1971).
3. DARNELL, J. E., PHILIPSON, L., WALL, R., ADESNIK, M.: Science **174**, 507 to 510 (1971).
4. EDMONDS, M., VAUGHAN, M. H., NAKAZOTO, N.: Proc. nat. Acad. Sci. (Wash.) **68**, 1336—1340 (1971).
5. FUJINAGA, K., GREEN, M.: Proc. nat. Acad. Sci (Wash.) **65**, 375—382 (1970).
6. FUJINAGA, K., MAK, S., GREEN, M.: Proc. nat. Acad. Sci (Wash.) **60**, 959 to 966 (1968).
7. GILEAD, Z., GINSBERG, H. S.: J. Virol. **2**, 7—14 (1968).
8. GILLESPIE, S., GILLESPIE, D.: Biochem. J. **125**, 481—487 (1971).
9. GREEN, M., PARSONS, J. T., PIÑA, M., FUJINAGA, K., CAFFIER, H., LANDGRAF-LEURS, I.: Cold Spr. Harb. Symp. quant. Biol. **35**, 803—818 (1970).
10. HODGE, L. D., SCHARFF, M. D.: Bact. Proc. **1968**, 173.
11. JELINEK, W., ADESNIK, M., SALDITT, M., SHEINESS, D., WALL, R., MOLLOY, G., PHILIPSON, L., DARNELL, J. E.: J. molec. Biol. **75**, 515—532 (1973).
12. KATES, J.: Cold Spr. Harb. Symp. quant. Biol. **35**, 743—752 (1970).
13. LEE, Y., MENDECKI, J., BRAWERMAN, G.: Proc. nat. Acad. Sci. (Wash.) **68**, 1331—1335 (1971).
14. LINDBERG, U., DARNELL, J. E.: Proc. nat. Acad. Sci. (Wash.) **65**, 1089 to 1096 (1970).
15. LINDBERG, U., PERSSON, T.: Europ. J. Biochem. **31**, 246—254 (1972).
16. LINDBERG, U., PERSSON, T., PHILIPSON, L.: J. Virol. **10**, 909—919 (1972).
17. LONBERG-HOLM, K., PHILIPSON, L.: J. Virol. **4**, 323—338 (1969).
18. LUCAS, J. J., GINSBERG, H. S.: J. Virol. **8**, 203—212 (1971).
19. MCGUIRE, P. M., SWART, C., HODGE, L. D.: Proc. nat. Acad. Sci (Wash.) **69**, 1578—1582 (1972).
20. ÖBERG, B., PHILIPSON, L.: Europ. J. Biochem. **11**, 305—315 (1969).
21. PARSONS, J. T., GREEN, M.: Virol. **45**, 154—162 (1971).
22. PARSONS, J. T., GARDNER, J., GREEN, M.: Proc. nat. Acad. Sci. (Wash.) **68**, 557—560 (1971).
23. PEACOCK, A. C., DINGMAN, C. W.: Biochemistry **7**, 668—674 (1968).
24. PENMAN, S., SCHERRER, K., BECKER, Y., DARNELL, J. E.: Proc. nat. Acad. Sci. (Wash.) **49**, 654—662 (1963).
25. PENMAN, S., ROSBASH, M., PENMAN, M.: Proc. nat. Acad. Sci. (Wash.) **67**, 1878—1885 (1970).
26. PHILIPSON, L., WALL, R., GLICKMAN, G., DARNELL, J. E.: Proc. nat. Acad. Sci. (Wash.) **68**, 2806—2809 (1971).
27. PRICE, R., PENMAN, S.: J. Virol. **9**, 621—626 (1972).
28. RASKAS, H. J., OKUBO, C. K.: J. Cell Biol. **49**, 438—449 (1971).
29. RASKAS, H. J., THOMAS, D. C., GREEN, M.: Virol. **40**, 893—902 (1970).
30. RUSSEL, W. C., KNIGHT, B. E.: J. gen. Virol. **1**, 523—528 (1967).

31. Russell, W. C., Skehel, J. J.: J. gen. Virol. **15**, 45—57 (1972).
32. Schochetman, G., Perry, R. P.: J. molec. Biol. **63**, 577—590 (1972).
33. Sheldon, R., Jurale, C., Kates, J.: Proc. nat. Acad. Sci. (Wash.) **69**, 417—421 (1972).
34. Wall, R., Philipson, L., Darnell, J. E.: Virol. **50**, 27—34 (1972).

Discussion

J. Paul (Glasgow): Thank you very much for your talk. We shall now enter the discussion. Since people tend to forget the first speaker, I would like to invite questions especially to Dr. Scherrer.

C. A. Thomas (Boston): Can I ask you what is the length of the RNA chains that you find in your polysomes, and may I ask Dr. Philipson how this agrees with his findings?

K. Scherrer (Lausanne): About 30 S is our maximum length. This corresponds to 1.5×10^6 daltons molecular weight.

L. Philipson (Uppsala): The molecular weight might be a little higher, in the region of two million; but 30—33 S is the maximum that can be observed for messenger RNA.

C. A. Thomas: We saw some electron micrographs taken by Oscar Miller last week at the Gatlinbury Meeting showing polysomes of incredible length, up to 6 to 10 μ.

K. Scherrer: We have in HeLa cells a population of many polyribosomes. A giant polysome would correspond to a message of 26 S. Such a messenger RNA of 26 S — *e.g.*, the myosin messenger — could make a polysome with 60 to 80 ribosomes.

H. Tiedemann (Berlin): What happens to the RNA during treatment with DMSO?

K. Scherrer: There are mainly three effects. First of all DMSO can degrade RNA if you do not take the proper precautions; this has been shown in Hofschneider's laboratory. Secondly DMSO reduces the sedimentation of molecules; the molecules are unfolded, and this increases the viscosity, so that the sedimentation becomes slower. And thirdly, it breaks down aggregates. 80% of the RNA hybridizing, *i.e.*, heavy RNA, breaks down in the DMSO gradient.

L. Philipson: In our experiments, RNA is completely intact in the DMSO gradient. The RNA was isolated without phenol treatment and under special precautions. I want to emphasize that DMSO gradients will always reveal nicks in your RNA, irrespective of how you make them.

Miller: I would like to remind you of some earlier experiments of Prof. Harris where he showed that heating of Hn-RNA produced molecules with a mean sedimentation coefficient of 16 S. These data suggest that Hn-RNA molecules would be aggregates [Bramwell, M.E., Harris, H.: Biochem. J. **103**, 816 (1967).

K. Scherrer: Heat treatment would always tend to break down RNA molecules. The decomposition of viral RNA by heat has been studied in detail by John Holland. Even treatment at 65° C, as was recommended in many isolation procedures, would lead to degradation of at least 10% of Hn-RNA.

C. A. Thomas: Since we have all these experts here, I cannot resist to ask another technical point. You know that Edström and his colleagues, in isolating the giant RNA from the Balbiani ring of chromosome IV, always fix the material with acetic acid-ethanol. In spite of this treatment, the RNA obtained seems to be unbroken. I wonder if any one of you have ever treated nuclei of cells with acetic acid-ethanol?

K. Scherrer: I have tried it, using the method of Cox to reveal hidden breaks. There you bring giant RNA to pH 3 and then up again. Moreover, if you use acetic acid in ethanol it is quite another thing as aqueous acetic acid.

G. R. Phillipp (Bonn): This question is addressed to both speakers. In your hybridization assays did you consider the presence of *poly A* stacks in mRNA since these may probably have an effect upon the percentage of RNA which is hybridized?

L. Philipson: In the hybridization of adenovirus mRNA we score hybrids with and without RNase treatment and furthermore the *poly* (*A*) is not hybridized to adenovirus DNA, as previously shown [Proc. Nat. Acad. Sci. **68**, 2806 (1971)]. Therefore the values given for RNase-treated hybrids are probably not containing poly (A) stacks.

K. Scherrer: May I comment on this? I think we must distinguish between the true poly A stretches of 150 to 200 nucleotides and that which I call oligo-A sequences of 20—30. Clusters of this kind have been demonstrated long ago.

L. Philipson: Yes. I meant that everything which is internal in the DNA is really transcribed; but the long stretch of poly A at the 3′-terminal is probably put on posttranscriptionally by a special enzyme.

P. Chambon (Strasbourg): My question is to Dr. Scherrer. Have you tried to translate your giant RNA?

K. Scherrer: Yes, we have tried to translate it in oocytes. We got some stimulation, but the question is that of a proper control. We have not yet looked for globin sequences in the translation product.

Scholtissek: I have a question on your poly U stretches in the adenovirus RNA. What is the evidence that it is on the 5′-end, and have you any idea about the function? And finally, does poly U also occur in m-RNA from non-infected cell?

L. Philipson: I think the work published is mostly done in non-infected cells. The evidence that it is at the 5′-end is not very strong; but it must be towards the 5′-side of the molecule, and it is not retained in the messenger.

Dube (Göttingen): Is there a difference between the size of the free messenger in the cytoplasm and messenger bound to polyribosomes? And secondly, is there any evidence that the messenger must be processed before it can be translated?

K. Scherrer: If you look at the whole spectrum, it is the same. However, in differentiated cells, especially in the erythroblasts, there are differences. All the mRNA falls in sizes between 4 and 30 S. But in the polyribosomes, you can have 30% 9 S, while free in the cytoplasm there might be between 5 and 25% of 9 S-size. As for processing, I do not believe that this is necessary, since we took 9 S messenger from the particle and translated it.

Miller: I would be inclined to say that all the polysomal heterodisperse-sedimenting RNA molecules which contain *poly A* stretches are messengers actually being translated. This results from experiments where HeLa cell polysomal RNA (labeled for messenger RNA) has been chromatographed on methylated bovine albumin kieselguhr coated (MAK) columns. This method also reveals the existence in the polysome region of the sucrose gradient of another class of RNA (present as 1.40 g/ml ribonucleoprotein particles) which does not have *poly A* sequences, which might correspond to the hidden form of mRNA awaiting translation. This fits recent data showing that for certain messengers the presence of the poly A fragment is necessary for initiation [Rosenfeld, M. G., Abrass, I. N., Perkins, L. A: Biochem. Biophys. Res. Comm. **49**, 230 (1972)].

I would agree with a previous remark that all the RNA retained on Sepharose-poly U columns may not necessarily be messenger RNA: a fraction may stick by virtue of short stretches of *oligo A* distributed internally and differ from active messenger where the *poly A* is the 3′-OH end. [Bramwell, M. E., Harris, H.: Biochem. J. **103**, 876 (1967)].

J. Paul: I think one point was misleading in this discussion, *i.e.*, that a *poly A* part is a prerequisite for translation. It is known that the histone messenger which has no *poly A* is readily translated.

Transcription II

Double-Stranded Viral and Cellular DNA's as Templates for Purified Mammalian DNA-Dependent RNA Polymerases

P. Chambon, J. L. Mandel, F. Gissinger, C. Kedinger, M. Gross-Bellard and P. Hossenlopp

Institut de Chimie Biologique, Unité de Recherche sur le Cancer de l'INSERM et Centre de Neurochimie du CNRS, Faculté de Médecine, Strasbourg France

Abbreviations

SV 40 DNA-Form I (FI)	supercoilded circular DNA of Simian Virus 40.
SV 40 DNA-Form II (FII)	open circular DNA of Simian Virus 40.
SV 40 DNA-Form III (FIII)	linear DNA of Simian Virus 40.

Introduction

The multiplicity and different intranuclear localizations of DNA-dependent RNA polymerases have suggested that gene expression in animal cells is regulated, at least in part, by distinct RNA polymerases with different template specificities [1]. One of the most direct ways to support this hypothesis would be to show *in vitro* that the various purified enzymes specifically transcribe different parts of the deproteinized chromosomal DNA. A first step in such a study is to ask whether the purified enzymes alone can in fact initiate RNA synthesis on an intact double-stranded DNA, or whether the presence of additional initiation factor(s), similar to the *E. coli* σ factor [2, 3] is (are) required. Previous studies involving normal calf thymus DNA preparations and the initiation inhibitor AF/013 [4–7] have suggested that purified AI and B enzymes initiate at different sites on calf thymus DNA, and that the B enzymes could in fact initiate on "true" double-stranded regions of the DNA while AI enzyme could lack an initiation factor. It is nevertheless clear that the possible specificity of the animal enzymes cannot be unequivocally demonstrated using these ordinary preparations of DNA's, since all of them contain a high number of single-stranded nicks ("high" compared to the expected number of true promoter sites), acting as efficient non-specific initiation sites [8, 9]. On the other hand, using intact phage DNA's, we have shown [9, 10] that the purified calf thymus enzymes can bind to phage DNA's, but do not initiate RNA synthesis and behave, in many respects, like the bacterial core enzyme which requires the σ initiation factor to initiate RNA synthesis on phage DNA's [11]. Since the phage DNA's are obviously non-physiological templates for the animal RNA polymerases, two interpretations could be proposed to explain these results. Either some general initiation factor(s) is (are) lost during the purification of the calf thymus RNA polymerases, or the animal enzymes are unable to recognize the initiation signals present in the DNA of these too-remote organisms. Studies of the transcriptability of intact DNA of animal viruses (SV 40 and adenovirus) which do not possess their own polymerase and therefore are transcribed, at least early after the infection, by one of the cellular DNA-dependent RNA polymerases should help to discriminate between these two alternatives. Analysis of the transcriptability of the high molecular weight chromosomal DNA, free of single-stranded nicks, which was recently prepared by Gross-Bellard et al. [12] should also be informatory. We have performed both types of studies and we have also characterized the RNA transcripts synthesized by AI and B enzymes on SV 40 DNA Form I, in the hope of revealing that the two types of enzyme exhibit different template specificities, and also to investigate whether the animal RNA polymerases exhibit the two main features of the prokaryotic transcription, namely selectivity and asymmetry [2, 3]. We discuss here the results of these studies which lead us to conclude that the purified mammalian RNA polymerases initiate RNA synthesis only very poorly, if at all, on unnicked "regular" double-stranded DNA's.

Transcription of a Viral Supercoiled Circular DNA

SV 40 DNA Form I (FI) is a supercoiled double-stranded circular DNA with a molecular weight of around 3.3×10^6 daltons. Study of its transcription *in vitro* by animal DNA-dependent RNA polymerases appealed to us for several reasons: 1. its circular structure resolves the problem of non-specific initiation at single-stranded nicks or at ends of DNA molecules; 2. *in vivo*, early viral m-RNA is complementary to about 30% of the sequences of one of the DNA strands, while late after infection additional mRNA appears, which is complementary to the remaining 70% of the other DNA strand, suggesting that *in vivo* transcription is selective and asymmetrical [13–19]; 3. *in vitro* transcription of SV 40 DNA-FI by *E. coli* holoenzyme is asymmetrical and primarily from the DNA strand which is transcribed early *in vivo* [16, 18–20].

Table 1 shows that SV 40 DNA-FI is readily transcribed by calf thymus AI and B RNA polymerases at low ionic strength in the presence of Mn^{++}. The most surprising finding was the very poor transcriptability of SV 40 DNA-FI by enzyme AI in the presence of Mg^{++}. This result markedly contrasts with the previous observations that enzyme AI is equally stimulated by Mn^{++} or Mg^{++} when the template is calf thymus DNA (Table 1 and [1, 21]).

Table 1. Transcription of allomorphic forms of SV40—DNA

Enzyme	Divalent cation	Salt (mM)	DNA (0.20 μg)			
			SV40—FI pmol	SV40—FII pmol	SV40—FIII pmol	N—CT pmol
CT—AI	Mn^{++}	—	32.4	13.0	7.2	45.7
		AS = 16	24.1	12.5	—	41.1
	Mg^{++}	—	5.6	9.6	5.5	31.6
CT—B	Mn^{++}	AS = 25	27.3	11.7	4.8	18.7
	Mg^{++}	AS = 16	4.1	4.1	1.1	10.0
E. coli holoenz.	Mg^{++}	KCl = 120	110.9	—	53.3	47.7
E. coli core enz.	Mg^{++}	KCl = 120	21.4	—	19.1	31.3

RNA synthesis was determined as previously described [52] and is expressed in pmoles [^{3}H]UTP incorporated in 10 min at 37° C. Mn^{++} (3 mM), Mg^{++} (8 mM), ammonium sulfate (AS) and KCl were added to the incubation medium as indicated in the table. CT—AI = purified calf thymus AI enzyme, CT—B = purified calf thymus B enzymes, N—CT = native calf thymus DNA [52] (for details, see MANDEL and CHAMBON [53]).

It is also interesting to note that SV 40 DNA-FII (open double-stranded circular form) and, even more so, SV 40 DNA-FIII (double-stranded linear form) are very poor templates for both AI and B enzymes and that SV 40 DNA-FI is efficiently transcribed by *E. coli* core enzyme, as previously reported by SUGDEN and SAMBROOK [22].

The size of the RNA's synthesized by calf thymus polymerases AI and B on SV 40 DNA-FI was analyzed by sedimentation through formaldehyde-sucrose gradients [57]. Most of the RNA made in 30 min by enzyme AI is much larger than 28 S rRNA, like the RNA made by *E. coli* holoenzyme on the same template [23, 24, 57]. RNA polymerase AI can therefore pass over its own initiation site on SV 40 DNA-FI and synthesize RNA which has at least twice the length of a DNA strand. On the other hand, the RNA made by calf thymus B enzymes has a definite maximum size [57] which is always less than a complete transcript of the viral genome, irrespective of the divalent cation in the incubation medium. As reported elsewhere [25] this result was not due to the presence of traces of RNAase in the B enzyme preparation. Furthermore, measurement of the incorporation of [^{32}P]-γ-labeled ATP and GTP and calculation of the average chain lengths of the RNA's synthesized by AI and B enzymes on SV 40 DNA-FI have confirmed the much larger size of the RNA transcripts made by enzyme AI [25]. These determinations indicate also that AI and B enzymes have multiple initiation sites on SV 40 DNA-FI, since both ATP and GTP were found at the 5′ end of the synthesized RNA chains.

The possible asymmetry of the transcription catalyzed by calf thymus enzymes AI and B on SV 40 DNA-FI was analyzed by self-annealing assay of the synthesized RNA's. As shown in Table 2, the transcription is largely symmetrical with both enzymes, irrespective of the incubation conditions and of the SV 40 DNA form. In contrast, the *E. coli* holoenzyme transcription is essentially asymmetrical as previously found by WESTPHAL [20]. On the other hand, it is interesting to note that *E. coli* core enzyme does not transcribe preferentially one of the DNA strands, a result in keeping with the symmetrical transcription of phage DNA's by this enzyme [2, 3]. Since the size of the RNA synthesized by enzyme AI is larger than the SV 40 genome size, these results demonstrate that enzyme AI transcribes the two strands of SV 40 DNA-FI over their entire length. In the case of the B enzymes, the smaller size of the RNA transcripts suggest that, although symmetrical, the *in vitro* transcription of SV 40 DNA-FI could nevertheless exhibit some selectivity, only a part of the genome being transcribed.

Table 2. Symmetric transcription of SV40 DNA by AI and B RNA Polymerases

Enzyme	Incubation conditions	SV40 DNA template	Ribonuclease resistance after self-annealing (in %)
CT—AI	Mn^{++}, no salt	FI	76.9
CT—B	Mn^{++}, AS 25 mM	FI	62.4
	Mg^{++}, AS 16 mM	FI	56.5
	Mn^{++}, AS 25 mM	FIII	66.6
RL—B	Mn^{++}, AS 25 mM	FI	69.5
	Mg^{++}, AS 16 mM	FI	50.7
E. coli holoenz.	Mg^{++}, KCl 120 mM	FI	20.0
	Mn^{++}, KCl 120 mM	FI	21.0
	Mg^{++}, KCl 120 mM	FIII	27.6
	Mg^{++}, KCl 120 mM	denatured FI	67.3
E. coli core enz.	Mn^{++}, no salt	FI	66.9

RNA's were synthesized and self-annealed as described elsewhere [25]. Mn^{++} (3 mM), Mg^{++} (8 mM), ammonium sulfate (AS) and KCl were added to the incubation medium as indicated in the table. CT—AI = purified calf thymus AI enzyme, CT—B = purified calf thymus B enzymes (for details, see MANDEL and CHAMBON [25]).

This alternative is clearly not the case since most of the AI or *E. coli* transcripts can be hybridized with the RNA synthesized by the B enzymes (Table 3). In these experiments the concentration of the AI and *E. coli* transcripts was low enough to prevent their self-annealing. One should therefore conclude that the maximum definite size of the RNA synthesized by the B enzymes on SV 40 DNA-FI does not reflect the selective transcription of some part of the genome. Other results, not shown [25], suggest that this maximum size is related to an inhibition of chain elongation by the RNA product.

Taken together, these results demonstrate that the *in vitro* transcription of SV 40 DNA by the two types of calf thymus RNA polymerases does not exhibit the two main features of *in vivo* and *in vitro* prokaryotic transcription, asymmetry and selectivity. Similar results were obtained with the rat liver B enzymes when the

Table 3. Hybridization of RNA's synthesized *in vitro* by calf thymus AI enzyme and *E. coli* holoenzyme with RNA made *in vitro* by calf thymus B enzymes on SV40 DNA—FI

Labeled RNA's present during the hybridization period and enzymes used for their synthesis		RNAase resistance of [^{3}H]-RNA after hybridization (in %)
[^{3}H]-RNA	[^{32}P]-RNA	
CT—AI	—	23.6
CT—AI	CT—AI	84.2
CT—AI	CT—B	74.9
E. coli holoenz.	—	11.2
E. coli holoenz.	CT—AI	76.6
E. coli holoenz.	CT—B	68.6

Concentrations of [^{3}H]-RNA and [^{32}P]-RNA were 3.3×10^{-3} µg/ml and 4×10^{-2} µg/ml, respectively. Hybridization was performed at 68 °C for 3 hours. CT—AI = purified calf thymus AI enzyme, CT—B = purified calf thymus B enzymes (for details, see MANDEL and CHAMBON [25]).

phosphocellulose chromatography step was omitted from the purification in order to prevent the loss of a possible animal σ-like factor [25, 26]. Our results are at variance with previous reports showing that SV 40 DNA-FI was asymmetrically transcribed *in vitro* by KB cell [27] and rat liver [28] RNA polymerases primarily from the DNA strand which is also transcribed *in vitro* by *E. coli* enzyme [27]. However, more recent results are in keeping with our observations since they indicate that RNA's synthesized *in vitro* by cellular DNA-dependent RNA polymerases on either SV 40 DNA-FI (SAMBROOK, personal communication) or polyoma DNA-FI (MONJARDINO, personal communication) are largely symmetrical. In this respect, it is important to stress that the lack of selectivity and the symmetry of the *in vitro* transcription of SV 40 DNA-FI by either types of animal DNA-dependent RNA polymerases does not necessarily mean that the *in vitro* situation is completely different from that *in vivo* since it has been recently shown that viral cytoplasmic mRNA's are probably derived from much larger nuclear precursors [15, 29–32] and that *in vivo* transcription is mostly symmetrical [33, 58].

That both AI and B RNA polymerases transcribe double-stranded circular DNA suggests that these enzymes can initiate on intact native double-stranded DNA and that no additional initiation factors are required. In fact, it was recently shown that supercoiled circular DNA's are not perfectly double-stranded and contain, due to conformational constraints, small loosely basepaired regions [34–37]. In this respect it is noteworthy that the *E. coli* core enzyme, which initiates RNA synthesis very poorly on linear unnicked double-stranded phage DNA's, efficiently transcribes SV 40 DNA-FI as well as other supercoiled circular DNA's [38]. If one assumes that the purified animal RNA polymerases are in fact unable to initiate on a regular double-stranded DNA, this peculiar structure of SV 40 DNA-FI would explain why it is more effectively transcribed than the linear Form III. This could also explain why Mn^{++} is much more efficient than Mg^{++} in stimulating SV 40 DNA-FI transcription since it is known that Mn^{++} has a destabilizing effect on double-stranded DNA [39, 40].

Transcription of Viral and Cellular Linear Double-Stranded DNA's

As discussed above, the ability of the purified animal enzymes to transcribe supercoiled double-stranded SV 40 DNA-FI does not rule out the possibility that, in fact, they lack some factor(s) necessary for the initiation of RNA synthesis on regular double-stranded DNA's.

We first chose another viral DNA – adenovirus-2 DNA – to answer this question since: 1. it has a well-defined unnicked linear double-stranded structure and its molecular weight (23×10^6) is high enough to minimize the importance of the initiations which could occur at the ends of the DNA molecules; 2. it was recently shown that the adenovirus genome is transcribed *in vivo* both by an amanitin-sensitive (B type) RNA polymerase [41–43] and an amanitin-resistant (A type) enzyme [41, 44]. Results presented in

Table 4 show that native adenovirus-2 DNA is very poorly, if at all, transcribed by purified calf thymus AI and B RNA polymerases, irrespective of the divalent cation present in the incubation medium. Nor is it transcribed by a partially purified calf thymus

Table 4. Transcription of adenovirus-2 DNA

Enzyme	Salt (mM)	Divalent cation	DNA (0.5 μg)	[^{32}P]-α-ATP incorporated (10 min, 37 °C) (pmol)
CT—AI	AS = 25	Mn^{++}	N-Ad2	0.7
			D-Ad2	28.8
			N-CT	36.2
		Mg^{++}	N-Ad2	1.2
			N-CT	29.3
CT—B (PC2)	AS = 25	Mn^{++}	N-Ad2	2.3
			D-Ad2	133
			N-CT	42
		Mg^{++}	N-Ad2	0.1
			N-CT	29.6
CT—B (DCB)	AS = 25	Mn^{++}	N-Ad2	3.5
			N-CT	19.6
		Mg^{++}	N-Ad2	1.5
			N-CT	11.2
RL—B	AS = 25	Mn^{++}	N-Ad2	2.3
			N-CT	30
		Mg^{++}	N-Ad2	0.0
			N-CT	15.6
E. coli holoenz.	KCl = 120	Mg^{++}	N-Ad2	52
			N-CT	104
E. coli core enz.	KCl = 120	Mg^{++}	N-Ad2	5.6
			N-CT	38.4

Calf thymus AI enzyme (CT—AI) was 0.35 μg of fraction GG [54], calf thymus B enzymes (CT—B) were either 0.70 μg of fraction PC2 [55] or 20 μg of fraction DCB [52], rat liver B enzymes (RL—B) were 0.6 μg of the glycerol gradient fraction [26]. *E. coli* holoenzyme (0.40 μg) and *E. coli* core enzyme (0.30 μg) were prepared as described in legend to Table 5. RNA synthesis was determined as previously described [52]. Mn^{++} and Mg^{++} were 4 and 8 mM, respectively. Ammonium sulfate (AS) or KCl was added to the incubation mixture as indicated in the table. N—Ad2 = intact native adenovirus-2 DNA, D—Ad2 = denatured adenovirus-2 DNA, N—CT= native calf thymus DNA [52].

B enzyme preparation (fraction DCB) or by rat liver B RNA polymerases which were purified without a phosphocellulose chromatography step. As expected, native adenovirus DNA is also very poorly transcribed by *E. coli* core enzyme. Transcription is blocked at the initiation step since denatured adenovirus-2 DNA (Table 4) and adenovirus DNA nicked with pancreatic DNAase (results not shown) are readily transcribed by the animal enzymes.

These results indicate either that the animal enzymes lack a general initiation factor required for initiation on any regular double-stranded DNA or that transcription of an animal viral DNA

Table 5. Transcription of high molecular weight DNA from CV1 cells

Enzyme	Salt (mM)	Divalent cation	DNA (0.5 μg)	[^{32}P]-α-UTP incorporated (10 min, 37 °C) (pmol)
CT—AI	AS = 25	Mn^{++}	CV1	19.3
			N—CT	95.7
		Mg^{++}	CV1	9.4
			N—CT	79.5
CT—B	AS = 25	Mn^{++}	CV1	7.35
			N—CT	91.4
		Mg^{++}	CV1	1.4
			N—CT	46.6
E. coli holoenz.	KCl = 120	Mg^{++}	CV1	77.7
			N—CT	78.0
E. coli core enz.	KCl = 120	Mg^{++}	CV1	14.0
			N—CT	111.3
E. coli core enz. plus sigma factor	KCl = 120	Mg^{++}	CV1	117.5
			N—CT	140.6

Calf thymus AI enzyme (CT—AI) was 0.7 μg fraction GG [54] and calf thymus B enzymes were 1.0 μg fraction PC2 [55]. *E. coli* holoenzyme (0.30 μg) was prepared according to Berg et al. [56] and further purified on a glycerol gradient. *E. coli* core enzyme (0.70 μg) and σ factor (0.20 μg) were separated as described by Burgess et al. [45], but phosphocellulose chromatography was repeated twice. RNA synthesis was determined as previously described [52]. Mn^{++} and Mg^{++} were 4 and 8 mM, respectively. Ammonium sulfate (AS) or KCl was added to the incubation mixture as indicated in the table. CV1 = DNA from CV1 cells (an established line of African green monkey kidney cells) prepared as described elsewhere [12]. N—CT = native calf thymus DNA [52].

requires some additional factor(s) which could be of cellular or viral origin. In an attempt to distinguish between these two possibilities we prepared and used as template a high molecular weight unnicked cellular DNA [12]. This DNA preparation, obtained from CV 1 cells, has an average molecular weight of 2×10^8 and does not contain any detectable single-stranded nicks. As shown in Table 5, this DNA is very poorly transcribed by the purified calf thymus AI and B enzymes. Determination of the incorporation of [^{32}P]-γ-labeled ATP and GTP as well as measurement of pyrophosphate exchange revealed that initiation is in fact blocked on such a template. The ready transcription of this DNA, when denatured or nicked with DNAase, confirmed that the initiation step was blocked (results not shown). It is interesting that *E. coli* core enzyme also very poorly transcribes this high molecular weight eukaryotic DNA and that the stimulation by σ factor is of the order of that seen with intact linear phage DNA's [38, 45].

Conclusion

Our results demonstrate that purified calf thymus AI and B RNA polymerases as well as the purified rat liver B enzymes are unable to transcribe an intact linear double-stranded DNA of viral or cellular origin efficiently. In fact, RNA synthesis was never higher than that achieved with *E. coli* core enzyme which is known to initiate very poorly, if at all, on intact linear double-stranded DNA's [2, 3]. Several hypotheses, some of them not necessarily mutually exclusive, could account for these observations: 1. a general initiation factor similar to the *E. coli* σ factor could be lost during the purification, the purified animal enzymes thus resembling *E. coli* core enzyme; 2. the animal enzymes could be highly species-specific and transcribe only their homologous DNA, like phage and mitochondrial RNA polymerases [3, 46]. This would explain the failure of calf thymus or rat liver enzymes to transcribe CV 1 DNA; 3. the very limited RNA synthesis on cellular DNA could be significant. This synthesis could correspond to initiations at a small number of specific sites which are laboriously located by the enzymes. This hypothesis would also imply that, unlike the bacterial enzymes, the animal RNA polymerases are highly specific; 4. the animal RNA polymerases could be intrinsically incapable of initiating RNA synthesis on a naked, intact double-stranded DNA.

In this case initiation would occur only when the secondary structure of DNA is altered. This could be achieved either a) by the introduction of single-stranded nicks [47], b) by the binding of some regulatory factors (protein or RNA [48–50]), or c) by the supercoiling of the DNA in the chromatin which would result in loops of single-stranded DNA [51]. In this respect, it is interesting that RNA synthesis is readily initiated by animal enzymes on supercoiled DNA Form-I of SV 40 or PM 2 phage [10]. Studies aimed at distinguishing between these various hypotheses are in progress.

Acknowledgements

We thank Dr. Doerfler (Köln) for a generous gift of adenovirus-2 DNA. The technical assistance of Mrs. M. Acker, K. Dott, C. Hauss and of Mr. G. Dretzen is gratefully acknowledged. F. Gissinger is Attaché de Recherche CNRS and P. Hossenlopp is Boursier de la Ligue Nationale Française contre le Cancer.

This investigation was supported by grants from the Centre National de la Recherche Scientifique (ATP differenciation), the Institut National pour la Santé et la Recherche Médicale and the Commissariat à l'Energie Atomique.

References

1. Cold Spring Harbor Symp. on Quant. Biology, Transcription of Genetic Material, Vol. 35, pp. 641—737. Cold Spring Harbor Laboratory, Long Island, N.Y. 1970.
2. Burgess, R. R.: An Rev. Biochem. **40**, 711 (1971).
3. Bautz, E. K. F.: In: Davidson, J. W., Cohn, W. E. (Eds.): Prog. in Nucleic Acid Res. and Mol, Biol, Vol. 12, pp. 129—160. New York: Academic Press 1972.
4. Chambon, P., Meilhac, M., Walter, S., Kedinger, C., Mandel, J. L., Gissinger, F.: In: Hollaender, A., gen. (Ed.): Gene Expression and its Regulation, pp. 75—90. New York: Plenum Press 1973.
5. Chambon, P., Gissinger, F., Kedinger, C., Mandel, J. L., Meilhac, M., Nuret, P.: Acta Endocrin. Supl. N° **168**, 222 (1972).
6. Meilhac, M., Tysper, Z., Chambon, P.: Europ. J. Biochem. **28**, 291 (1972).
7. Meilhac, M., Chambon, P.: Europ. J. Biochem. **35**, 454 (1973).
8. Vogt, V.: Nature (Lond.) **223**, 854 (1969).
9. Gniazdowski, M., Mandel, J. L., Gissinger, F., Kedinger, C., Chambon, P.: Biochem. Biophys. Res. Commun. **38**, 165 (1970).
10. Chambon, P., Gissinger, F., Mandel, J. L., Kedinger, C., Gniazdowski, M., Meilhac, M.: Cold Spr. Harb. Symp. quant. Biol. **35**, 693 (1970).
11. Travers, A. A., Burgess, R. R.: Nature (Lond.) **222**, 537 (1969).

12. Gross-Bellard, M., Oudet, P., Chambon, P.: Europ. J. Biochem. **36**, 32 (1973).
13. Aloni, Y., Winocour, E., Sachs, L.: J. molec. Biol. **31**, 415 (1968).
14. Oda, K., Dulbecco, R.: Proc. nat. Acad. Sci. (Wash.) **60**, 525 (1968).
15. Tonegawa, W., Walter, G., Bernardini, A., Dulbecco, R.: Cold Spr. Harb. Symp. quant. Biol. **35**, 823 (1970).
16. Khoury, G., Byrne, J.C., Martin, M.A.: Proc. nat. Acad. Sci. (Wash.) **69**, 1925 (1972).
17. Khoury, G., Martin, M.A.: Nature (Lond.) New Biol. **238**, 4 (1972).
18. Lindstrom, D.M., Dulbecco, R.: Proc. nat. Acad. Sci. (Wash.) **69**, 1517 (1972).
19. Sambrook, J., Sharp, P.A., Keller, W.: J. molec. Biol. **70**, 57 (1972).
20. Westpahl, H.: J. molec. Biol. **50**, 407 (1970).
21. Chambon, P., Gissinger, F., Kedinger, C., Mandel, J.L., Meilhac, M.: In: Busch, H. (Ed.): The Cell Nucleus, in press. New York: Academic Press 1973.
22. Sugden, B., Sambrook, J.: Cold Spr. Harb. Symp. quant. Biol. **35**, 663 (1970).
23. Westphal, H., Kiehn, E.D.: Cold Spr. Harb. Symp. quant. Biol. **35**, **819** (1970).
24. Fried, A.H., Sokol, F.: J. gen. Virol. **17**, 69 (1972).
25. Mandel, J.L., Chambon, P.: Submitted for publication (1973a).
26. Mandel, J.L., Chambon, P.: Febs Letters **15**, 175 (1971).
27. Keller, W., Goor, R.: Cold Spr. Harb. Symp. quant. Biol. **35**, 671 (1970).
28. Herzberg, M., Winocour, E.: J. Virol. **6**, 667 (1970).
29. Jaenisch, R.: Nature (Lond.) New Biol. **235**, 46 (1972).
30. Weinberg, R.A., Warnaar, S.O., Winocour, E.: J. Virol. **10**, 193 (1972).
31. Sokol, F., Carp, R.I.: J. gen. Virol. **11**, 177 (1971).
32. Rozenblatt, S., Winocour, E.: Virology **50**, 558 (1972).
33. Aloni, Y.: Proc. nat. Acad. Sci. (Wash.) **69**, 2404 (1972).
34. Maestre, M.F., Wang, J.C.: Biopolymers **10**, 1021 (1971).
35. Dean, W.W., Lebowitz, J.: Nature (Lond.) **231**, 5 (1971).
36. Delius, H., Mantell, M.J., Alberts, H.: J. molec. Biol. **67**, 341 (1972).
37. Hossenlopp, P., Oudet, P., Chambon, P.: Submitted for publication (1973).
38. Sugiura, M., Okamoto, T., Takanami, M.: Nature (Lond.) **225**, 598 (1970).
39. Anderson, J.A., Kuntz, G.P.P., Evans, H.H., Swift, T.J.: Biochemistry **10**, 4368 (1971).
40. Luck, G., Zimmer, C.: Europ. J. Biochem. **29**, 528 (1972).
41. Price, R., Penman, S.: J. Virol. **9**, 621 (1972).
42. Wallace, R.D., Kates, J.: J. Virol. **9**, 627 (1972).
43. Ledinko, N.: Nature (Lond.) **233**, 247 (1971).
44. Price, R., Penman, S.: J. molec. Biol. **67**, 433 (1972).
45. Burgess, R.R., Travers, A.A., Dunn, J.J., Bautz, E.K.F.: Nature (Lond.) **221**, 43 (1969).

46. KUNTZEL, H., SCHAFER, K. P.: Nature (Lond.) New Biol. **231**, 265 (1971).
47. CASCINO, A., RIVA, S., GEIDUSCHEIK, E. P.: Cold Spr. Harb. Symp. quant. Biol. **35**, 213 (1970).
48. BRITTEN, R. J., DAVIDSON, E. H.: Science **165**,349 (1969).
49. BRAM, S.: Biochimie **54**, 1005 (1972).
50. PAUL, J.: Nature (Lond.) New Biol. **238**, 444 (1972).
51. CRICK, F.: Nature (Lond.) **234**, 25 (1971).
52. KEDINGER, C., GISSINGER, M., GNIAZDOWSKI, M., MANDEL, J. L., CHAMBON, P.: Europ. J. Biochem. **28**, 269 (1972).
53. MANDEL, J. L., CHAMBON, P.: Submitted for publication (1973b).
54. GISSINGER, F., CHAMBON, P.: Europ. J. Biochem. **28**, 277 (1972).
55. KEDINGER, C., CHAMBON, P.: Europ. J. Biochem. **28**, 283 (1972).
56. BERG, D., BARRETT, K., CHAMBERLIN, M.: In: GROSSMAN, L., MOLDAVE, K. (Eds.): Methods in Enzymology, Vol. XXI, p. 506.
57. MANDEL, J. L., KEDINGER, C., GISSINGER, F., CHAMBON, P., FRIED, A. H.: FEBS Letters **29**, 109 (1973).
58. ALONI, Y.: Nature (Lond.) New Biol. **243**, 2 (1973).

Discussion

KLAMERTH (Heidelberg): 1. How do the different eukaryotic polymerases behave in an electrophoretic focusing system in Sephadex gel?

2. Does the eukaryotic polymerase, besides its specificity, accept poly C or the polynucleotides as templates?

P. CHAMBON (Strasbourg): 1. I don't know. I am not aware of such experiments.

2. Yes. The relative template efficiency of various polynucleotides was recently reported by D. LENTFER [LENTFER, D., LEZIUS, A. G.: Eur. J. Biochem. **30**, 278 (1972)].

H. KERSTEN (Erlangen): Is it possible to exchange certain subunits of the A and B forms of the RNA polymerases and, if so, are these hybrid enzymes inactive, less active, or even as active as the normal enzymes?

P. CHAMBON: I can't answer your question since unfortunately up to now we were unable to dissociate and to efficiently renature the animal RNA polymerases.

J. PAUL (Glasgow): In view of the claims made by TOCCHINI-VALENTINI and GRIPPA that injections of bacterial σ factor stimulated RNA synthesis in frog oocytes, has anyone studied the effect of adding *E. coli* σ factor to the eukaryotic enzymes?

P. CHAMBON: Yes, it was shown in 1970 [see Cold Spring Harbor Symposium **35**, 693 (1970)] that σ factor does not stimulate the transcription of phage DNA's by calf thymus RNA polymerases. Similarly, we recently found that the transcription of the high molecular weight cellular DNA is not stimulated by the σ factor either. The reason for the discrepancy between the *in vivo* and *in vitro* experiments is unknown.

E. Wintersberger (Würzburg): Dr. Ponta in our laboratory has prepared antibody against the yeast B enzyme which in our hands does not inhibit activity of the yeast A enzyme to any significant extent. Our results therefore complement those of P. Chambon.

Studies on *in vitro* RNA Synthesis by RNA Polymerases from Rat Liver Tissue

K. H. Seifart, A. Ferencz and B. Benecke

Institut für Physiologische Chemie, Philipps-Universität Marburg/Lahnberge, W. Germany

With 13 Figures

DNA-dependent RNA polymerase (E.C.2.7.7.6.) has in recent years been isolated from a number of different eukaryotic organisms and tissues and in all examined cases this enzyme has been found to occur in multiple forms [1]. With the exception of mitochondrial RNA polymerase [2–5], these enzymes are located in the cell nucleus although recent reports describe the existence of a cytoplasmically located enzyme [6, 7]. The biological significance and possible biosynthetic interelationships between the cytoplasmic and nuclear enzymes as well as the question concerning the role which these enzymes may play in controlling rate and specificity of RNA synthesis are incompletely understood at present.

The present report will be addressed to the following three questions:

1. multiplicity and structural localization of RNA polymerase species in rat liver tissue;
2. studies concerning the biological half-life of these enzymes and questions related to this parameter; and,
3. the formation of initiation complexes between RNA polymerase A and DNA.

Certain parts of this account have previously been published [6, 8].

1. Multiplicity and Structural Localization of RNA Polymerase Species from Rat Liver Tissue

The multiple forms of RNA polymerase have initially been designated by Roeder and Rutter [9] as I, II and III for the enzymes from sea urchin and rat liver according to their elution order from DEAE-Sephadex and, by Chambon and his collaborators, as A and B for the calf thymus enzymes according to their sensitivity toward α-amanitin [10]. This toxin inhibits nuclear RNA synthesis as was initially shown by Stirpe and associates [11]. This inhibition is a specific phenomenon involving polymerase B as was initially shown for rat liver by Seifart and Sekeris [12] and calf thymus by Kedinger et al. [10].

Recently it has been shown that the B enzyme consists of more than one subclass [13, 14] and that enzyme A may likewise represent a multicomponent group [15–17] although the functional significance of this finding is unclear at present. It is generally accepted that enzyme A (or I) is located in the nucleolus and is chromatographically distinct and functionally different from enzyme B (or II) which is associated with the extranucleolar chromatin. Since both nuclear enzymes are at least in part structurally bound, most of the techniques which have been devised for their quantitative extraction employ ultrasonic treatment in media of high ionic strength. For a review of this aspect, see [18].

It has nevertheless been observed recently that the particulate-free cytosol of the liver cell [6] contains an RNA polymerizing activity also reported to occur in thymus and ascites cells [7] as well as in developing *Xenopus* oocytes [19]. For the sake of differentiation, this activity has been designated as enzyme C. Initially is was suspected that the activity observed in the cytosol either (i) represents a mixture of nuclear enzymes A and B which leaked into the cytoplasm during the preparation or (ii) is of mitochondrial origin. The former alternative was obviously most prevalent since current preparation methods cause a certain amount of nuclear disruption. The striking observations is, however, that if the phenomenon is to be reduced to mere nuclear leakage, only one of the polymerases can be involved since only a single peak of activity in the chromatographic region of enzyme A is recovered after DEAE-cellulose chromatography of the cytosol (Fig. 1). By this criterion, very little if any B activity is present in the cytosol. It should be mentioned that the crude cytoplasmic supernatant fraction before ultracentrifugation contains RNA polymerase activity which also bears the properties normally associated with nuclear polymerase B. This activity is, however, not DNA-dependent, can be removed by ultracentrifugation, and is by all probability chromatin-bound or structurally associated in some other fashion.

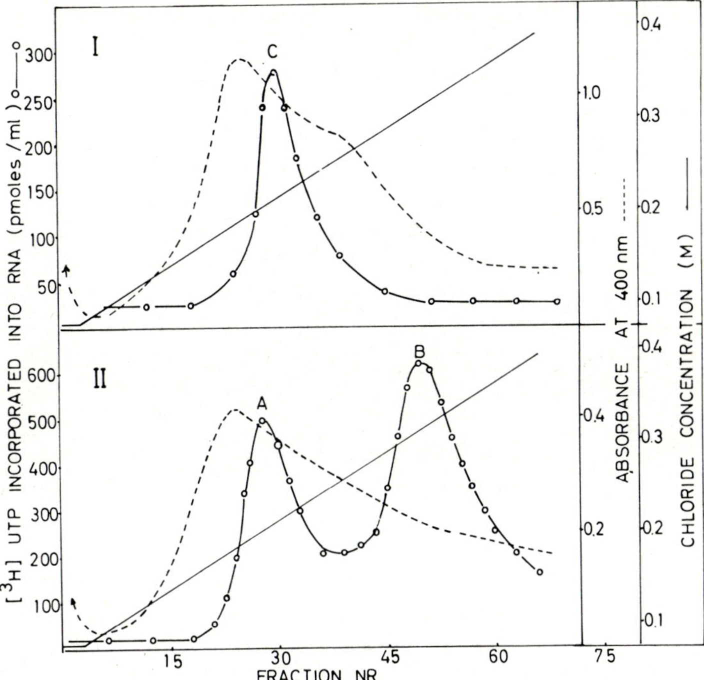

Fig. 1. DEAE-cellulose chromatography of the cytosol on a 4 × 18 cm column (part I) and the nuclear extract on a 2.5 × 13 cm column (part II). Elution of approximately 8.0 ml fractions was achieved with a linear gradient (550 + 500 ml; 0.0—0.3 M NH_4Cl in buffer II) pumped from an ultrograd mixer at a flow rate of about 60 ml/hr. RNA polymerase was assayed as described in [6]

It would be readily conceivable that polymerase A has a lesser degree of structural association and is much more prone to cytoplasmically directed diffusion, implying that the cytoplasmic polymerase would emanate from enzyme A. This would be in accord with the indiscernible chromatographic behavior of A and C on DEAE-cellulose. However, several points argue against this assumption. Simultaneous chromatography of all three enzymes on individual phosphocellulose columns under very stringent conditions seems to clearly differentiate enzyme A and C [6].

Since the initial distinction between enzymes A and B was achieved by their sensitivity toward α-amanitin, titration curves were conducted for this inhibitor with each of the enzymes A, B and C. In agreement with previously published results [12], amanitin very specifically inhibits the B enzyme at a stochiometry of 1:1. Enzyme A is not effected at all whereas the C enzyme is intermediate between the two (Fig. 2), indicating that the enzyme in question is neither A nor B. This differential amanitin sensitivity has been observed with highly purified preparations of enzyme C and it cannot be attributed to contaminating, amanitin-binding compounds. The amanitin concentration to achieve 50% inhibition of enzyme C (±50 μg/ml) is rather high. Since enzyme A is not effected by this concentration, however, this cannot be a non-specific,

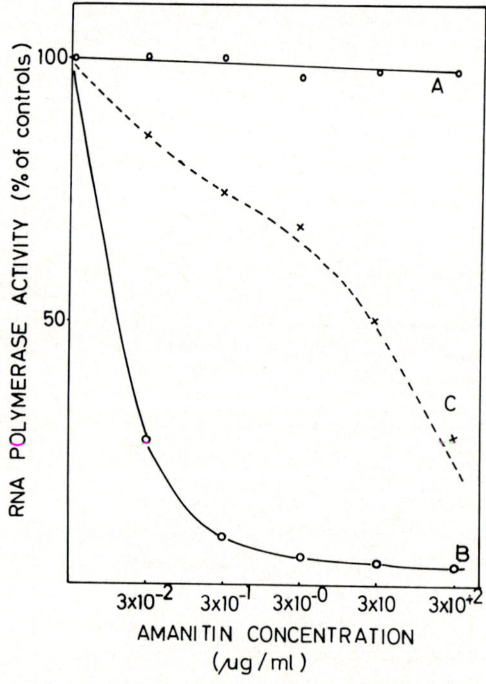

Fig. 2. Effect of α-amanitin at increasing concentrations on the activity of RNA polymerase species *A*, *B*, and *C*. Amanitin concentrations reflect final concentrations in the assay

toxic effect. It is also interesting to note that the sensitivity of polymerase B from yeast for amanitin is in the same order of magnitude [20–23]. If enzyme C is incubated with radioactively labeled γ-amanitin at a concentration required to yield approximately 50 % inhibition and the excess of non-bound inhibitor is subsequently removed by gel filtration through Sephadex G 75 (Fig. 3), the recovered enzyme remains inhibited approximately to the same extent (44 %) although the actual inhibitor concentration is drastically reduced. Therefore the enzyme-inhibitor complex must possess a high stability although high concentrations of in-

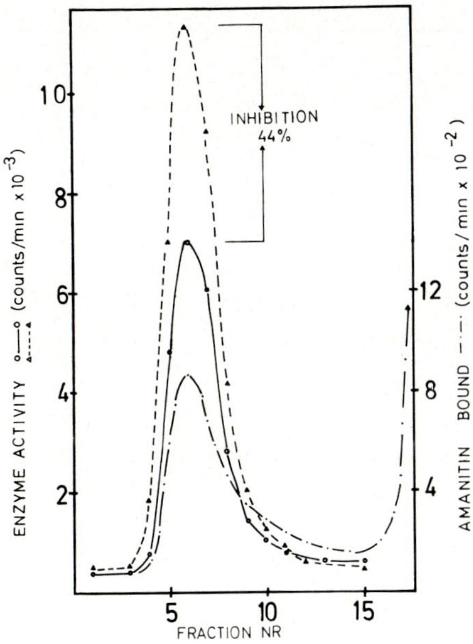

Fig. 3. Binding of radioactive [^{14}C] methyl-γ-amanitin to enzyme C obtained after phosphocellulose. 5 μC Amanitin at a concentration of 50 μg/ml were incubated with the enzyme for 20 min at 37°. Controls were conducted by incubating an identical RNA polymerase fraction with buffer in the same manner. Separation of free and protein-bound amanitin was achieved by gel filtration of 1.0 ml of the incubate on a 1×30 cm Sephadex G 75 column, and counting a 250 μl aliquot of each fraction in 10.0 ml Bray's scintillation fluid. Concurrently RNA polymerase activity was assessed in the fractions of the columns obtained from control (▲- - - -▲) and amanitin-treated (○——○) samples. From [6]

hibitor are required to form this complex. The amanitin titration curve for enzyme C (Fig. 2) also shows quasi-sigmoidal characteristics, possibly indicating cooperative binding although additional experiments are required to clarify this interesting point.

Even more convincing evidence minimizing the possibility that enzyme C represents nuclear leakage is put forth by an experiment in which amanitin was applied to rats *in vivo* at a concentration of 100 μg/100 g body weight. It is known that amanitin, when applied *in vivo*, will not bind to enzyme A but will form an exceedingly stable complex with enzyme B which is maintained through subsequent extraction and purification steps [24]. Therefore this ap-

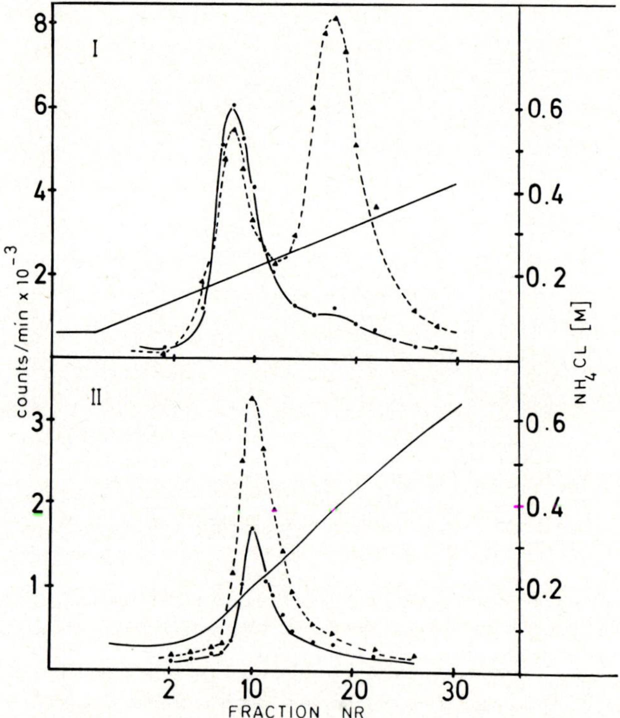

Fig. 4. DEAE-cellulose chromatography of RNA polymerase isolated from nuclei (part I) or cytoplasm (part II) from control (▲- - - - -▲) or amanitin-treated (●———●) animals. Details of the chromatography are as in the legends of Figs. 6 and 7

proach enables one to freeze the *in situ* situation and allows certain conclusions concerning the distribution of the enzymes before the preparation disrupts the cellular integrity. If the enzymes are subsequently extracted from such amanitin-treated animals and analyzed by DEAE-cellulose chromatography, the results in Fig. 4, Part I, concerning the nuclear enzymes very clearly show that amanitin treatment completely eliminates enzyme B, but leaves enzyme A virtually uneffected. In contrast, enzyme C (Fig. 4, Part II) is partly inhibited, but is clearly present. Consequently, the enzyme cannot represent B, since this component was completely eliminated before the preparation was started and it is not A since that component is not inhibited.

In addition, enzyme C can be differentiated functionally with respect to template and ionic requirements from both enzymes A

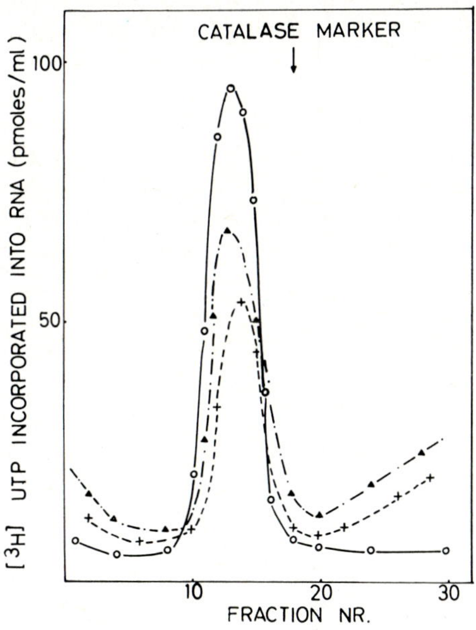

Fig. 5. Density gradient centrifugation of enzyme activities A (+ - - - - - +), B (○—○) and C (▲—·—·▲). Enzyme solutions (1.1 ml) were layered over 11.0 ml linear 5—20% sucrose gradients in buffer containing 10% glycerol + 80 mM $(NH_4)_2SO_4$ throughout and centrifuged for 15 h at 210000 g (max) in the SW 41 rotor of the Beckman centrifuge. Fractions (0.4 ml) were collected through a needle from the bottom of the tube. From [6]

and B [6] although it is questionable whether these criteria are very stringent and meaningful. The enzyme does not support homopolymer formation nor the end-addition of nucleotides to preexisting ribopolymers (rRNA, poly rA, rU, rC, rG). The incorporation of UTP is dependent on the presence of the remaining 3 nucleoside triphosphates and DNA, and is inhibited by actinomycin.

RNA polymerase C has a molecular weight of 500000 and sediments in a single peak in density gradients containing 0.1 M $(NH_4)_2SO_4$ (Fig. 5). The enzyme is completely rifampicin-resistant in all stages of purification rendering it unlikely to be of mitochondrial origin [2, 3, 25] although some discrepancy apparently exists with respect to the general sensitivity of the mitochondrial enzymes to rifampicin ([5], and discussion by WINTERSBERGER and KÜNTZEL, this volume).

If obvious artifacts have been excluded and the existence of a cytoplasmically located RNA polymerase can be substantiated by additional work from this and other laboratories, the rather perplexing question concerning the significance of this finding and its biological relationship to the known function of the enzyme, *i.e.*, the transcription of DNA, has to be dealt with. It is well documented that in the case of DNA polymerase a large portion of this enzyme is located in the cytoplasm [26–29]. Attempts to differentiate the nuclear and cytoplasmic DNA polymerase have thus far not provided a conclusive answer. These findings are very similar to the results reported here, although the implications are not clear in either case. Precursor relationships or transformation phenomena between these enzymes are, of course, possible although no, or very few, data are available to solidify this concept. It is generally accepted that the cytoplasm is the site of synthesis for most, if not all, of the nuclear proteins. It could therefore be possible that the RNA polymerase found in the cytoplasm partly represents a population of newly synthesized molecules.

2. Studies Concerning the Biological Half-Life of RNA Polymerases

In order to obtain an estimate of the biological half-life and thus to gain insight into possible biosynthetic interrelationships between these enzymes, cycloheximide was administered to rats *in vivo* for time periods between 5 min and 24 h before sacrifice. The

Table 1. Inhibition of general hepatic protein biosynthesis by the application of cycloheximide *in vivo* to rats at a concentration of 400 μg/100 g body weight. Thirty minutes before sacrifice the animals received L-[^{14}C]-tyrosine intraperitoneally (22 μCi/100 g body weight). Radioactivity was measured in the acid-precipitable (10% TCA, 95°, 10 min) fraction of the cytosol. From [8]

Time after inhibitor application	5 min	4 h	12 h	24 h
Inhibition of protein synthesis (cpm/mg protein) (% of controls)	0.5	86.9	76.0	33.7

inhibitor was administered at a concentration of 400 μg/100 g body weight which resulted in an inhibition of overall protein biosynthesis of ±90% at 4 h and 75% at 12 h. It was not attempted to achieve complete inhibition of protein synthesis since pilot experiments had shown that higher concentrations of cycloheximide (3 mg/100 g body weight) which completely blocked peptide synthesis killed the animals within 6 h, thus preventing an assessment of slowly turning over proteins. For each treatment time, a control and cycloheximide-treated group was killed simultaneously, and RNA polymerase was extracted from the nuclei and cytoplasm as described previously [6]. The enzyme extracts were chromatographed on DEAE cellulose and all steps of extraction and particularly purification of the enzymes from the inhibitor-treated and corresponding control group were conducted simultaneously under identical conditions, thus ensuring that the obtained curves for enzyme activities are directly comparable. These values were deliberately not corrected for protein content since protein concentrations could vary as a function of cycloheximide treatment and a correction for this variable could then possibly eliminate an effect which the experiment was designed to measure.

An analysis of the results reveals no marked differences for the activities of RNA polymerase extractable by this technique from nuclei (Fig. 6) or cytoplasm (Fig. 7) of control and cycloheximide-treated animals, although overall protein synthesis was markedly inhibited (Table 1).

Therefore the enzyme molecules seem to have extended half-life periods, representing a fairly stable population. Among others, one of the aims of these experiments was to possibly determine the biosynthetic interrelationships between enzymes A, B, and particularly C. It is questionable, however, whether this approach is

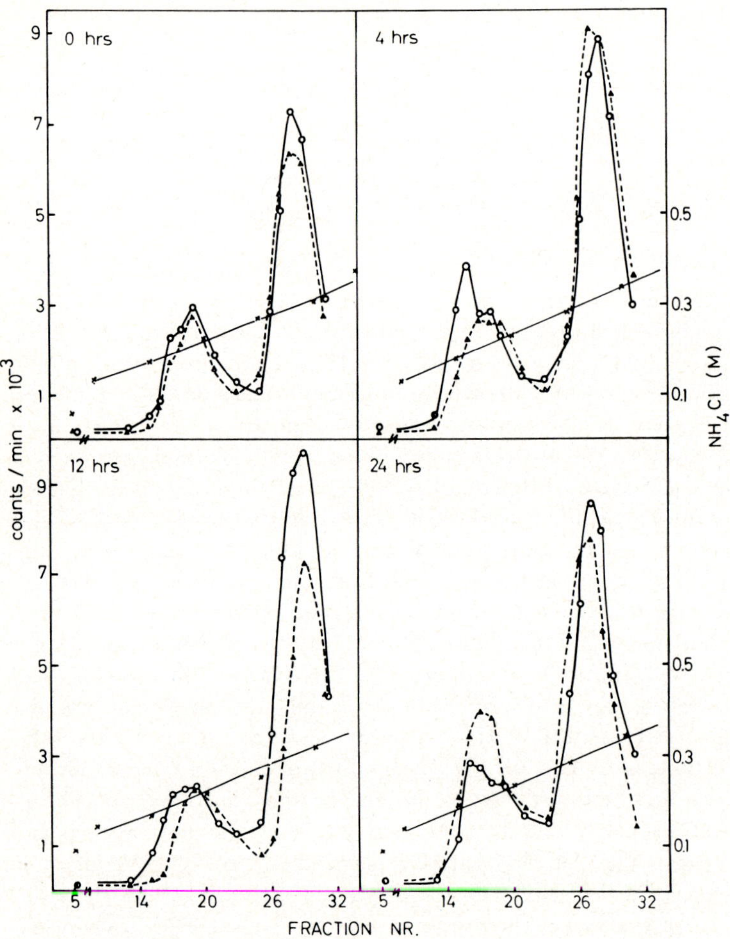

Fig. 6. Simultaneous chromatography of nuclear RNA polymerase activities from control (▲-----▲) and cycloheximide-treated (○——○) animals on identical 2.5 × 5 cm DEAE-cellulose columns. Elution of approx. 3.5 ml fractions was achieved with a linear gradient from 0.0—0.35 NH_4Cl in buffer (*tris*-Cl, 0.05 M, pH 7.9, 20% glycerol, 5 mM mercaptoethanol, 0.25 mM EDTA) pumped from an ultrograd mixer at a flow rate of 40 ml/h. In order to achieve identical conditions all columns were packed from the same batch of exchanger and were developed in pairs at identical elution rates from one pump (with multiple outlets) from the same elution medium. From [8]

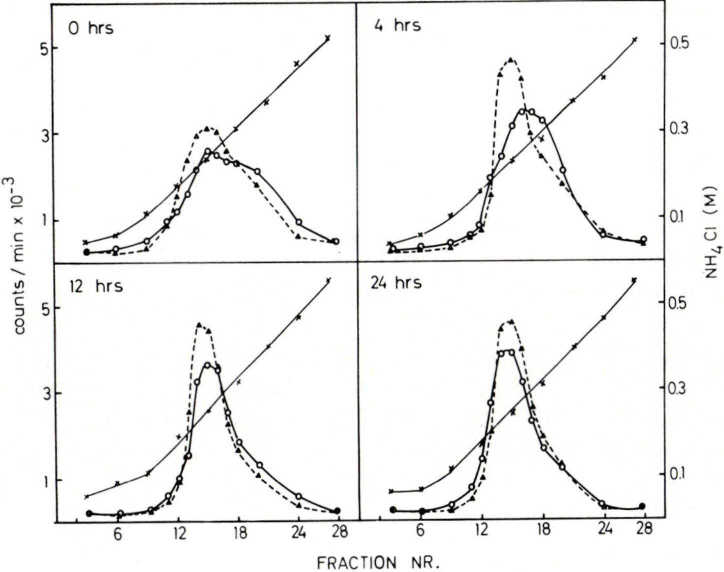

Fig. 7. Simultaneous chromatography of the cytosol from control (▲- - - - -▲) and cycloheximide-treated (○——○) animals on identical 2.5 × 5 cm DEAE-cellulose columns. Conditions were as described in Fig. 6. In this case the developing gradient was from 0.0 to 0.60 M NH_4Cl. From [8]

feasible with proteins of very long half-life values The conclusions drawn from these experiments are valid under the assumption that the inhibition of protein synthesis by cycloheximide is random for all proteins and that the level of residual protein synthesis (±10 %) required to keep the animals alive for periods over 6 h does not account for completely normal synthesis of RNA polymerase. No such selective mechanism of escape-synthesis after cycloheximide has been described, however, and experiments conducted with 3.0 mg cycloheximide/100 g for 4 h showed complete blockage of peptide synthesis but maintenance of enzyme levels exceeding 85 % of the control values.

It is quite clear that cycloheximide will at some point in time and concentration inhibit the formation of RNA polymerase which, as a central biomolecule, is obviously turning over. The data presented here lead to the conclusion, however, that this may be a rather slow process and that *short-term* fluctuations in the rate of RNA synthesis may not be controlled through the factor of enzyme concentration.

These findings are in apparent disaggreement with results reported by YU and FEIGELSON [30] who found a rapid decrease of nucleolar RNA polymerase activity in rat liver after cycloheximide application, measured as UTP incorporation of intact nucleoli under certain experimental conditions. These workers concluded from their findings that polymerase A has a rapid turnover time and that control of RNA synthesis, in particular of ribosomal RNA, can effectively be achieved by a variation in the amount of enzyme synthesized. This assumption would be reconcilable with experiments stating an enhanced amount of nucleolar RNA polymerase after hormone treatment [31]. This conclusion is, however, not supported by data presented here and results of SEKERIS and coworkers [32, 33].

It is beyond any doubt that cycloheximide *in vivo* will dramatically inhibit both total RNA synthesis and the incorporation of precursors into RNA *in vitro* by nuclei isolated from such animals [34]. It is questionable, however, whether this inhibition of *RNA synthesis* in fact reflects a disappearance of the enzyme – defined as the catalytic unit responsible for the DNA-dependent incorporation of nucleoside triphosphates – or whether it much more reflects dependence on continued protein synthesis for the enzyme to express its biological activity in the nucleus or nucleolus. Analogous observations have been made in cultures of *Chlorella* [35] where cycloheximide inhibits DNA synthesis, probably without altering the amount of enzyme present. In addition, it is known that some of the developmental changes of *Xenopus* oocytes are characterized by dramatic flutuations in the rate of rRNA synthesis although the amount of RNA polymerase A (or I) does not vary accordingly [36]. It is possible that under the influence of factors which grossly effect the pattern of RNA synthesis, *e.g.*, tissue regeneration after partial hepatectomy, preexisting enzyme levels which are normally present in excess no longer suffice to cope with the increased demand, thus leading to enhanced *de novo* synthesis of enzyme. These are long-term effects related to cellular growth, however, which may be different from rapid modulations in the rate of RNA synthesis.

It is also of importance to note that synthesis of ribosomal and of heterogeneous nuclear RNA are possibly interrelated as has recently been concluded by Schmid and Sekeris [37], and it is possible that a modulation of RNA synthesis, in particular of ribosomal RNA by hormones [38–40] and other agents [41], have to be seen in this context, *i.e.*, that the synthesis of ribosomal RNA requires the continued production of certain proteins. Therefore regulation of RNA synthesis in the eukaryotic nucleus is possibly achieved through other factors controlling the rate of this process. As has been postulated by Muramatsu [34], it is possible that short-lived and inducible proteins which could be synthesized on short-lived RNA messages function as such specifiers of RNA synthesis. Collectively, these observations are reminiscent of the phenomenon of stringent control in bacteria and, although this assumption is by no means proven for the eukaryotic cell, indications exist supporting this hypothesis [42].

3. Formation of Initiation Complexes by RNA Polymerase A

The complicated process of RNA synthesis can be divided into several steps which have been well characterized for the bacterial RNA polymerase (*e.g.*, [43]) and more recently for the enzyme from animal cells [44]. The initial steps of the reaction involve binding of the enzyme to DNA and the formation of an initiation complex. Once bound into such a true initiation complex, the enzyme from *E. coli* becomes resistant to compounds like rifampicin [45, 46], poly I [47] or heparin [48] which are known to act as inhibitors of free polymerase, *i.e.*, enzyme which is either not bound at all or attached to DNA only very loosely and, therefore, with limited specificity. Similar studies have been conducted for several rifamycin derivatives (AF/O13, AF/O5, PR19) in their action on eukaryotic RNA polymerases [44, 49] although the question of specificity of their action is not completely clear at present.

Recently conducted experiments with heparin have shown that purified RNA polymerase A from rat liver is strongly inhibited by minute quantities of this compound (Fig. 8). The interesting point is, however, that the enzyme is inhibited to a comparable extent if heparin is added to enzyme A which is actively engaged in the transcription of purified DNA (Fig. 9). Therefore no heparin-resistant complexes are formed by this enzyme although RNA chains have been initiated.

Since the action of heparin on mammalian RNA polymerases may be entirely different from that of *E. coli*, experiments were

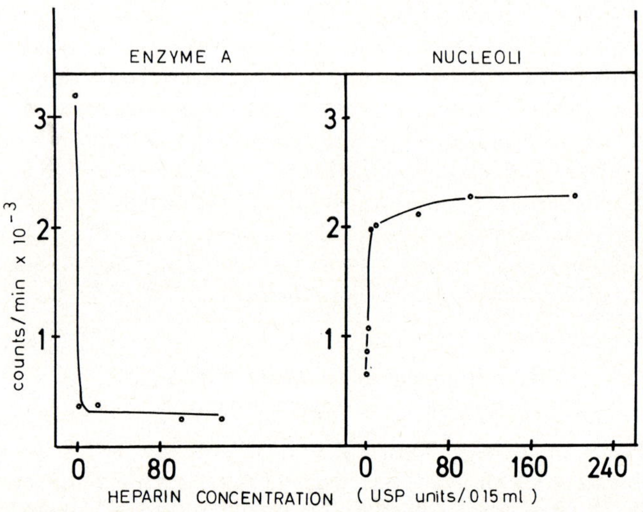

Fig. 8. Effect of heparin *in vitro* on RNA synthesis catalyzed by enzyme A and isolated nucleoli, respectively. Heparin was added to the enzyme or nucleoli before the reaction was started

conducted with isolated rat liver nucleoli known to contain the same enzyme [6, 50] which is, however, functionally initiated *in vivo*. The results show an apparently paradoxical stimulation by heparin of nucleolar RNA synthesis (Fig. 8). This stimulation is, possibly among others, due to two investigated effects:

1. An inhibition of RNAse described previously. Experiments outlined in Fig. 10 demonstrate that nucleoli contain an RNA degrading enzyme (or enzymes) which is sensitive to heparin inhibition.

2. A general loss of nucleolar structure after heparin treatment can be observed directly by phase contrast microscopy and by indirect measurement of a decreased light-scattering at 400 nm.

This effect can likewise contribute toward enhanced RNA synthesis by physically facilitating this process.

Whatever the explanation for the observed stimulation of RNA synthesis may be, the interesting point remains that heparin does

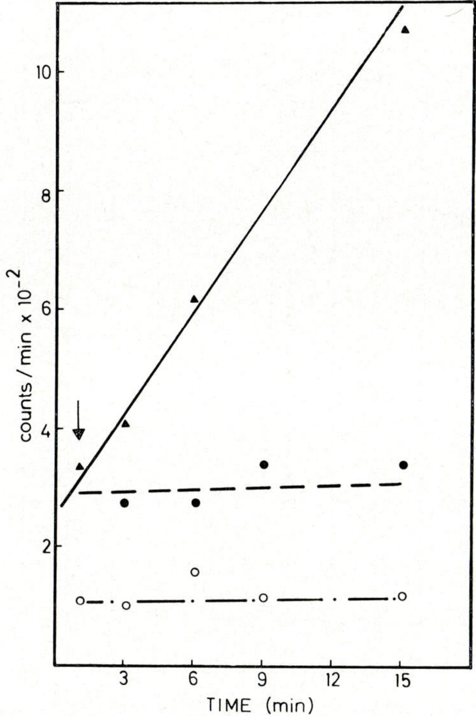

Fig. 9. Effect of heparin on transcription of isolated DNA by purified enzyme A. Control synthesis was conducted for 15 minutes (▲——▲). Heparin (40 USP units/0.15 ml) was added either to the free enzyme (○—·—·○) or 1 minute after the reaction mixture had been incubated at 37° in the presence of all the reaction components (●-----●)

not inhibit the *in vitro* completion reaction of RNA chains correctly initiated by the nucleolar enzyme A *in vivo*. Therefore this enzyme must be in a different state – conformational or otherwise – rendering it heparin-resistant in contrast to the purified A enzyme transcribing isolated DNA. Employing the definition of heparin resistance, the latter enzyme cannot form a true initiation complex on DNA.

An interesting aspect concerns the apparent abundance of the enzyme within the nucleolus since a portion of the polymerase does not seem to be engaged in transcription. Addition of exogenous

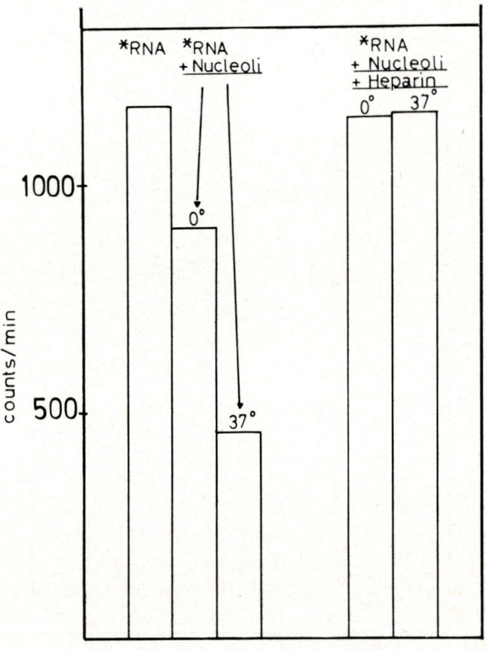

Fig. 10. Effect of heparin (40 USP units/0.15 ml) on the degradation of exogeneously added labeled RNA by a nuclease contained in isolated nucleoli. [^{3}H]-labeled heterogeneous nuclear RNA, purified by classical phenol extraction and ethanol precipitation, was incubated alone or in the presence of nucleoli at 0° and 37° for 20 min (left side). Addition of heparin to nucleoli prevented the degradation at both temperatures for 20 min (right side)

DNA to nucleoli results in an increase in RNA synthesis possibly indicating that a population of previously non-transcribing polymerase molecules are now bound to the exogenously added template (Fig. 11). Alternatively, this result could of course be interpreted in terms of a derepression of nucleolar chromatin through the transfer of inhibitory proteins to the new template. This latter interpretation is rendered unlikely since DNA does not stimulate RNA synthesis of isolated nucleoli if these have been pretreated with heparin (Fig. 11).

Qualitatively similar results are obtained from time-kinetic experiments which show that *in vitro* RNA synthesis by isolated

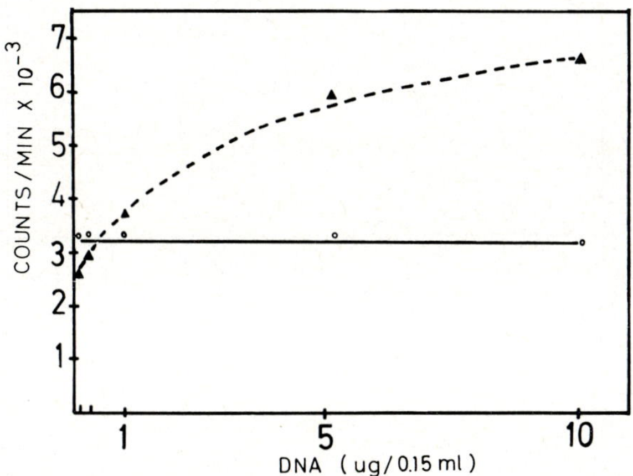

Fig. 11. Effect of exogenously added DNA on nucleolar RNA synthesis *in vitro* in the absence (▲-----▲) or presence (○——○) of heparin (40 USP units/0.15 ml)

nucleoli reaches a plateau after approximately 20 min. If exogenous DNA is added to this plateau, synthesis is resumed (Fig. 12, Part II), possibly because free enzyme in the nucleolus can utilize this template. If nuclei are preincubated with heparin for 20 min, exogenously added DNA will have absolutely no effect (Fig. 12, Part I) indicating that free enzyme was complexed by heparin during the preincubation period, thus preventing it from transcribing the exogenous template.

Moreover, the stimulatory effects of DNA and heparin are nonadditive as shown in Fig. 13. If added separately to nucleoli before the reaction is started, each of these compounds will enhance RNA synthesis *in vitro*. If combined, the stimulation is no greater than that oberved with heparin alone (Fig. 13), supporting the previously drawn conclusion that the nucleolus contains an excess of free enzyme which is inhibitable by heparin if added from the beginning of the reaction. The addition of heparin at 24 min to nucleoli preincubated with DNA (Fig. 13) results in an additional enhance-

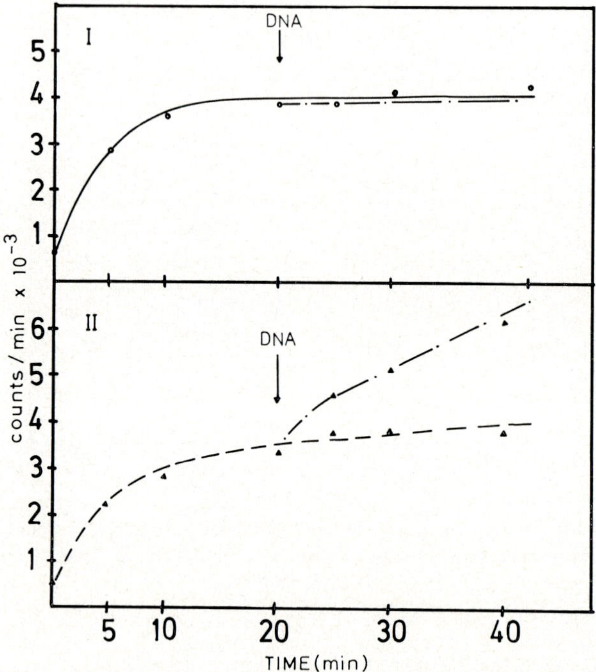

Fig. 12. Effect of exogenously added DNA on nucleolar RNA synthesis *in vitro* in the presence (Part I) or absence (Part II) of heparin. Where appropriate, heparin was added to nucleoli before start of the reaction at a concentration of 40 *m*SP units/0.15 ml. In both cases DNA (10 µg/0.15 ml) was added after 20 minutes when synthesis had reached a plateau

ment of RNA synthesis which greatly exceeds the value obtained when both DNA and heparin are present from the start of the reaction. Therefore we are certainly not dealing with a simple additive phenomenon implying the possible interpretation that the excess enzyme present in nucleoli can form initiation complexes with added DNA which are subsequently resistant to heparin. Additional studies, among others involving the direct measurement of the initiation reaction, are obviously required to substantiate this point.

Collectively the results of this latter series of experiments allow the following conclusions. The A enzyme contained in the nucleolus and initiated *in vivo* is able to form heparin-resistant complexes. Such initiation complexes are not formed between isolated DNA and purified A enzyme extracted from nuclei. The nucleolus ap-

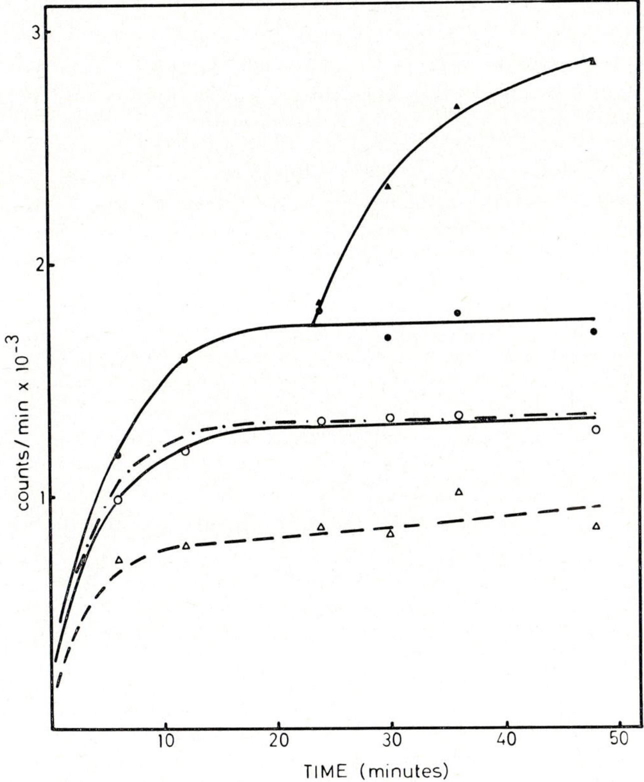

Fig. 13. Time course of nucleolar RNA synthesis in vitro. Before starting the reaction, nucleoli were mixed with either buffer (△——△), DNA (●——●; 10 μg/0.15 ml), heparin (○——○; 40 USP units/0.15 ml) or heparin + DNA (—·—·—). At 24 min, when synthesis had reached a plateau, heparin (40 USP units/0.15 ml) was added to the system in which nucleoli had been preincubated with DNA (▲——▲)

pears to contain an excess of A enzyme which can transcribe exogenously added DNA and preliminary experiments do not rule out that *this* enzyme is in fact able to form heparin-resistant initiation complexes on DNA. It is therefore possible that the nucleolus contains a factor which is required for the formation of the true initiation complexes. This factor need not be an integral part of the enzyme and may not be co-extracted and purified with the catalytic unit. Future research will have to clarify this interesting point.

Investigations concerning a low molecular weight, basic protein factor acting on enzyme B [51], presumably by stimulating the elongation process of enzyme B on native templates, have deliberately not been dealt with since this will be the topic of Dr. Stein's lecture

Acknowledgements

We gratefully acknowledge financial support of the Deutsche Forschungsgemeinschaft and excellent technical assistance from Miss DORIS SCHWARTZ. We thank Prof. PETER KARLSON for providing research fascilities and DIETER GALLWITZ, CONSTANTIN SEKERIS and WOLFGANG SCHMID for valuable discussions. Thanks are extended to Prof. TH. WIELAND, Heidelberg, for having generously donated α-amanitin and [^{14}C]-γ-amanitin. One of us, A.F., is the recipient of a fellowship from the Alexander von Humboldt Stiftung.

References

1. An excellent collection of articles is found in: Cold Spr. Harb. Symp. quant Biol. **35**, 641—742 (1970).
2. KÜNTZEL, H., SCHÄFER, K. P.: Nature (Lond.) New Biol. **231**, 265 (1971).
3. REID, B. D., PARSONS, P.: Proc. nat. Acad. Sci. (Wash.) **68**, 2830 (1971).
4. WINTERSBERGER, E., WINTERSBERGER, U.: FEBS Letters **6**, 58 (1970).
5. WU, G. J., DAWID, I. B.: Biochemistry **11**, 3589 (1972).
6. SEIFART, K. H., BENECKE, B. J., JUHASZ, P. P.: Arch. Biochem. Biophys. **151**, 519 (1972).
7. AMALRIC, F., NICOLOSO, M., ZALTA, J. P.: FEBS Letters **22**, 67 (1972).
8. BENECKE, B., FERENCZ, A., SEIFART, K. H.: FEBS Letters **31**, 53 (1973).
9. ROEDER, R. G., RUTTER, W. J.: Nature (Lond.) **224**, 234 (1969).
10. KEDINGER, C., GNIAZDOWSKI, M., MANDEL, J. L., GISSINGER, F., CHAMBON, P.: Biochem. Biophys. Res. Commun. **38**, 165 (1970).
11. STIRPE, F., FIUME, L.: Biochem. J. **105**, 779 (1967).
12. SEIFART, K. H., SEKERIS, C. E.: Z. Naturforsch. **24**b, 1538 (1969).
13. KEDINGER, C., NURET, P., CHAMBON, P.: FEBS Letters **15**, 169 (1971).
14. WEAVER, R. F., BLATTI, S. P., RUTTER, W. J.: Proc. nat. Acad. Sci. (Wash.) **68**, 2994 (1971).
15. CHESTERTON, C. J., BUTTERWORTH, P.: FEBS Letters **12**, 301 (1971).

16. Smuckler, E. A., Tata, J. R.: Nature (Lond.) **234**, 39 (1971).
17. Sajdel, E., Jakob, S. T.: Biochem. Biophys. Res. Commun. **45**, 707 (1971).
18. Jakob, S. T.: Progr. Nucl. Acid Res. molec. Biol. (Davidson, J. N., Cohn, W., Eds.). New York-London: Academic Press **13**, 93 (1973).
19. Personal communication by Williams, R. A., Crippa, M., Institut Biologique Moleculaire, Geneva, Switzerland.
20. Dezelee, S., Sentenac, A., Fromageot, P.: FEBS Letters **7**, 220 (1970).
21. Adman, R., Schultz, L., Hall, B. D.: Proc. nat. Acad. Sci. (Wash.) **69**, 1702 (1972).
22. Brogt, Th., Planta, R.: FEBS Letters **20**, 47 (1972).
23. Ponta, H., Ponta, U., Wintersberger, E.: FEBS Letters **18**, 204 (1971).
24. Chambon, P., Gissinger, F., Kedinger, C., Mandel, J. L., Meilhac, M., Nuret, P.: Acta Endocrinologica (Kbh.) Suppl. **168**, 222 (1972).
25. Gadaleta, M., Greco, M., Saccone, C.: FEBS Letters **10**, 54 (1970).
26. Schlabach, A., Fridlender, B., Bolden, A., Weissbach, A.: Biochem. Biophys. Res. Commun. **44**, 879 (1971).
27. Roychoudhury, R., Bloch, D.: J. biol. Chem. **244**, 3359 (1969).
28. Chang, L. M., Bollum, F. J.: J. biol. Chem. **246**, 5385 (1971).
29. Haines, M. E., Johnston, I. R., Mathias, A. P.: Fed. Europ. Biochem. Soc. Letters **10**, 113 (1970).
30. Yu, F. L., Feigelson, P.: Proc. natl. Acad. Sci. (Wash.) **69**, 2833 (1972).
31. Blatti, S. P., Ingles, C. J., Lindell, T. J., Morris, W. P., Weaver, R. F., Weinberg, F., Rutter, W. J.: Cold Spr. Harb. Symp. quant. Biol. **35**, 649 (1970).
32. Doenecke, D., Marmaras, V., Sekeris, C. E.: Insect Biochemistry 1973 (in press).
33. Sekeris, C. E., Schmid, W.: Proc. IV. Internat. Congr. of Endocrinology Washington, 1972 (in press).
34. Muramatsu, M., Shimada, N., Higashinakagawa: J. molec. Biol. **53**, 91 (1970).
35. Wanka, F., Moors, J.: Biochem. Biophys. Res. Commun. **41**, 85 (1970).
36. Roeder, R., Reeder, R., Brown, D. D.: Cold Spr. Harb. Symp. quant. Biol. **35**, 727 (1970).
37. Schmid, W., Sekeris, C. E.: Biochim. biophys. Acta **312**, 549 (1973).
38. Tata, J. R.: Progr. Nucl. Acid Res. **5**, 191 (1966).
39. Baulieu, E., Wira, C., Milgrom, E., Raynaud-Jammet: Acta Endocrinologica (Kbh.) Suppl. **169**, 383 (1972).
40. Gorski, J., Toft, D., Shyamala, G., Smith, D., Notides, A.: Rec. Progr. Hormone Res. **24**, 45 (1968).
41. Cooke, A., Kay, J. E.: Exp. Cell Res. **79**, 179 (1973).
42. Frarye-Fernández, M., Pogo, A. O.: Proc. nat. Acad. Sci. (Wash.) **68**, 3040 (1971).
43. Burgess, R.: Ann. Rev. Biochem. (Snell, R., Ed.), Ann. Rev. Inc. Palo Alto, Calif. USA **40**, 711 (1971).
44. Meilhac, M., Chambon, P.: Europ. J. Biochem. **35**, 454 (1973).
45. Sippel, A., Hartmann, G.: Europ. J. Biochem. **16**, 162 (1970).

46. BAUTZ, E. K. F., BAUTZ, F. A.: Nature (Lond.) **226**, 1219 (1970).
47. Personal communication by BAUTZ, E., Institut für molekulare Genetik, Heidelberg.
48. SCHÄFER, R., ZILLIG, W., ZECHEL, K.: Europ. J. Biochem. **33**, 207 (1973).
49. JUHASZ, P., BENECKE, P., SEIFART, K. H.: FEBS Letters **27**, 30 (1972).
50. ROEDER, R., RUTTER, W. J.: Proc. nat. Acad. Sci. (Wash.) **65**, 675 (1970).
51. SEIFART, K. H., JUHASZ, P. P., BENECKE, B. J.: Europ. J. Biochem. **33**, 181 (1973).

Discussion

R. K. ZAHN (Mainz): You have shown that in nucleoli heparin can stimulate the enzyme. This could be the effect of polyanions on RNase, but couldn't it be the setting free and making available of DNA template?

K. H. SEIFART (Marburg): The reasons for the stimulation of RNA synthesis by heparin can be manifold and the one you mentioned is one possibility which we discussed (see p. 152). The important point is, however, that the enzyme in the nucleolus is able to form a heparin-resistant complex and this is in contrast to the purified enzyme. Moreover, it could be argued that the addition of DNA has a similar effect to the one you mentioned. However, the non-additivity of heparin and DNA, if added at different times (Fig. 13), argues against this assumption.

E. BAUTZ (Heidelberg): It may be difficult extrapolating from *E. coli* to mammalian enzymes with regard to resistance of initiation complexes to heparin. In the case of T3 enzyme, *e.g.*, this enzyme can perfectly well initiate but even up to synthesizing the first dinucleotide it remains heparin-sensitive.

K. H. SEIFART: What we know for certain is that the enzyme in the nucleolus forms a heparin-resistant complex. Under these conditions one probably measures mainly the completion reaction of chains initiated *in vivo*, and since we don't know the exact chain-length of the initiated RNA, we obviously don't know how many nucleotides we need in the complex to acquire heparin resistance. Therefore, your comment is appreciated. However, we do not see heparin resistance with the purified enzyme even after one minute of full synthesis (Fig. 9) which should be enough time to initiate the enzyme and form the first few phosphodiester bonds. This question requires further investigation, however.

PH. FEIGELSON: Perhaps some previously established facts should be mentioned to place Dr. SEIFART's present contribution in a broad perspective to enable the audience to understand the present state of knowledge concerning the regulation of nucleolar RNA polymerase activity. These published facts from our and other laboratories include:

1. A variety of *in vivo* and cell culture tracer studies demonstrating a rapid inhibition of rRNA synthesis following cessation of protein synthesis by translational inhibitors or amino acid deprivation.

2. Isotope precursor studies showing that within hours after administration of several hormones, hepatic rRNA synthesis is accelerated *in vivo*.

3. Nucleoli isolated from livers of glucocorticoid-treated animals manifest *in vitro* enhanced rates of nucleotide incorporation into RNA. This hormonally enhanced polymerase A activity is evident with both endogenous DNA or exogenous synthetic deoxynucleotide templates.

4. *In vivo* administration of cycloheximide or actinomycin D or α-amanitin results in progressive diminuation in nucleolar polymerase A activity with a $t_{1/2} \cong 1.5$ h, whether assayed *in vitro* with endogenous or exogenous deoxynucleotide templates.

Our recent unpublished studies indicate that pretreatment of animals with α-amanitin or cycloheximide prevents the *in vivo* glucocorticoidal induction of nucleolar polymerase A activity. This body of experimental evidence suggests the existence of one or more rapidly turning over polypeptide component(s) essential for the functioning of polymerase A; furthermore, the mRNA coding for the synthesis of this polypeptide(s) seems to be itself short-lived and hormonally modulatable.

Dr. Seifart has reported today that *in vivo* cycloheximide pretreatment does not lead to any decreased activity in a "polymerase A" fraction extracted from hepatic nuclei and isolated by ion exchange chromatography. He infers from this that RNA polymerase A is metabolically stable and not hormonally inducible. I will not dispute here his experimental findings and interpretations although in several respects they are not in accord with published reports from the laboratories of Rutter and Jacob. The comment I do wish to interpose here is that Dr. Seifart not leave the audience with the unqualified statement that RNA polymerase A activity is metabolically stable and uninducible but rather to make the more operational conclusion that these seem to be attributes of the polymerase A *core enzyme* that he isolates and assays under his unique set of conditions. It seems likely from all the aforementioned studies that one or more metabolically labile and hormonally inducible polypeptide entities exist which may possibly be somewhat analogous to the sigma factor of microbial polymerase which may participate *in vivo* and in the isolated nucleolus in mediating RNA polymerase A activity.

K.H. Seifart: I appreciate your comments and since I knew that you were here to present them, I have not explicitly discussed your results in my talk. I have not dealt with the phenomenon of hormonal induction of RNA synthesis because it was not central to the point, but we are of course aware of your and other people's work in this connection. I would, however, like to make some comments in connection with the cycloheximide experiments which you mentioned. It is beyond doubt that cycloheximide will inhibit the synthesis of ribosomal RNA and the question is, as I have outlined in my talk, whether this is due to the disappearance of the enzyme itself or some factor necessary for enzyme activity as you indicated in your comment, as our data seem to suggest and as was initially discussed by Muramatsu [34]. Such a factor is as yet hypothetical and remains to be isolated. In general, I think one has to be extremely careful to extrapolate findings obtained for the microbial RNA polymerase to the animal enzymes before direct experimental proof has been obtained. Moreover, nothing is known whether such a factor is a part of the enzyme (as in *E. coli*) or is a completely separate entity. The definition of "core enzyme" is therefore only an assumption, tempting as this may be, of course. At any rate, if you interpret your results [Proc. Nat. Acad. Sci. (USA) **69**, 2833 (1972)] to mean that the half-life of $1^1/_2$ h refers to a factor which is not necessarily the polymerase itself, I think we would completely agree with you. The open question which then still remains is why polymerase A requires such an activating factor to transcribe poly dAT which is the template which you employed since poly dAT is, as far as I know, transcribed very efficiently by all enzymes — even "core enzymes".

A Protein Factor Influencing Mammalian RNA Polymerase B

HANS STEIN and HORST HAMEISTER

Max-Planck-Institut für Biologie, Tübingen, W. Germany

and CLAUDE KEDINGER

Centre de Neurochimie du CNRS et Institut de Chimie Biologique, Faculté de Médecine, Strasbourg, France

With 4 Figures

The current analysis of transcription in eukaryotes is faced with the serious problem that the functional unit of the transcriptional machinery has not yet been identified. This leads to the operational problem that we do not know how extensively the chromatin may be broken down and still allow a clear functional analysis. Moreover, a large part of the chromatin seems to consist of elements, such as repeated DNA sequences, whose function remains unknown. In the light of these difficulties, it seemed promising to investigate the eukaryotic RNA polymerase as part of the transcriptional machinery because its function is relatively well known. Furthermore, studies with the prokaryotic RNA polymerase suggested that perhaps one protein, or a limited number of enzymes, would express this function in eukaryotes. In fact as we now know, RNA polymerase activity of eukaryotes consists of several different enzymes which may be distinguished from one another by their chromatographic behavior [1, 2], their subunit composition [3], and in part by their sensitivity to the toxin amanitin [2]. From a functional point of view, the different sensitivity of two species of RNA polymerase to amanitin was of special importance in the localization of the two species within the nucleus. The amanitin-resistant enzyme has been shown to be associated with the nucleolus, the amanitin-sensitive enzyme with the chromatin [4, 5]. The question then arises as to whether each of the several distinct enzymes performs different functions. This question implies a role for RNA polymerases in the regulation of gene transcription. On the surface, such a concept has little attraction because the number of identified RNA polymerases does not currently exceed four to five [5–8]. However, each of these enzymes may be extensively modified by additional cofactors as has been shown for the prokaryotic RNA polymerase [9]. If similar modifications exist in eukaryotes, it would be possible to generate an appreciable number of combinations that could serve as the basis for an effective gene controlling system.

For this reason, we are presently extensively analyzing a group of cellular proteins which stimulate the amanitin-sensitive RNA polymerase of calf thymus *in vitro* [10, 11]. The specificity of this stimulating effect in one case suggests that these factors may play a special role for *in vivo* RNA synthesis as well. We are attempting to determine the chemical nature of these factors, how they act in the *in vitro* system, and what their physiological significance might be.

The Nature of the Stimulating Agent

Several substances such as salts [12] and polycations [13, 14] as well as proteins [15, 16] have previously been shown to have a stimulatory effect on the RNA polymerase reaction. To determine whether the stimulatory effect found within calf thymus crude extracts was due or related to one of these substances, the stimulating agent was characterized. First, it was found that the effect of the stimulatory factor found in calf thymus homogenates was completely lost upon digestion of the cell lysates with proteolytic enzymes [10]. Thus the stimulating agent appeared to be a protein. The factor was then purified 500–1000-fold by a procedure which included a DEAE-cellulose batch-treatment, ammonium sulfate precipitation, Sp-Sephadex and hydroxyapatite chromatography. (Further details of this purification procedure will be published elsewhere.) The final product was shown to have a molecular weight of around 20000 by electrophoresis with SDS-gels and appeared as one band (Fig. 1). When the factor was subjected to chromatography on Sephadex G-100, a good correlation of the stimulating activity with the protein concentration was obtained in each of the several fractions [17]. The molecular weight of the factor as determined by gel chromatography was again found to be around 20000. Thus, it may be assumed that the single protein band found within the SDS-polyacrylamide gels in fact represented the stimulating agent. On this basis, the molar ratio of factor to RNA polymerase that was necessary for optimal stimulation was estimated to be about 2. The fact that preparations of homogeneous factor may be achieved after a roughly 500-fold purification further suggests that

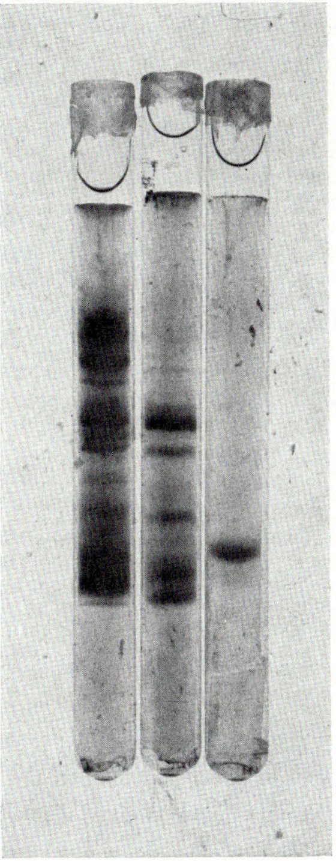

Fig. 1. SDS-polyacrylamide electrophoresis of factor fractions obtained at intervals throughout the purification of factor. The gel at the left shows the protein pattern of a factor fraction obtained after DEAE-treatment and ammonium sulphate precipitation. The gel in the middle shows the factor preparation after further SP-Sephadex chromatography. On the right, the final stage of purity of the factor preparation after hydroxyapatite chromatography is demonstrated

factor may be present within the cell in relatively high amounts. More rigorous tests are being performed to provide further proof of the identity of the purified protein with the stimulating agent.

It should be mentioned that the stimulating agent within the calf thymus appeared to be heterogeneous. A limited heterogeneity in charge was observed when the stimulating activity of calf thymus was subjected to ion-exchange chromatography on DEAE-cellulose or to isoelectrofocusing [17]. In the latter case, three distinct peaks could be clearly distinguished with isoelectric points between 7.5 and 8.5. This result characterized the stimulating factor(s) as (a) basic protein(s). Material derived from the three peaks activated the enzyme reaction to the same extent. However, the stimulatory effect of these fractions was not additive.

Upon chromatography of the factor on Sephadex G-100, there appeared beside the main fraction an additional fraction with a molecular weight of 40–50000. Indirect evidence suggests that this heavy fraction resulted from an aggregation of the lighter form. It may further be added that the factor activity was relatively insensitive to heat, and storage of the factor between pH 9 and 2 did not affect its activity. Treatment of the factor with sulphydryl reacting reagents, such as *p*-mercuribenzoate, could destroy its effect.

The Effect of the Stimulating Agent on the RNA Polymerase Reaction *in vitro*

To obtain an indication of the physiological role of the factor within the cell, we investigated the properties of the factor stimulated reaction *in vitro*. Our primary concern has been to determine what steps in the RNA polymerase reaction are affected by the factor. Furthermore, results will be summarized which show that the factor mediates remarkable changes in the enzymatic reaction in addition to its stimulatory effect. Some preliminary experiments will be also cited in which we investigated the site of interaction of the factor *in vitro*.

In regard to the modification of the enzyme reaction by the factor *in vitro*, we have previously reported that the factor shifted the salt optimum of the reaction [10]. The ratio of the activity in the presence of Mn^{++} or Mg^{++} ions was drastically changed as well. Furthermore, the enzyme was more active on native DNA in the presence of factor, whereas optimal activity in the absence of factor was observed with denatured DNA. Finally, the kinetics of the reaction were altered [10, 17]. Under the conditions of low ionic strength, the unstimulated reaction was completed within 30 min of incubation. In the presence of factor, however, the same initial rate of incorporation was maintained throughout this period [17]. This result implies that there is a delay in the stimulating effect when the factor is present from the beginning of the reaction. However, addition of the factor to a preincubated sample results in an immediate stimulation of RNA synthesis.

We have therefore attempted to determine whether the factor affects the initiation, elongation, or the termination-reinitiation step of the reaction. A first hint that initiation was not stimulated by the factor was provided by the observation that the initial rate of RNA synthesis was not increased by the factor (see above). Further evidence against an initiation promoting effect was provided by experiments in which the enzyme reaction was allowed to proceed under conditions which were highly inhibitory for RNA synthesis. In these experiments the enzyme reaction was depressed either by addition of increasing amounts of actinomycin D, thereby reducing mainly the elongation reaction [18], or by lowering the concentration of UTP keeping the other three ribonucleoside triphosphates constant. As UTP is likely not used for the initiation [19], its reduction must mainly affect the elongation. A secondary reduction in the rate of the termination-reinitiation step may also occur under these conditions. Experiments performed with actinomycin D revealed that the factor-stimulated reaction was more sensitive to the drug than the unstimulated reaction. Thus at high inhibitor concentrations the residual activity was nearly the same in the presence and absence of factor (Fig. 2). Upon depression of the rate of RNA synthesis by lowering the UTP level, a rapid decrease of the factor effectivity was observed (Fig. 3). At very low substrate concentrations the factor had only little stimulatory effect. Both types of experiments support the hypothesis that the number of initiating RNA polymerase molecules is not increased by the factor.

Finally, an experiment analogous to those performed with *E. coli* RNA polymerase and the sigma factor was carried out to test for an initiation promoting capacity [20]. In the presence of only two triphosphates, enzyme B was allowed to form an initiation complex with native DNA in the presence and absence of factor. The RNA synthesizing capacity of this complex was then measured in the presence of the rifamycin derivative AF/013 to prevent reinitiation [21]. Equivalent incorporation in a subsequent elonga-

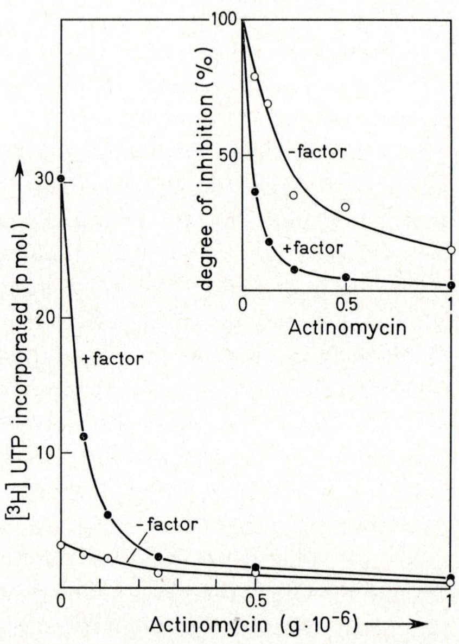

Fig. 2. Inhibition of the enzyme reaction in the presence and absence of factor by actinomycin D. RNA synthesis was carried out in a reaction system [10] with RNA polymerase B, with (1 μg) and without factor and differing amounts of actinomycin D. The insert shows the percent inhibition as a function of actinomycin D

tion assay was found whether factor was present or absent during the initiation period [17]. This again allows the conclusion that the factor does not increase the number of initiating RNA polymerase molecules. The validity of this conclusion is restricted, however, because AF/013 may interact with the factor as well as with the RNA polymerase (see below). If this is the case, then initiation complexes which might require factor for their stabilization would not be detected in the presence of AF/013. With this one reservation, the three experiments discussed definitely demonstrate that the factor does not positively influence initiation. Thus the elongation step or the termination-reinitiation events remain as possible functions affected by the factor. To differentiate between these two possibilities, the effect of the factor on the enzyme reaction was observed in the presence of AF/013, heparin and poly I. These sub-

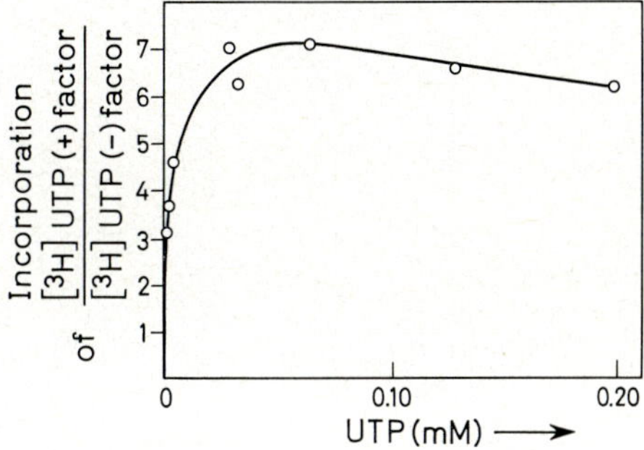

Fig. 3. Dependency of the stimulatory effect of the factor on the UTP concentration. Reactions were carried out in reaction systems [10] containing different concentrations of UTP in the presence (0.5 μg) and absence of factor

stances have each been shown to inhibit preferentially the initiation or reinitiation step [21–23]. Furthermore, denatured DNA was used in combination with ribonuclease H to compete with native DNA for the template-free RNA polymerase. The combination of dDNA with RNAse H was necessary as denatured DNA by itself serves as template for the RNA polymerase. However, the product of this reaction which is hybridized to DNA can be specifically broken down by RNAse H [24]. In all experiments, inhibitor and factor were subsequently added after the enzyme reaction had been allowed to start with RNA synthesis by 5–30 min preincubation. Under these conditions, a nearly complete block of the factor effect was seen in the presence of heparin (Fig. 4). The results obtained for AF/013 (not shown) under the same conditions were analogous to those obtained with heparin: again a more than 95 % reduction in the factor effect was found. With poly I only a partial reduction of the factor stimulation could be observed. In the case of the denatured DNA-ribonuclease H combination, no impairment of the factor effect could be demonstrated (Fig. 4). The effect of heparin and AF/013 suggests that the reinitiation step is an essential although not necessarily a sufficient requirement for the action of the factor. The incomplete block of the factor effect by poly I is consistent with this conclusion if one assumes that some re-

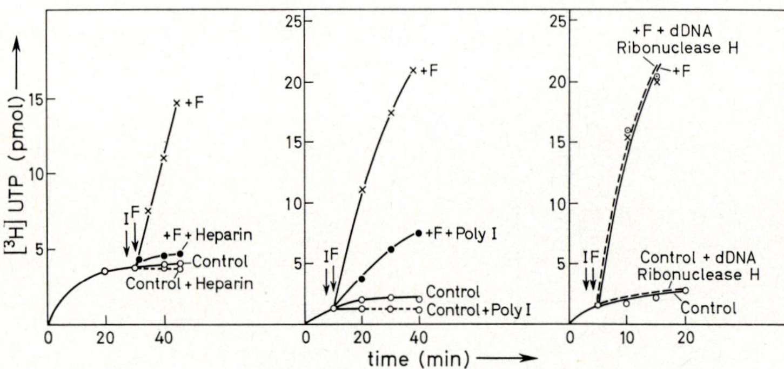

Fig. 4. Effect of heparin, poly I and a denatured DNA-ribonuclease H mixture on the enzyme reaction in the absence and presence of factor. At the times indicated, either 50 μg of heparin or 1 μg of poly I or 20 μg of denatured DNA together with an excess of ribonuclease H were added to reaction systems [10] containing 1 μg of DNA. Factor (1 μg) was added subsequent to the inhibitors to half of the samples. Processing of the samples and measurement of radioactivity was performed as previously described [10]

initiating RNA polymerase molecules can escape the poly I inhibition. From the fact that denatured DNA together with RNAse H shows no inhibitory effect at all, we can finally deduce that prior to or during the action of the factor no release of the enzyme from its template occurs.

A prediction can be made concerning the number of RNA chains present in the factor-mediated system. If reinitiation is necessary for the stimulatory effect of the factor as suggested by the AF/013 and heparin experiment, then at least a two-fold increase in the number of RNA chains should result from the action of the factor. However, a direct determination of the 5′-terminal nucleotides of RNA revealed that the factor increased their number

Table 1. Incorporation of labeled nucleotides

Incubation conditions	[^{3}H]-UTP pmol	[^{32}P]-γ-GTP pmol	[^{32}P]-γ-ATP pmol	Average chain length (nucleotides)
30 min, without factor	73	0.177	0.064	1,090
30 min, plus factor	236	0.213	0.079	2,900
40 min, without factor	89	0.180	0.076	1,241
40 min, factor added at 30 min	199	0.201	0.095	2,401

Incubation mixture contained in 0.125 ml : 100 mM Tris HCl pH 7.9, 4 mM thioglycerol, 0.1 mM dithiothreitol, 0.1 mM EDTA, 12% glycerol, 25 μg bovine serum albumin, 3 mM Mn^{++}, 0.5 mM GTP or ATP (one of them labeled with [^{32}P] in the γ position), 0.16 mM CTP, 0.16 mM [^{3}H]-UTP (40 cpm/pmol), calf thymus DNA 5 μg, purified calf thymus B enzymes (fraction PC 2) 1.3 μg. Factor (10 μl) was added where indicated. After incubation at 37 °C, as indicated in the table, the incorporation of the labeled nucleotides was determined as described by M. Meilhac and P. Chambon (Eur. J. Biochem., in press).

only by a factor of 1.2 (Table 1). Thus, in contradiction to the results presented above, the factor appears mainly to affect the length and not the number of the newly synthesized RNA chains. Control experiments have clearly shown that the γ-^{32}P-(ATP) and -(GTP) incorporation used to measure the 5'-termini is dependent on RNA synthesis and that it is RNAse A-sensitive. Furthermore, the factor itself does not appear to contain a phosphorylase activity as measured by the conversion of ATP to ADP. Such an activity could have resulted in an underestimate of the number of synthesized RNA molecules. One reservation with regard to this experiment is, however, the fact that we monitored only purine termini. This limitation is not likely to affect the conclusion because the polymerase has been shown to initiate preferentially with purines. [19]. With regard to the discrepancy in the conclusion between the endgroup labeling experiments and the inhibitory experiments, we tend to favor the results of the more direct endgroup labeling experiments. At present, relatively unspecific effects of the heparin, poly I and AF/013 upon the factor cannot be ruled out. Heparin and poly I, as negatively charged molecules, might directly bind to the positively charged factor. AF/013 on the other hand may directly inactivate the factor protein.

Thus, in the light of these experiments the factor appears mainly to increase the performance of the RNA polymerase molecules in chain propagation. The small increase in the number of RNA chains influenced by the factor (Table 1) may indicate an additional effect on the termination-reinitiation reaction, but it may also represent a secondary phenomenon. Although the factor is a likely candidate for an elongation – promoting factor, fundamental questions concerning its *in vitro* action remain open. For example, it is not known why the reaction rate of the control reaction decreases within 30 minutes of incubation and how the factor can reactivate an apparently inactive reaction system. Exhaustion of substrate or of template in the absence of factor does not account for this phenomenon (unpublished results).

Previous experiments demonstrated a strong binding of factor to enzyme B under standard assay conditions [11]. Thus, the RNA polymerase appears to be one site of interaction for the stimulating agent. Based on the observation that factor effectively adsorbs to nitrocellulose filters, binding studies between native DNA and purified factor preparations were recently carried out. The results demonstrate that up to 10^4 factor molecules are required for retention of one fdmolecule on the filters under standard salt conditions. This takes into account a 80% purity of the factor. Lowering the salt concentration increases significantly the binding capacity for DNA. The binding forces thus may be mainly electrostatic in nature. A similar weak binding between RNA and factor was found in analogous studies in which *in vitro*-synthesized RNA was employed. As a final conclusion we would suggest that neither DNA nor RNA may be able to compete with the RNA polymerase for binding the factor under the conditions of the standard assay. Thus enzyme B may provide the essential site of interaction between the factor and the components of the *in vitro* assay.

With regard to the general concept of this symposion we will now consider the possible biological meaning of the stimulating factor. The high degree of specificity encountered in the interaction of the factor with the responding enzyme, as shown by the *in vitro* experiments, substantiates a possible function of this factor in cellular RNA synthesis. There are two major alternatives for its function: 1. The factor could act as a regulatory element in RNA synthesis either in a quantitative or a qualitative way. 2. The factor could represent an essential component for enzyme B.

The extensive stimulatory effect leading to a more than 5-fold stimulation of the RNA synthesis by the factor in the *in vitro* system would be consistent with a *quantitative* control function of the stimulatory proteins. However, it must be kept in mind that the degree of stimulation is highly dependent on the ionic environment [10]. The several-fold stimulation is only observed at low salt concentrations. At salt concentrations presumably closer to physiological values, *e.g.*, 0.10 M KCl, the stimulatory effect in the *in vitro* system does not exceed a factor of about two. Thus, the factor may not be essential for regulating the rate of RNA synthesis. On the other hand, whether the factor effects *qualitative* changes in the cellular RNA synthesis mediated by enzyme B cannot be decided on the basis of the existing *in vitro* experiments. These experiments, although not totally consistent, were interpreted to suggest that the factor mainly affects elongation. One interpretation of this effect is that the factor enables enzyme B to synthesize special RNA classes, *e.g.*, as an "anti-termination" factor. Alternatively, other mechanisms such as the reversion of a product inhibition could lead to the same elongation effect. In this case, a change in the template specificity of enzyme B would probably not be involved. So far, the only finding which supports the idea that factor can induce an alteration of the template requirement of enzyme B are changes in the activity ratio of native versus denatured DNA in the presence of enzyme B [10]. However, other explanations for this effect, which do not suppose a change in the template specificity of enzyme B, may be considered as well.

A change in the template specificity of enzyme B on native DNA alone has not yet been demonstrated as introduced by the factor. No differences could be detected in terms of total base composition between the reaction product synthesized on native calf thymus DNA in the presence and absence of factor. Moreover, the base composition of the RNA synthesized in reaction mixtures with or without factor was nearly identical with that of the template DNA, suggesting that in both cases the DNA was completely transcribed (unpublished results). Further experiments with less complex templates and thus better characterizable transcription products will be required to prove or disprove an effect of the factor on the template specificity of enzyme B for native DNA.

An estimation of the factor concentration in crude calf thymus homogenates suggests that the molar ratio of factor to enzyme B [25, 26] is roughly one or greater than one. In view of the observation that a two to one relationship of factor to enzyme may be enough for optimal stimulation of the enzyme assay *in vitro*, we tentatively conclude that the factor is not limiting within this tissue. This calls in question a regulatory role of the factor, either in qualitative or quantitative respect.

The factor thus could well represent a normal constituent of the transcriptional machinery required for the performance of enzyme B. Such a hypothesis would be consistent with the observation that a similar or identical stimulating activity has recently been found in a variety of cells and tissues including bovine testes (unpublished observation), rat liver [27, 28] and mouse myeloma cells [29].

The factor activity isolated from mouse myeloma tissue shows a particularly close relationship to the factor extracted from calf thymus tissue in regard to its effect on the enzyme reaction as well as on its chemical properties. Both factors have the following properties in common with the activity from rat liver: 1. they are basic, 2. they have relatively small molecular weights, 3. they appear to restore the preference of enzyme B for native DNA and 4. they stimulate *in vitro* mainly by an effect on chain propagation as measured by γ-^{32}P incorporation. Differences between the mouse liver factor on the one hand and the mouse myeloma and calf thymus factor on the other hand may be seen in their salt dependencies and reaction kinetics.

As to our knowledge, no related factors have so far been detected in prokaryotes. Even in yeast where an amanatin-sensitive RNA polymerase similar to enzyme B of higher organisms is present [30], no comparable factor activity has been detected so far.

Acknowledgement

The generous support of this work by Prof. Dr. W. BEERMANN is gratefully acknowledged. We are greatly indebted to Prof. P. CHAMBON for many helpful discussions and suggestions. Thanks are due to I. Böttcher (Tübingen) and G. Dretzen (Strasbourg) for expert technical assistance.

This work was supported by the Deutsche Forschungsgemeinschaft.

References

1. Roeder, R. G., Rutter, W. J.: Nature (Lond.) **224**, 234 (1969).
2. Kedinger, C., Gniazdowski, M., Mandel, J. L., Gissinger, F., Chambon, P.: Biochem. Biophys. Res. Commun. **38**, 165 (1970).
3. Mandel, J. L., Chambon, P.: FEBS-Letters **15**, 175 (1971).
4. Jacob, S. T., Sajidol, E. M., Munro, H. N.: Biochem. Biophys. Res. Commun. **38**, 765 (1970).
5. Lindell, T. J., Weinberg, J., Morris, P. W., Roeder, R. G., Rutter, W. J.: Science **170**, 447 (1970).
6. Kedinger, C., Nuret, P., Chambon, P.: FEBS-Letters **15**, 169 (1971).
7. Chesterton, C. J., Butterworth, P.: FEBS-Letters **12**, 301 (1970).
8. Seifart, K. H., Benecke, B. J., Juhasz, P. P.: Arch. Biochem. **151**, 519 (1972).
9. Burgess, R.: Ann. Rev. Biochem. **40**, 711 (1971).
10. Stein, H., Hausen, P.: Europ. J. Biochem. **14**, 270 (1970).
11. Stein, H., Hausen, P.: Cold Spr. Harb. Symp. quant. Biol. **35**, 709 (1970).
12. Fuchs, E., Millette, R. C., Zillig, W.: Europ. J. Biochem. **3**, 183 (1967).
13. Fox, C. F., Weiss, S. B.: J. biol. Chem. **239**, 175 (1964).
14. Petersen, E. E., Kröger, H., Hagen, U.: Biochim. biophys. Acta **161**, 325 (1968).
15. Vogt, V., Nature (Lond.) **223**, 854 (1969).
16. Burgess, R. R., Travers, A. A., Dunn, J. J., Bautz, E. K. F.: Nature (Lond.) **221**, 43 (1969).
17. Hameister, H., Wilson, M., Stein, H.: Stud. Biophys. **31/32**, 33 (1972).
18. Sentenac, A., Simon, E. J., Fromageot, P.: Biochim. biophys. Acta **161**, 299 (1968).
19. Meilhac, M., Chambon, P.: Europ. J. Biochem. (in press).
20. Sippel, A. E., Hartmann, G. R.: Europ. J. Biochem. **16**, 152 (1970).
21. Meilhac, M., Tysper, Z., Chambon, P.: Europ. J. Biochem. **28**, 291 (1972).
22. Walter, G., Zillig, W., Palm, P., Fuchs, E.: Europ. J. Biochem. **3**, 194 (1967).
23. Mandel, J. L., Chambon, P.: Personal communication.
24. Hausen, P., Stein, H.: Europ. J. Biochem. **14**, 278 (1970).
25. Gissinger, F., Chambon, P.: Europ. J. Biochem. **28**, 277 (1972).
26. Roeder, R. G., Rutter, W. J.: Proc. nat. Acad. Sci. (Wash.) **65**, 675 (1970).
27. Seifart, K. H.: Cold Spr. Harb. Symp. quant. Biol. **35**, 719 (1970).
28. Seifart, K. H., Juhasz, P. P., Benecke, B. J.: Europ. J. Biochem. **33**, 181 (1973).
29. Lentfer, D., Lezius, A. G.: Europ. J. Biochem. **30**, 278 (1972).
30. Ponta, H., Ponta, U., Wintersberger, E.: Europ. J. Biochem. **29**, 110 (1972).

Discussion

J. Drews: Since the factor allows the formation of longer chains, is it possible that its main effect is just a stabilization of the polymerase? Does the polymerase become more heat stable in the presence of the factor?

H. Stein (Tübingen): The decrease of the rate of RNA synthesis observed under our reaction conditions in the absence of factor is probably not due to an heat inactivation of RNA polymerase B in the course of RNA synthesis. This is suggested by the fact that the factor can completely reactivate a reaction system which was nearly inactive in terms of RNA synthesis after a 40 min preincubation at 37°. The thermal inactivation profile of rat liver RNA polymerase B, recently published by Tata and Shields [FEBS-Letters **31**, 209 (1973)], furthermore demonstrates a 80% residual activity of RNA polymerase B after a 15 min pre-heating at 45 °C.

D. Lentfer (Göttingen): Did you ever look for the intracellular localization of your stimulating activity?

H. Stein: This has not yet been done.

Starlinger (Köln): 1. Could the effect of the factor be similar to the effect of the N-product of phage λ?

2. Are the same sequences formed (by hybridization competition) in the presence and in the absence of factor?

H. Stein: Regarding the first question we currently envisage mainly two alternatives as interpretations for the stimulatory effect of the factor: (a) the factor could allow the enzyme to overcome certain termination signals of the DNA template, or (b) abolish a product inhibition of the RNA polymerase by the newly formed RNA. In the first case, the factor effect may be similar to the Nproduct effect of phage λ.

Hybridization experiments, which could give an answer to the second question, were not performed up to now. In terms of total base composition, there is no difference to be detected between the RNA synthesized either in the presence or absence of factor with calf thymus DNA as template.

R.K. Zahn (Mainz): Did you look at amount and/or activity of your factors under the action of hormones — oestrogens and androgens?

H. Stein: The activity and or amount of stimulating factor after hormone application to hormone-sensitive tissues has not yet been investigated. However, it may be of interest that in phytohemagglutinin-stimulated lymphocytes, no quantitative correlation between the rate of RNA synthesis and the factor activity was found.

Chr. Scholtissek (Gießen): By isoelectric focusing, three fractions were separated. Is it possible that your factor embodies different amounts of sialic acid molecules which leads to this kind of separation as was found, *e.g.*, for interferon? Treatment with sialidase might lead to a single band which might help to purify the factor.

H. STEIN: The reasons for the "microheterogeneity" found within factor fractions are not at all clear at the moment. One possible explanation would in fact be that factor embodies different amounts of sialic acid molecules.

A. HILDEBRANDT (Heidelberg): I would like to add two pieces of information to Dr. CHAMBON's talk. Firstly, he found no immunological crossreaction between enzymes A and B from mammalian system; we found such crossreaction with the corresponding yeast enzymes. Secondly, we found in yeast that PMSF is able to inhibit only two out of three protekinases.

H.G. ZACHAN: Has the stimulating factor already been used on P. CHAMBON's 200×10^6 MW DNA or on adenovirus DNA?

H. STEIN: Yes, the very low transcription of these DNA's was stimulated by the factor, but not more than the transcription of calf thymus DNA. It follows that the relative transcription of the high molecular weight cellular RNA and of adenovirus DNA remains very low even in the presence of the factor.

E. BAUTZ (Heidelberg): I have a comment to the previous question. It might be worth while to try this factor on circular DNA of SV40. The factor is supposed not to act on denatured DNA.

H. STEIN: This might not give a solution to the problem. Once the enzyme has initiated on a non-denatured piece of DNA, it will continue to act also on the denatured part.

Transcription in Mitochondria

E. WINTERSBERGER

Department of Physiological Chemistry,
University of Würzburg, W. Germany

Introduction

Extensive studies in the past 10 years have shown that mitochondria of all eukaryotic cells contain a unique DNA which has been well characterized physically and chemically in a large number of cases. (See [1–5] for recent reviews.) This DNA is genetically active; it is replicated and transcribed within the organelle and it carries information for the RNA of mitochondrial ribosomes, for at least some mitochondrial tRNAs and for a few mitochondrial membrane proteins. Its presence and function are required for the biogenesis of mitochondria.

Studies on the transcription of DNA in mitochondria have been initiated almost ten years ago [6–8]; they have since been carried out at several levels of complexity starting from *in vivo* experiments using intact growing cells to experiments with isolated mitochondria or solubilized and partially purified mitochondrial RNA polymerases. All these investigations have provided interesting and valuable results which will be useful for the understanding of the subtle regulatory mechanisms which couple the genetic activities of mitochondria and nuclei during cell growth and development.

Synthesis of Mitochondrial RNA in Intact Cells

One of the major advances in our understanding of transcription in mitochondria was brought about by the comprehensive experiments of ATTARDI and coworkers [9–12] with HeLa cells. There are several reasons which make this system such a useful one for this kind of investigations: 1. Mitochondrial DNA of HeLa, like of most animal cells, can be readily isolated as an intact, circular double-stranded molecule by CsCl gradient centrifugation in the presence of ethidium bromide. Most important, this DNA can be denatured after introduction of one single-strand break and the complementary strands separated and isolated in pure form. This allows the determination of the strand of mitochondrial DNA coding for the various transcription products. 2. The interference with the labeling of mitochondrial RNA by the rapid synthesis of large amounts of RNA of cytoplasmic ribosomes can be prevented by carrying out the labeling in the presence of low concentrations (0.04 μg/ml) of actinomycin D. 3. Mitochondrial RNA synthesis,

Table 1. Stable transcription products of mitochondrial DNA

Type of RNA	Number of cistrons on mitochondrial DNA	Coding strand of animal mtDNA
Ribosomal RNA present in the two subunits of mitochondrial ribosomes	1	H-strand
Transfer RNA	11	8 on H-strand 3 on L-strand

but not the transcription of nuclear DNA, is sensitive to ethidium bromide [13, 14]; this provides a useful control to prove the origin of the transcription product studied. Aloni and Attardi [10] first established that total mitochondrial RNA, synthesized in HeLa cells during long labeling periods, hybridizes almost exclusively to the heavy (H) strand of mitochondrial DNA; less than 2% of the light (L) strand forms hybrids with this RNA. In hybridization saturation experiments it was found that the heterogeneously sedimenting RNA saturates 100% of the H-strand. This RNA includes the 16 S and 12 S mitochondrial rRNA components and mitochondrial 4 S RNA as well as other RNA species consisting in part of precursors of these stable RNA's (Tables 1 and 2). The only RNA species hybridizing to the L-strand under these labeling conditions are about 3 mitochondrial tRNA's. Annealing experiments using purified rRNA and 4 S RNA from HeLa mitochondria revealed the presence on mitochondrial DNA of one cistron each for the 16 S and the 12 S RNA and about eleven for tRNA of which eight are localized on the H-strand and three on the L-strand

[11]. (See Table 1.) In a most elegant continuation of this work, Wu et al. [15] were able to directly localize the relative positions of the 4 S RNA genes and of the rRNA genes on the H- and L-strands of mitochondrial DNA by electron microscopy. In order to visualize the small regions on the DNA coding for 4 S RNA, this RNA was covalently coupled to the electron-opaque label, ferritin, which is visible in the electron microscope. Mitochondrial 16 S, 12 S and 4 S RNA-ferritin-conjugate were then hybridized to the H- and L-strand HeLa cell mitochondrial DNA and the relative positions of the duplex regions were examined. With this

Table 2. Turning over transcription products of mitochondrial DNA

Characteristics of RNA	Coding strand of animal mtDNA
A. Heterogeneously sedimenting RNA representing in part precursor molecules of mitochondrial ribosomal RNA (observed under pulse labeling as well as steady state labeling conditions)	H-strand (100%)
B. Heterogeneously sedimenting RNA which is rapidly degraded or exported and observed only under pulse labeling conditions	L-strand (100%)

method 9 reproducible binding sites for 4 S RNA are found on the H-strand and 3 on the L-strand. While the presence in mitochondria of a 5 S rRNA has not yet been proven, it is possible that, in keeping with the small size of the other species of mitochondrial rRNA, an rRNA with the function of 5 S RNA might be present in the 4 S fraction. One of the binding sites for 4 S RNA found on the H-strand could therefore represent a region coding for such an rRNA.

Under conditions of pulse labeling, Aloni and Attardi [12] obtained quite different results (Table 2): in that case, the transcript hybridizes to the L-strand of mitochondrial DNA as well as to the H-strand. This was confirmed by the observation that the mitochondrial RNA synthesized during short labeling periods is self-complementary. Transcription of HeLa mitochondrial DNA therefore appears initially to be symmetric, but most of the L-strand product is rapidly degraded or exported.

Although analogous extensive studies have not yet been carried out in other systems, there is good evidence that pulse labeled mitochondrial RNA of hamster cells is also rapidly turning over [16]; the situation in hamster cells may then be quite similar to that in HeLa. Synthesis of mitochondrial RNA *in vivo* was also studied in a number of other animal cells including sea urchin embryos [17] and *Xenopus levis* eggs, ovaries and kidney cells [18]. In the latter case it was again shown that the stable RNA species hybridizes largely to the H-strand of mitochondrial DNA.

In vivo experiments on the synthesis of mitochondrial RNA in lower eukaryotes (yeast, *Neurospora crassa*, *Tetrahymena*) have not yet been carried out as extensively. Although these unicellular eukaryotes should be very useful for such investigations and in fact are extremely valuable for many studies on mitochondrial biogenesis, they have some severe disadvantages for this type of *in vivo* experiments: 1. Mitochondrial DNA is much larger and difficult to isolate in intact form. 2. DNA strand separation has not yet been achieved.

In vivo experiments on mitochondrial RNA synthesis in *N. crassa* have been initiated by Kuriyama and Luck [19] and have already provided good evidence for the existence of a precursor molecule of mitochondrial rRNA which sediments at about 32 S. In the process of maturation of this molecule to yield the 2 rRNA's present in the mitochondrial subunits, approximately 22% of the precursor is lost. Thus, there appears to be some similarity between *Neurospora crassa* and HeLa cells in that mitochondrial RNA is synthesized in the form of larger precursor molecules. In this respect, the biosynthesis of mitochondrial rRNA resembles the production of cytoplasmic rRNA in nucleoli [20].

Synthesis of RNA in Isolated Mitochondria

Isolated mitochondria retain at least in part the capacity to synthesize RNA. In fact, mitochondrial RNA synthesis was first demonstrated almost 10 years ago using isolated organelles [6–8]. It was suggested that mitochondria contain an RNA polymerase which is active in the transcription of their DNA. Since then, extensive studies of this reaction have been carried out in various systems. In many cases the product of the reaction was isolated and examined by sedimentation in sucrose gradients. It is found that, in addition to some heterogeneously sedimenting materials, the stable RNA's are labeled [8, 21, 25]. Incorporation of radioactively labeled UTP or ATP generally requires all four ribonucleoside triphosphates and magnesium ions (contrast [26, 27]). The incorporation process is considerably faster if the mitochondria are swollen, probably due to faster penetration of the ribonucleoside triphosphates under these conditions. Mitochondrial RNA synthesis is then inhibited by actinomycin D, proving its dependence on DNA. A large difference was observed in some cases between the extent of incorporation of ATP vs. UTP [28]. This is most likely due to the presence in mitochondria, in addition to the DNA-dependent RNA polymerase, of a poly(A)-polymerase [29] which is probably responsible for the synthesis of the poly (A) stretches recently found in mitochondrial messenger RNA [30].

The product of the *in vitro* incorporation reaction hybridizes with mitochondrial DNA. In rat liver, where separation of the complementary strands of mitochondrial DNA is possible, it was found that the labeled material again forms hybrids with the H-strand [31]. This supports the contention that isolated mitochondria are capable, at least for some time, of carrying out RNA synthesis just as do mitochondria within the living cell, except that the process appears to be much slower *in vitro* than *in vivo*.

Mitochondrial RNA Polymerase

For a more detailed characterization of the transcription process and its regulation, it would of course be desirable to have a complete *in vitro* system consisting of isolated mitochondrial DNA and purified mitochondrial RNA polymerase. To this end, many attempts have been made to isolate these components in a form suitable for such studies. While the isolation of intact circular mitochondrial DNA from animal cells is readily possible, this is not the case for the 5 times larger mitochondrial DNA molecules of the simple eukaryotes such as yeast or *N. crassa.* In fact, there is as yet no reliable method available for the isolation of intact mitochondrial DNA from yeast.

As far as solubilization and isolation of mitochondrial RNA polymerase is concerned, the difficulties are equally large. They are due to several facts. Mitochondrial RNA polymerase appears to be bound to membrane and/or DNA within the organelle [32–40]. This calls for the necessity to disrupt the membrane and to free the enzyme from the membrane and the DNA. This can be achieved by several techniques; however, in many cases the enzyme then proved to be very labile and to become rapidly inactivated during purification (see later). Despite great efforts, no procedure has as yet been worked out which could be applied generally. In most cases a procedure that can be successfully used with one type of mitochondria does not work with mitochondria from another organism or even tissue. In the light of these problems it is not surprising that the field of mitochondrial RNA polymerase is still confusing and controversial. Upon inspection of whatever solid data are available, there are, however, a number of observations which seem to be rather general and which therefore might provide a first clue as to some of the properties of this enzyme. One major concern in purifying mitochondrial polymerases is to exclude contamination by nuclear or cytoplasmic enzymes; mitochondrial RNA polymerase is present in the organelle in rather low amounts only and contamination of the mitochondrial preparation by nuclear material will therefore cause additional problems. No fast and easy way is at hand to test for contamination by nuclear RNA polymerases. There is general agreement that mitochondrial RNA polymerase is not inhibited by α-amanitin, thus nuclear type *B* RNA polymerases can readily be excluded. However, it is more difficult to exclude contamination by nuclear A enzymes. In some but not all cases, mitochondrial RNA polymerase has been found to be rifampicin-sensitive. As no rifampicin-sensitive nuclear RNA polymerase is known, this is a good criterion to exclude contamination, but it can be applied only to those cases where the mitochondrial enzyme is sensitive to the drug. Still, it seems to me that enough data on various distinct properties of mitochondrial RNA polymerases are available now to rigorously prove the mitochondrial origin of an enzyme preparation. Before discussing differences found in various systems and sometimes even in the same system by different authors, let me first summarize the generalities that can be extracted from the work of a number of groups (Table 3).

1. Starting from isolated and purified mitochondria, everyone encounters difficulties in extracting the enzyme. Therefore, mitochondrial RNA polymerase must exist in a particle-bound form within the organelle. It is not really known whether this means binding to mitochondrial inner membrane directly or via DNA. 2. The polymerase, once solubilized, appears to be quite labile. Sometimes it is inactivated by the common techniques, such as column chromatography, used in enzyme purification. Lability of

Table 3. Properties common to mitochondrial RNA polymerases isolated from various organisms and tissues

1. Solubilization requires destruction of mitochondrial membrane.
2. Soluble enzymes usually are very labile.
3. Requirement for all four ribonucleoside triphosphates and Mg^{++} for activity.
4. Salt inhibits activity drastically.
5. DNA-dependent; native mitochondrial DNA is good template but in most cases calf thymus DNA is also accepted. Denatured calf thymus DNA is preferred over native DNA.
6. Elution from DEAE-cellulose or DEAE-Sephadex columns at 0.15 to 0.25 M salt concentration.
7. Insensitive to even high concentrations of α-amanitin.
8. Aggregates form very readily.
9. Are composed of polypeptide chains of rather low molecular weight.

the RNA polymerase may be induced by the means applied for solubilization; it cannot be excluded that the enzyme as extracted from mitochondria is no longer intact and is missing a protein subunit whose absence renders it labile. 3. Mitochondrial RNA polymerases require magnesium ions for activity. In many cases 10 mM Mg^{++} was found to be optimal. Sometimes, but not in every system, Mg^{++} can be replaced by Mn^{++}; the optimal concentration of Mn^{++} is much lower (about 1 mM) and higher concentrations of Mn^{++} are inhibitory. 4. One very typical property of mitochondrial RNA polymerases is their sensitivity towards addition of salt. Concentrations of NaCl, KCl or $(NH_4)_2SO_4$ exceeding about 25 mM strongly inhibit the enzyme. 5. Mitochondrial RNA polymerases use native mitochondrial DNA as template, but except for one case [36] are about equally active with denatured calf thymus DNA. Denatured calf thymus DNA usually is approximately twice as active a template as native thymus DNA. 6. Upon chromatography on DEAE-cellulose or DEAE-Sephadex, mitochondrial RNA polymerases elute at salt concentrations between 0.15 and 0.25 M. 7. All mitochondrial RNA polymerases studied so far are insensitive towards α-amanitin, even if high concentrations of the drug are used. As outlined above, this distinguishes mitochondrial RNA polymerase from nuclear B enzymes, but also from the newly discovered cytoplasmic RNA polymerase [42]. 8. Mitochondrial RNA polymerases have a tendency to form aggregates in solutions of low ionic strength. This property is not typical for mitochondrial RNA polymerase, as bacterial or nuclear enzymes likewise aggregate quite readily; it can, however, be conveniently used for the purification of the enzyme (for instance [36]). When enzyme preparations were obtained pure enough for determination of the molecular weight of subunits by polyacrylamide gel electrophoresis in the presence of sodium dodecyl sulfate, it is found that the subunits are of low molecular weight. 64000 was found by KÜNTZEL and SCHÄFER [36] for the enzyme from *N. crassa*, 46000 with a shoulder at 50000 by WU and DAWID [38] for the RNA polymerase from *Xenopus laevis* mitochondria and three polypeptides with molecular weight of about 97000, 69000 and 59000 by ROGALL and WINTERSBERGER [43] in a not-quite-pure preparation of yeast mitochondrial RNA polymerase (Table 4). Where only one subunit is found by gel electrophoresis in presence of sodium dodecyl sulfate, it is not known whether this represents an oligomer of identical polypeptide chains as no separation of subunits according to charge has been tried. When comparing the as yet few data on subunit molecular weights, it is interesting to note that the subunits of the animal mitochondrial RNA polymerases are definitely smaller than those of the lower eukaryotes. Even if the yeast enzyme must still be considered impure, it is already clear that there is no subunit present with a molecular weight as low as those found in *Xenopus*. This might indicate that with respect to the size of mitochondrial RNA polymerase, the situation is similar to the one found for the mitochondrial ribosome which also is smaller in higher eukaryotes than in yeast or *Neurospora* [5]. In any case, the subunit structure of mitochondrial RNA polymerase also clearly distinguishes this enzyme from the nuclear polymerases. (See article by CHAMBON in this volume.)

Taken together, it seems to me that there are now sufficient criteria known which have been established for mitochondrial RNA polymerases from quite different sources and which indicate some sort of similarity among these enzymes. It is the more surprising that with respect to one property there is still absolute disagreement between different groups and this concerns the sensitivity or insensitivity of the polymerase to rifampicin. This problem has concerned investigators in the field ever since mitochondria or mitochondrial extracts were for the first time tested for their sensitivity against this drug four years ago. In intact mitochondria, insensitivity to

Table 4. Molecular weights of polypeptide chains of mitochondrial RNA polymerases

Source of enzyme	Subunit molecular weight
Neurospora crassa	64000
Yeast (probably impure)	95—98000; 66—69000; 56—59000
Xenopus laevis	46000; 50000

rifampicin could, of course, simply be due to a permeability problem and was frequently interpreted that way, although in some cases the same preparation of mitochondria was sensitive to actinomycin D which means that the particles must have been permeable. There are many reasons why rifampicin could or could not inhibit transcription in whole mitochondria and I therefore will only discuss results obtained with soluble and DNA-dependent preparations of mitochondrial RNA polymerase. The number of papers in the literature reporting isolation of a rifampicin-sensitive RNA polymerase (*e.g.* [33, 36, 37, 41]) approximately equals that on insensitive enzymes (*e.g.* [32, 34, 35, 38–40]). Sensitive and insensitive polymerases need not necessarily be isolated from different cells as in some cases mitochondrial RNA polymerase has been isolated from the same organism in a sensitive or an insensitive form. In the case of the liver enzyme, for instance, one group consistently finds a sensitive polymerase [33, 41], in another report the enzyme is said to be inconsistently inhibited by the drug depending on the purity of the preparation [37], while in a third case a completely insensitive polymerase was isolated [39]. Except for the mitochondrial RNA polymerase isolated from *N. crassa* [36], rather high concentrations of rifampicin, far exceeding those causing complete inhibition of bacterial RNA polymerases, are necessary to inhibit the enzyme and complete inhibition has never been observed. The discrepancies could be interpreted to mean that there are two different RNA polymerases in mitochondria, one being inhibited by and the other resistant to rifampicin, one or the other enzyme then being isolated by the various workers. Except for the rather conclusive evidence for the presence in mitochondria of a poly(A)-polymerase [29] (with properties which clearly distinguish this enzyme from a DNA-dependent RNA polymerase), nothing is known which would confirm such an assumption. On the contrary, this possibility does not appear very attractive in view of the small size of the mitochondrial genome. Still, the presence in mitochondria of two different DNA-dependent RNA polymerases cannot be dismissed at present. Considering the many similarities between mitochondrial RNA polymerases of different sources, regardless of whether they are sensitive or insensitive to rifampicin, the latter property may only depend on some particular state or conformation the enzyme has assumed during the solubilization process; this could expose or cover binding sites for rifampicin. It has been found by Wu and Dawid [38] that the mitochondrial RNA polymerase of *Xenopus laevis* ovaries, though resistent to even high concentrations of rifampicin, is inhibited by other rifamycin derivatives such as AF/05 or AF/013, and the same was observed by ourselves with the yeast enzyme [44]. There might therefore exist binding sites for these drugs on the enzyme molecule. Oxime derivatives of rifamycin are of course also well known to inhibit nuclear RNA polymerases (*e.g.* [45]) and recently their specificity has in fact been questioned [46]. Another possibility for which there is again no experimental evidence as yet would be that preparations of insensitive enzymes contain, in addition to the polypeptide responsible for the catalytic activity, some bound protein subunit which prevents rifampicin to reach its binding site on the enzyme. That this is in fact feasible has been shown recently by Di Mauro, Hollenberg, and Hall [47] who found that a small protein isolated from yeast chromatin can protect eukaryotic RNA polymerases against inhibition by rifamycin AF/013.

Since a rifampicin-insensitive RNA polymerase was obtained from mitochondria of ascites tumor cells, the possibility was considered [40] that rifampicin insensitivity of mitochondrial RNA polymerase may be typical for transformed cells or for cells with a low respiratory capacity. A close inspection of the literature reveals, however, that this explanation is rather unlikely.

No doubt, much effort is still needed to bring light into this controversial issue.

The Site of Synthesis and the Genetic Origin of Mitochondrial RNA Polymerase

In considering the genetic autonomy of mitochondria, it is of particular interest to determine the site of synthesis (mitochondrial or cytoplasmic) and the genetic origin (mitochondrial or nuclear) of those mitochondrial components which are intimately involved in the maintenance and expression of the mitochondrial genetic information. The results of such studies could also provide the foundation for investigations into control mechanisms of mitochondrial biogenesis and into interrelations between nuclei and mitochondria. There are in principle two ways to solve this problem; one is the application of specific inhibitors of mitochondrial or cytoplasmic macromolecular synthesis, another is the use of a class of cytoplasmic mutants of yeast, the petite mutants. Petite mutants do not carry out mitochondrial protein synthesis [48, 49], therefore any protein present in mitochondria of such mutants must be synthesized on cytoplasmic ribosomes. Petite mutants have an altered mitochondrial DNA which causes their inactivity in protein synthesis (see [5]). In some cases, however, mitochondrial DNA is no longer detectable and such mutants can be used to prove not only cytoplasmic synthesis but also nuclear coding of a particular mitochondrial protein because in this case even the unlikely possibility of an export of some mitochondrial messenger RNA that still may be synthesized on petite mitochondrial DNA can be excluded. Using petite mutants of yeast, I have shown some time ago that RNA polymerase can still be found in mitochondria, and is even more readily solubilized from such mitochondria due to their less rigid membrane system [34]. Similar results were reported by Tsai et al. [35]. This has been taken as proof that mitochondrial

RNA polymerase is synthesized on cytoplasmic ribosomes [34, 50]. Using the first mentioned approach, the specific inhibition by chloramphenicol or ethidium bromide of macromolecular synthesis in *N. crassa* mitochondria, BARATH and KÜNTZEL [51] came to the same conclusion. They found that these inhibitors did not diminish the synthesis of mitochondrial RNA polymerase which therefore must be cytoplasmically synthesized and coded for by nuclear DNA. An unexpected result of these studies was that mitochondrial RNA polymerase was not only undiminished in presence of the inhibitors, but in contrast accumulated in the cytoplasm of the so-treated cells. This was interpreted to mean that inhibition of mitochondrial macromolecular synthesis resulted in a breakdown of some regulatory mechanism assuring mitochondrial control over nuclear activity. As such an induction of mitochondrial RNA polymerase may be very useful for increasing the amounts of the enzyme in the cell for purpose of isolation and purification, we have examined this process in yeast [52]. Cells were grown in the presence of ethidium bromide (10 μg/ml) for several generations and then tested for the presence in cytoplasmic extracts of a Mg^{++}-activated, salt-sensitive RNA polymerase. Unfortunately we did not find any significant increase of such an activity in ethidium bromide-treated cells. Also a chromatographic separation of cellular RNA polymerases on DEAE-sephadex columns did not reveal any effect of ethidium bromide on the number, absolute quantity, and relative amount of the RNA polymerases known to be separable under these conditions [53]. Only the nuclear enzymes were obtained in the usual manner (Table 5). Petite mutants whose mitochondria cannot carry out protein synthesis would provide another case of simulating the condition which leads to induction of mitochondrial RNA polymerase in *Neurospora*. I have therefore looked for an enzyme that might be present in the cytoplasm of such petite mutants but largely absent or at least present in much lower amounts in normal, respiring cells. However, no significant difference was found between wild type and mutant cells (Table 5). While these experiments have not fulfilled our hope to find conditions under which mitochondrial RNA polymerase from yeast can be obtained in larger quantites, they clearly show that the control mechanism proposed by BARATH and KÜNTZEL [51] from their studies with *N. crassa* is unlikely to work in yeast. This is not surprising as yeast, quite in contrast to most other eukaryotic cells, is a facultative anaerobe which can dispense with active mitochondria; hence nuclear-mitochondrial regulation may in fact be different in this organism.

Table 5. Ratio of yeast RNA polymerases A, B and C extracted from cells grown under different conditions

	A (in %)	B (in %)	C (in %)
Wild type cells grown to late log phase in medium containing 0.8% glucose (derepressed)	16	69	15
Wild type cells grown to late log phase in medium containing 5% glucose (repressed)	18	70	12
Wild type cells grown in the presence of 10 μg/ml of ethidium bromide	14	71	13
Petite mutant grown to late log phase	16	70	14

Conclusions

The process of mitochondrial RNA synthesis, though far from being understood at the cellular and molecular level, has been much enlightened in the last few years and it is to be expected that more insight will be gained in the near future. As mitochondrial DNA is of extremely small size compared with nuclear DNA of the same cell, transcription of mitochondrial DNA and the regulation of this process should provide a good model system for studies on control mechanisms in eukaryotes. Since the number of mitochondria is kept rather constant in a particular type of cell, there is tight coordination of cell division and rate of formation of mitochondria which most likely is governed by a strict regulation of mitochondrial DNA replication and transcription. The unraveling of these regulatory mechanisms will keep workers in this field busy for many years to come.

Discussion, see p. 207.

References

1. NASS, M. M. K.: Science **165**, 25 (1969).
2. BORST, P., KROON, A. M.: Int. Rev. Cytol. **26**, 107 (1969).
3. ASHWELL, M., WORK, T. S.: Ann. Rev. Biochem. **38**, 251 (1970).
4. SCHATZ, G.: In: RACKER, E. (Ed.): Membranes of mitochondria and chloroplasts, p. 251. New York: Von Nostrand-Reinhold 1970.

5. Borst, P.: Ann. Rev. Biochem. **41**, 333 (1972).
6. Wintersberger, E.: Hoppe Seylers Z. physiol. Chem. **336**, 285 (1964).
7. Luck, D.J.L., Reich, E.: Proc. nat. Acad. Sci. (Wash.) **52**, 931 (1964).
8. Wintersberger, E.: In: Tager, J.M., Papa, S., Quagliariello, E., Slater, E.C. (Eds.): Regulation of metabolic processes in mitochondria, p. 435. Amsterdam: Elsevier Publ. Comp. 1966.
9. Attardi, B., Attardi, G.: J. molec. Biol. **55**, 231 (1971).
10. Aloni, Y., Attardi, G.: J. molec. Biol. **55**, 251 (1971).
11. Aloni, Y., Attardi, G.: J. molec. Biol. **55**, 271 (1971).
12. Aloni, Y., Attardi, G.: Proc. nat. Acad. Sci. (Wash.) **68**, 1757 (1971).
13. Zylber, E., Vesco, C., Penman, S.: J. molec. Biol. **44**, 195 (1969).
14. Knight, Jr., E.: Biochemistry **8**, 5089 (1969).
15. Wu, M., Davidson, N., Attardi, G., Aloni, Y.: J. molec. Biol. **71**, 81 (1972).
16. Dubin, D.T.: J. biol. Chem. **247**, 2662 (1972).
17. Chamberlain, J.P., Metz, C.B.: J. molec. Biol. **64**, 593 (1972).
18. Dawid, I.B.: J. molec. Biol. **63**, 201 (1972).
19. Kuriyama, Y., Luck, D.J.L.: J. molec. Biol. **73**, 425 (1973).
20. Attardi, G., Amaldi, F.: Ann. Rev. Biochem. **39**, 183 (1970).
21. Suyama, Y., Eyer, J.: J. biol. Chem. **243**, 320 (1968).
22. South, D.J., Mahler, H.R.: Nature (Lond.) **218**, 1226 (1968).
23. Saccone, C., Gadaleta, M.N., Gallerani, R.: Europ. J. Biochem. **10**, 61 (1969).
24. Fukimachi, S., Bartoov, B., Mitro, R.S., Freeman, K.B.: Biochem. biophys. Res. Commun. **40**, 852 (1970).
25. Fukimachi, S., Bartoov, B., Freeman, K.B.: Biochem. J. **128**, 299 (1972).
26. Gamble, J.G., McCluer, R.M.: J. molec. Biol. **53**, 557 (1970).
27. Grant, W.D., Poulter, R.T.M.: J. molec. Biol. **73**, 439 (1973).
28. Saccone, C., Gadaleta, M.N., Gimigliano, A.F.: In: Slater, E.C., Tager, J.M., Papa, S., Quagliariello, E. (Eds.): Biochemical aspects of the biogenesis of mitochondria, p. 265. Bari: Adriatico Meditrice 1968.
29. Jacob, S.T., Schindler, D.G.: Biochem. biophys. Res. Commun. **48**, 126 (1972).
30. Perlman, S., Abelson, H.T., Penman, S.: Proc. nat. Acad. Sci. (Wash.) **70**, 350 (1973).
31. Aaij, C., Saccone, C., Borst, P., Gadaleta, M.N.: Biochem. biophys. Acta (Amst.) **199**, 373 (1970).
32. Wintersberger, E., Wintersberger, U.: FEBS-Letters **6**, 58 (1970).
33. Gadaleta, M.N., Greco, M., Saccone, C.: FEBS-Letters **10**, 54 (1970).
34. Wintersberger, E.: Biochem. biophys. Res. Commun. **40**, 1179 (1970).
35. Tsai, M.-J., Michaelis, G., Criddle, R.S.: Proc. nat. Acad. Sci. (Wash.) **68**, 473 (1971).
36. Küntzel, H., Schäfer, K.P.: Nature (Lond.) New Biol. **231**, 265 (1971).
37. Reid, J.D., Parsons, P.: Proc. nat. Acad. Sci. (Wash.) **68**, 2830 (1971).
38. Wu, G.-J., Dawid, I.B.: Biochemistry **11**, 3589 (1972).
39. Wintersberger, E.: Biochem. biophys. Res. Commun. **48**, 1287 (1972).

40. Jackisch, R., Jung, A., Schlegel, W., Mayer, D.: Hoppe Seylers Z. physiol. Chem. **353**, 1705 (1972).
41. Gallerani, R., Saccone, C., Cantatore, P., Gadaleta, M. N.: FEBS-Letters **22**, 37 (1972).
42. Seifart, K. M., Benecke, B. J., Juhasz, P. P.: Arch. biochem. biophys. **151**, 519 (1972).
43. Rogall, G., Wintersberger, E.: unpublished.
44. Kraft, V., Wintersberger, E.: unpublished.
45. Melihac, M., Tysper, Z., Chambon, P.: European J. Biochem. **28**, 291 (1972).
46. Riva, S., Fietta, A., Sivestri, L. G.: Biochem. biophys. Res. Commun. **49**, 1263 (1972).
47. Di Mauro, E., Hollenberg, C. P., Hall, B. D.: Proc. nat. Acad. Sci. (Wash.) **69**, 2818 (1972).
48. Kuzela, S., Grecna, E.: Experienta (Basel) **25**, 776 (1969).
49. Schatz, G., Saltzgaber, J.: Biochem. biophys. Res. Commun. **37**, 996 (1969).
50. Wintersberger, U., Wintersberger, E.: In: Harris, R., Allin, P., Viza, D. (Eds.): Cell Differentiation, p. 281. Copenhagen: Munksgaard 1972.
51. Barath, Z., Küntzel, H.: Nature (Lond.) New Biol. **240**, 195 (1972).
52. Wintersberger, E., Rogall, G.: Unpublished.
53. Ponta, H., Ponta, U., Wintersberger, E.: FEBS-Letters **18**, 204 (1971).

Expression of the Mitochondrial Genome in Wild Type and in an Extranuclear Mutant of *Neurospora Crassa*

H. KÜNTZEL, Z. BARATH, I. ALI, J. KIND, H. H. ALTHAUS and H. C. BLOSSEY

Max-Planck-Institut für experimentelle Medizin, Göttingen, W. Germany

With 8 Figures

1. Mitochondrial Genes

In some aspects mitochondria behave like endosymbiontic bacterial cells because they proliferate by division, growth and distribution to the progeny [1, 2] and because they contain a circular genome which is replicated and expressed by a mitochondrion-specific genetic apparatus of bacterial nature [3]. However, this genome is at least ten times smaller than the genome of the smallest free-living cells like *Mycoplasm*. The mitochondrion is therefore far from being autonomous in the genetic sense but depends heavily on nuclear genes of its "host" cell for growth and reproduction. The proteins of the outer membrane and the matrix seem all to be determined by nuclear genes and to be synthesized on cytoplasmic ribosomes whereas the biosynthesis of the inner membrane is controlled by both nuclear and mitochondrial genes [4].

The components of the mitochondrial genetic apparatus like DNA, ribosomal and transfer RNA, ribosomes, initiation and elongation factors, DNA polymerase, RNA polymerase, and amino-acyl-tRNA synthetases are now fairly well characterized [3, 5]. However, little is known about mitochondrial genes. We have therefore set out to learn more about the informational role of mitochondrial DNA and its regulatory interaction with nuclear DNA.

The mitochondrial genome of the fungus *Neurospora crassa* consists of a covalent circular DNA [6, 7] with a molecular weight ranging between 40×10^6 and 60×10^6 [6–9].

The only mitochondrial gene products which have been identified with certainty are ribosomal RNA's and transfer RNA's [5]. In *Neurospora* [10] as well as in several other organisms [5] it has been shown that each mitochondrial DNA circle contains one gene for each of the two rRNA's.

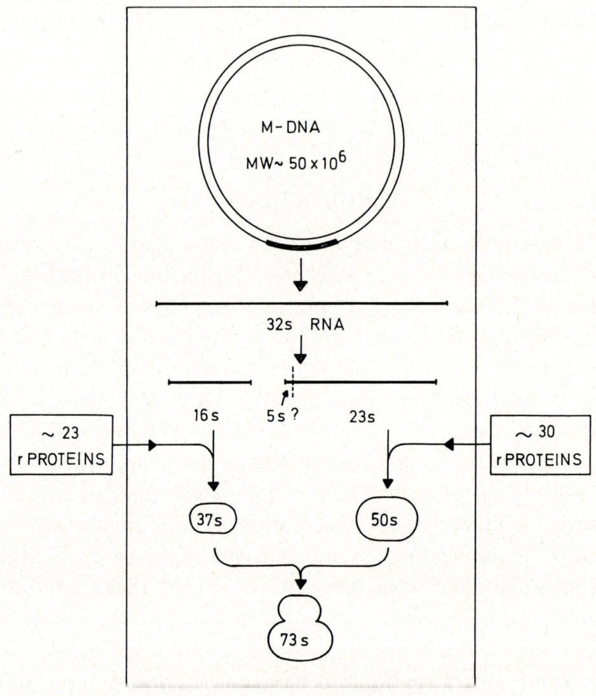

Fig. 1. Biosynthesis of mitochondrial ribosomes in *Neurospora*

Figure 1 shows schematically the biosynthesis of mitochondrial ribosomes in *Neurospora*. The tandem cistron for rRNA which occupies about 10 % of the coding strand is transcribed into a 32s RNA [11]; this precursor RNA is then processed into 23s and 16s rRNA which combine with the imported ribosomal proteins [12–14] to form the 50s and 37s subunits of the mitochondrial ribosome [15].

There are also a few mitochondrial genes for transfer RNA (the number varies in different organisms between 11 and 20 genes per genome) [5] which at least in HeLa cells are scattered all over the genome [16].

Unfortunately, no genetic data are available to prove the existence of mitochondrial structural genes coding for proteins. Possible candidates for such gene products are the proteins which are synthesized on cycloheximide-resistant mitochondrial ribosomes (Table 1).

Table 1. Proteins synthesized on mitochondrial ribosomes in *Neurospora:* possible candidates for mitochondrial gene products

MW	Function	Ref.
36000	subunit of cytochrome oxidase	[26]
28000	subunit of cytochrome oxidase	[26]
18000	subunit of cytochrome oxidase	[26]
32000	protein associated with cytochrome b	[27]
15000	lipoprotein component of the inner membrane	unpublished
15000	lipoprotein component of virus-like particles in the mutant abn-1	[24]
?	hypothetical regulatory protein	[18, 19]

Although one cannot exclude the possibility that mitochondrial ribosomes translate imported messenger RNA, we consider this to be rather unlikely because we have recently demonstrated that an *in vitro* transcription product of mitochondrial DNA stimulates in a sub-mitochondrial system or in an *E. coli* cell-free system the synthesis of a few large proteins [17], indicating that mitochondrial DNA from *Neurospora* codes for translatable RNA.

2. Regulatory Interaction between Mitochondrial and Nuclear Genes

To decide whether mitochondrial enzymes like RNA polymerase, ribosomal translocase, and methionyl-tRNA transformylase are produced by the mitochondrial or by the nuclear/cytoplasmic genetic system, we have studied the biosynthesis of these enzymes *in vivo* in the presence of selective inhibitors of mitochondrial transcription (ethidium bromide) and translation (chloramphenicol) [18, 19].

Both drugs led to an overproduction of the mitochondrial enzymes studied under conditions where the expression of mitochondrial DNA was blocked. An example to illustrate this effect is the induction of mitochondrial RNA polymerase [20] by ethidium bromide as shown in Fig. 2. In this case the appearance of the

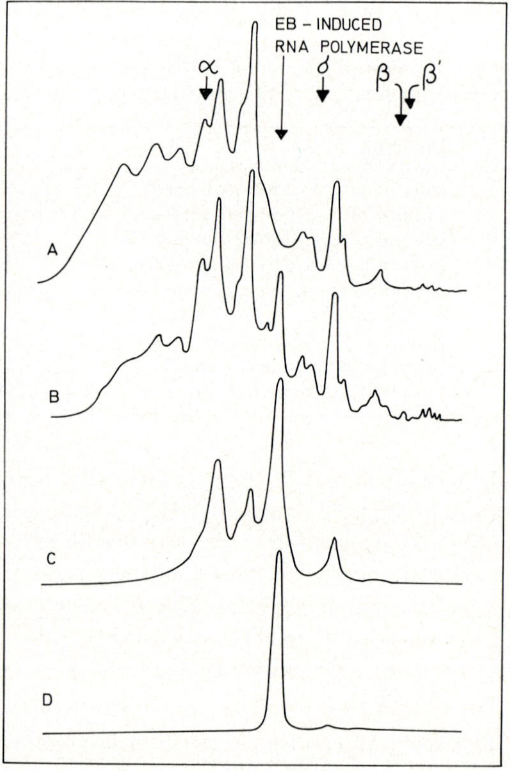

Fig. 2. Induction of mitochondrial RNA polymerase by ethidium bromide [19]. The proteins of postmitochondrial supernatant fractions from cells grown in the absence (*A*) and presence (*B*) of ethidium bromide were separated by SDS polyacrylamide gel electrophoresis, stained with Coomassie blue and recorded with a densitometer. *B* Partially purified preparation of EB-induced RNA polymerase. *C* Pure mitochondrial RNA polymerase from untreated cells. The markers are subunits of *E. coli* RNA polymerase

typical 64000 MW subunit in the cytoplasm of EB-treated cells can easily be visualized by SDS gel electrophoresis of crude S-30 fractions. Other experiments have shown that the appearance of the polymerase protein is paralleled by a dramatic increase of rifampicin-sensitive RNA polymerase activity in the mitochondria-free supernatant [19].

The unexpected finding that the biosynthesis of mitochondrial enzymes is stimulated by agents blocking transcription and translation of mitochondrial DNA suggests that these enzymes are coded by nuclear genes, synthesized on cytoplasmic ribosomes, and transported into the mitochondrion.

It also suggests that the nuclear genes coding for mi ochondrial enzymes are controlled by mitochondrial protein synthesis. The mechanism of this control, which could be negative or positive, is not yet known. There are two general possibilities: an indirect mechanism which would involve a metabolic control chain connecting respiration and ATP production in the mitochondrion with nuclear gene expression, or a direct control by a repressor-like protein which is coded by mitochondrial DNA, synthesized on mitochondrial ribosomes, and exported to the nucleus where it controls the nuclear partner genes for mitochondrial proteins (Fig. 3).

At the moment there is not enough experimental evidence available to decide between the two possibilities. However, we favor the second mechanism because there seems to be no correlation between the extent of respiratory deficiency (cytochrome composition) and cellular concentration of rifampicin-sensitive RNA polymerase in several extranuclear mutants.

3. Expression of Mitochondrial DNA in the Extranuclear Mutant "Abnormal-1" (abn-1)

It is generally believed that the respiratory-deficient phenotype of maternally inherited mutants of *Neurospora* is caused by a mutational alteration of mitochondrial DNA. One of these mutants, the slow growth variant "abnormal-1" (Table 2) is of special interest because it has been demonstrated that the "abnormal" character can be transmitted to normal cells by hyphal injection of "abnormal" mitochondria and that "abnormal" mitochondria are dominant over normal mitochondria in heteroplasmons [21]. This dominant or suppressive effect can be explained either by a replicative advantage of "abnormal" mitochondria or by an extramitochondrial exchange of mitochondrial gene products [22]. The

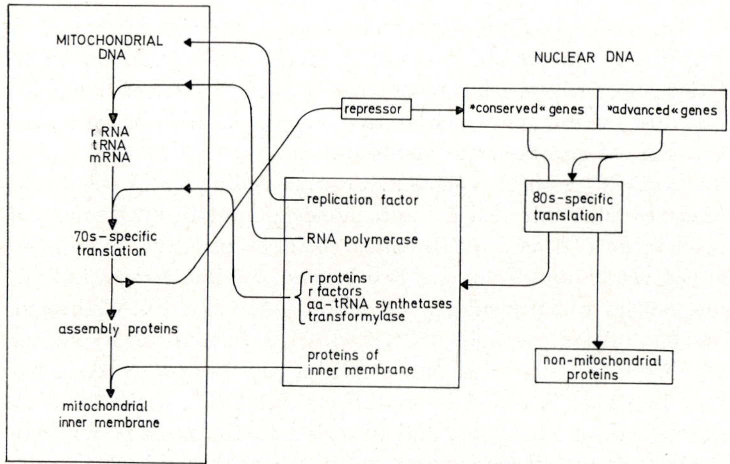

Fig. 3. Cooperation of mitochondrial and nuclear genes specifying the mitochondrial genetic apparatus [18]

recent discovery of intracellular virus-like particles (VLP) in abn-1 and "poky" mutants of *Neurospora* [23] raises the interesting possibility that these particles are responsible for the dominant charac-

Table 2. Properties of the extranuclear mutant "abnormal-1" of *Neurospora crassa* [21, 22]

Phenotype:	Slow, irregular growth ("stopper growth"). Respiratory deficiency: abnormal mitochondrial profiles, low content of cytochromes a and b, high content of cytochrome c. Female sterile.
Inheritance:	Not transmitted by the male parent in sexual crosses. Transmitted by heterocaryosis or by hyphal injection of mitochondria from abn-1. Dominance of "abnormal" character in heteroplasmons.

ter of the abn-1 genotype by infecting wild type mitochondria in heteroplasmons.

We have studied the structure and biosynthesis of the abn-1 particles which are not detectable in wild type cells [24]. Figure 4 shows a thin section of purified VLP after OsO_4 fixation. The particles appear as polymorphic vesicles of 250–400 nm diameter

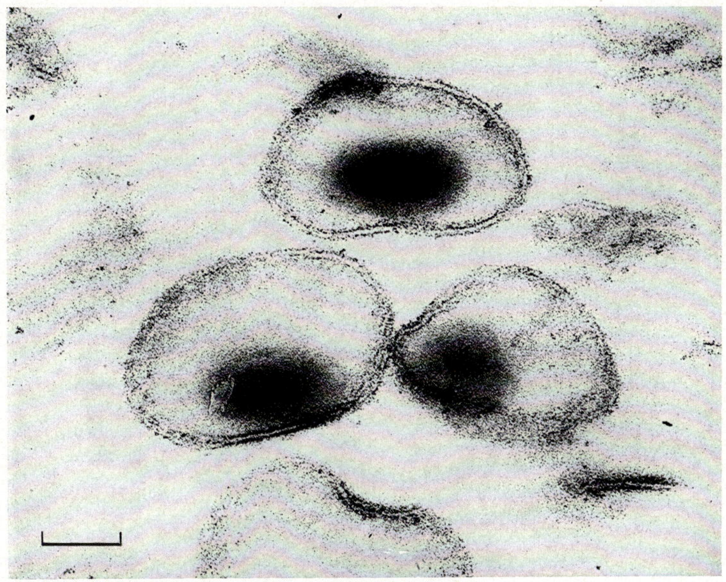

Fig. 4. Electron micrograph of a thin section of purified VLP from abn-1 after OsO_4 fixation and staining with uranyl acetate [24]. The bar represents 100 nm

containing an electron dense "nucleoid" of 120–170 nm diameter. This core region is surrounded by a trilamellar "unit" membrane. Chemical analysis has shown that VLP are composed of protein, RNA and phospholipids.

The RNA sediments at 33s and after heat denaturation at 7–9s (Fig. 5). From its base composition and its sensitivity to RNAse at high ionic strength, the RNA appears to be single-stranded; this is in marked contrast to other fungal virus-like particles which have double-stranded RNA [25].

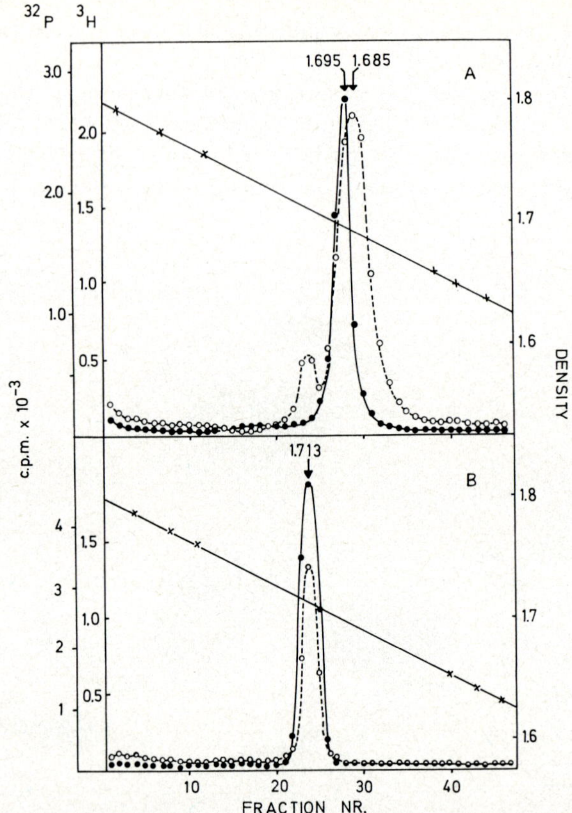

Fig. 5. Analysis of RNA from VLP by isokinetic sucrose gradient centrifugation [24]. (^{32}P)-labeled VLP RNA (●—●) was mixed with (^{3}H)-labeled rRNA and tRNA from *E. coli* (○- - -○) and analyzed either directly (A) or after heating at 80° for 2 minutes in the presence of 0.2% SDS (B)

From Figure 6 it is obvious that the abn-1 particles have a remarkably simple protein composition: they contain as major component a lipoprotein of MW 15000 and as minor component a glycoprotein of MW 95000.

Several observations suggest that VLP originate within mitochondria:

1. More than 50% of VLP are associated with mitochondria and can be released only by lysis of mitochondria with Triton X.

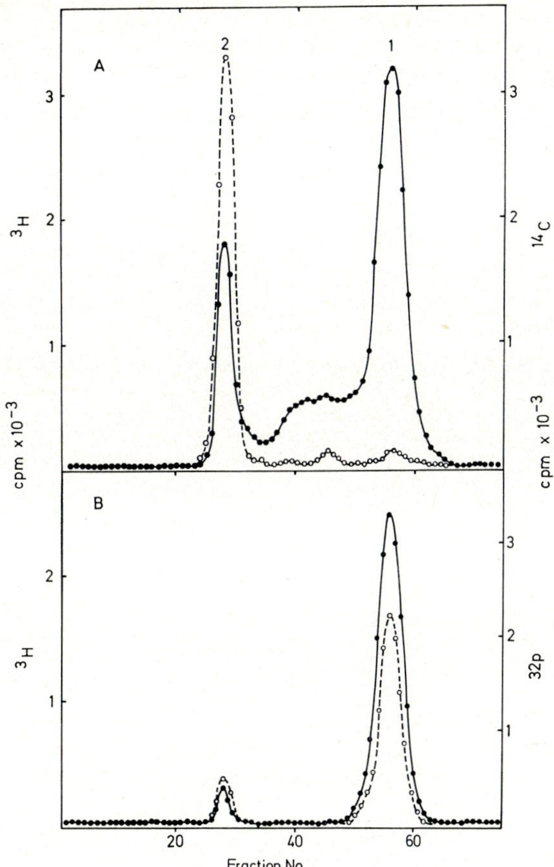

Fig. 6. Analysis of VLP proteins by filtration through Sephadex G 200 in the presence of SDS [24]. A VLP proteins labeled with (^{14}C)-leucine (● — ●) and (^{3}H)-glucosamine (○- - -○). B Methanol-soluble VLP proteins labeled with (^{3}H)-leucine (● — ●) and (^{32}P)-sodium phosphate (○- - -○)

The mitochondria-associated VLP cannot be distinguished in their morphology and chemical structure from the particles isolated from the mitochondria-free supernatant.

2. VLP contain cardiolipin, the marker lipid of the mitochondrial inner membrane.

3. The major membrane protein component of VLP, the lipoprotein of MW 15000, is synthesized on chloramphenicol-sensitive mitochondrial ribosomes and is possibly identical with a lipoprotein of the same molecular weight which is a product of mitochondrial protein synthesis in wild type.

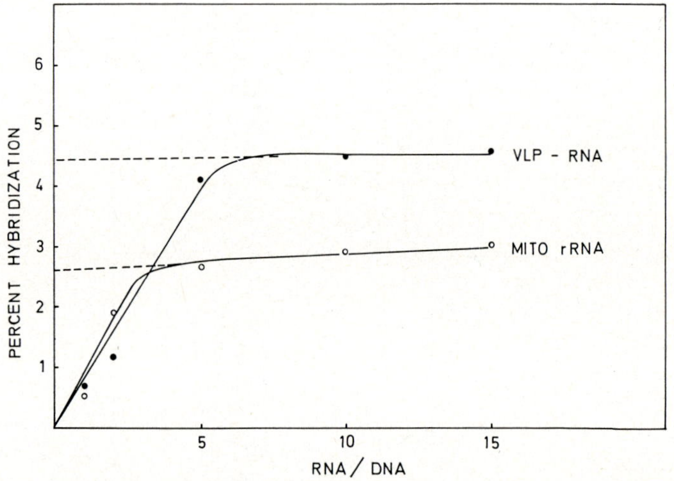

Fig. 7. Hybridization of ^{32}P-labeled RNA from VLP and from mitochondrial ribosomes with ^{3}H-labeled DNA from wild type mitochondria [10]

4. The 33s RNA of VLP hybridizes with mitochondrial DNA from both abn-1 and wild type cells (see Fig. 7).

5. VLP production is one of the stable phenotypic properties of the abn-1 mutant which are inherited exclusively by mitochondrial DNA.

Preliminary studies indicate that there is no sequence homology between mitochondrial rRNA and VLP RNA, and that the 33s RNA of VLP is not identical with the 32s precursor of mitochondrial rRNA. From hybridization studies (Fig. 7) we have to conclude that wild type mitochondrial DNA contains the gene for VLP RNA. However, this gene seems either to be repressed or its transcription product is rapidly degraded since we failed to detect 33s RNA of VLP containing this RNA in wild type cells.

At the moment it is entirely open as to what mutational event has triggered VLP production in the abn-1 mutant. We have compared the buoyant density of mitochondrial and nuclear DNA from wild type and abn-1 mutant and found that the mutation has caused a decrease in the density of mitochondrial DNA whereas the density of nuclear DNA has not changed (Fig. 8). This not

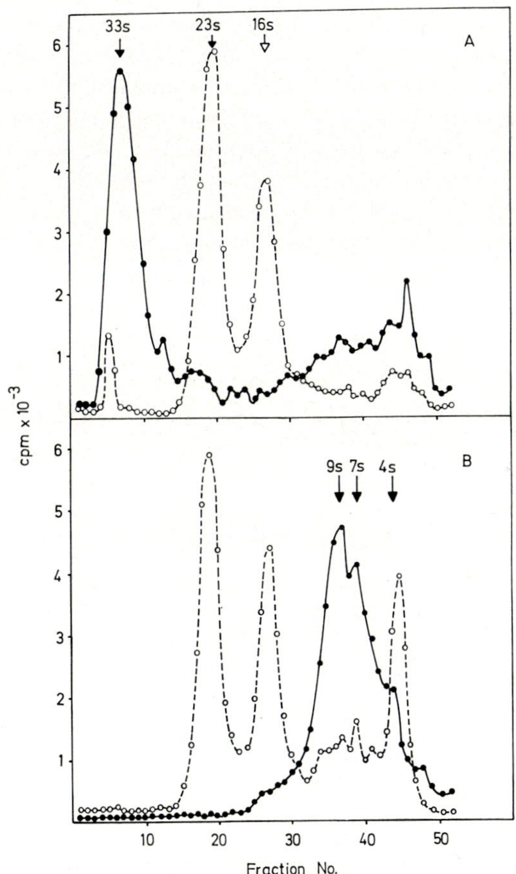

Fig. 8. Buoyant density of mitochondrial (A) and nuclear (B) DNA from *Neurospora* wild type (● — ●, (^{3}H)-labeled) and abn-1 (○- - -○, (^{32}P)-labeled) in a CsCl gradient

only confirms the mitochondrial inheritance of the abn-1 character but also suggests a major alteration of mitochondrial DNA, for instance a deletion.

Another open question is concerned with the possible viral function of VLP. Generally, infectivity of purified fungal virus-like particles cannot be demonstrated [23]; in the case of the abn-1 particles we would not expect infectivity from the early experiments by TATUM's group [22] showing that the abn-1 phenotype can only be transmitted by hyphal injection of abnormal mitochondria (which we know now to contain VLP) into wild type cells but not by coincubation. Although an intracellular infection of wild type mitochondria by VLP in heteroplasmons cannot be ruled out as an explanation of the dominant character of the abn-1 mutation, we better consider VLP as aberrant cellular structures derived from mitochondria. Nevertheless, we feel the term "virus-like particles" used by the discoverers to describe the abn-1 particles [23] is justified because structurally (not functionally) VLP resemble certain enveloped RNA viruses much more than cellular organelles like mitochondria.

Acknowledgement

We thank the Deutsche Forschungsgemeinschaft for support. Z. B. (present address: Slovak Academy of Sciences, Bratislava) and I. A. are recipients of DAAD fellowships.

References

1. LUCK, D. J. L.: J. Cell. Biol. **16**, 483 (1963).
2. HAWLEY, E. S., WAGNER, R. P.: J. Cell Biol. **35**, 489 (1967).
3. KÜNTZEL, H.: Curr. Top. Microbiol. Immunol. **54**, 94 (1971).
4. SCHATZ, G.: In: RACKER, A. (Ed.): Membranes of Mitochondria and Chloroplasts, p. 251. New York: Van Nostrand Reinhold Corp. 1969.
5. BORST, P.: Ann. Rev. Biochem. **41**, 333 (1972).
6. CLAYTON, D. A., BRAMBL, R. M.: Biochem. Biophys. Res. Commun. **46**, 1477 (1972).
7. AGSTERIBBE, E., KROON, A. M., VAN BRUGGEN, E. F. J.: Biochem. Biophys. Acta **269**, 299 (1972).
8. WOOD, D. D., LUCK, D. J. L.: J. molec. Biol. **41**, 211 (1969).
9. SCHÄFER, K. P., BUGGE, G., GRANDI, M., KÜNTZEL, H.: Europ. J. Biochem. **21**, 478 (1971).
10. SCHÄFER, K. P., KÜNTZEL, H.: Biochem. Biophys. Res. Commun. **46**, 1312 (1972).
11. KURIYAMA, Y., LUCK, D. J. L.: J. molec. Biol. **73**, 425 (1973).

12. KÜNTZEL, H.: Nature (Lond.) **222**, 142 (1969).
13. NEUPERT, W., SEBALD, W., SCHWAB, A. J., MASSINGER, P., BÜCHER, T.: Europ. J. Biochem. **10**, 589 (1969).
14. LIZARDI, P. M., LUCK, D. J. L.: J. Cell Biol. **54**, 56 (1972).
15. KÜNTZEL, H.: J. molec. Biol. **40**, 315 (1969).
16. WU, M., DAVIDSON, N., ATTARDI, G., ALONI, Y.: J. molec. Biol. **71**, 81 (1972).
17. BLOSSEY, H. C., KÜNTZEL, H.: FEBS Letters **24**, 335 (1972).
18. BARATH, Z., KÜNTZEL, H.: Proc. nat. Acad. Sci. (Wash.) **69**, 1371 (1972).
19. BARATH, Z., KÜNTZEL, H.: Nature (Lond.) New Biol. **240**, 195 (1972).
20. KÜNTZEL, H., SCHÄFER, K. P.: Nature (Lond.) New Biol. **231**, 265 (1971).
21. GARNJOBST, L., WILSON, J. F., TATUM, E. L.: J. Cell Biol. **26**, 413 (1965).
22. DIACUMAKOS, E. G., GARNJOBST, L., TATUM, E. L.: J. Cell Biol. **26**, 427 (1965).
23. TUVESON, R. W., PETERSON, J. F.: Virology **47**, 527 (1972).
24. KÜNTZEL, H., BARATH, Z., ALI, I., KIND, J., ALTHAUS, H. H.: Proc. nat. Acad. Sci. (Wash.), **70**, 1574 (1973).
25. ELLIS, L. F., KLEINSCHMIDT, W. J.: Nature (Lond.) **215**, 649 (1967).
26. WEISS, H., SEBALD, W., BÜCHER, T.: Europ. J. Biochem. **22**, 19 (1971).
27. WEISS, H.: Europ. J. Biochem. **30**, 469 (1972).

Discussion

HOFSCHNEIDER (München): Thank you very much for this interesting communication. We have some time left for discussion, and I would suggest that we first discuss transcription and then translation. To start the discussion, I have a question to Dr. WINTERSBERGER. If I understood you correctly, the DNA of mitochondria is totally transcribed, but the fate of part of the transcription product is not known. Is that correct?

E. WINTERSBERGER (Würzburg): This has been shown for HeLa cells, but for no other systems as yet. I mentioned it to demonstrate that mitochondrial DNA *can* be completely transcribed.

TH. STAEHELIN (Basel): I have a question to Dr. KÜNTZEL. When you centrifuged the DNA from the mutant and wild type with the different densities, did you notice that the band of the mutant DNA was considerably wider than that of wild type DNA? And further, did you calculate the molecular weight? You mentioned that it might be a deletion. Is it known what is missing?

H. KÜNTZEL (Göttingen): The broadening of the peak might be due to some heterogeneity. As for the deletion, we cannot say very much at the moment, but hope to learn more in the near future.

K. SCHERRER (Lausanne): Can you compete out the labeled 33 S RNA from the "virus-like" particle with total wild type mitochondrial RNA in a hybridization-competition assay?

H. KÜNTZEL: This experiment has not yet been done. All we can say is that ribosomal RNA from wild type mitochondria does not compete at all with 33 S RNA, indicating that ribosomal and virus-like RNA are transcribed from different regions of the mitochondrial genome.

L. PHILIPSON (Uppsala): 1. Please explain differences or similarities betwenn Dr. SEIFART's polymerase C and the mitochondrial polymerase.

2. Can the poly A-containing mitochondrial RNA help to elucidate transcription of mitochondrial DNA?

H. KÜNTZEL: 1. Dr. SEIFART is probably more competent to answer this question. From what I know of the literature it seems that mitochondrial RNA polymerase from rat liver is more sensitive to rifampicin and less sensitive to α-amanitin than RNA polymerase C from the same tissue.

2. The fact that mitochondrial messenger RNA contains poly A tracts may be very useful to separate this RNA from ribosomal and precursor RNA of mitochondria, which is otherwise difficult to achieve. It would also be interesting to study the attachment of poly A to the symmetrically transcribed mitochondrial RNA, in analogy to SV 40 transcription.

KERSTEN: Do you experts agree on the effect of rifampicin on the various polymerases?

H. KÜNTZEL: All I can say is that our purified preparation is sensitive to rifampicin, while our crude preparations are intensive to rifampicin. The enzyme isolated from *Xenopus laevis* in pure form is rifampicin-resistant. Thus, the enzymes from different species behave differently.

TH. STAEHELIN: It should be defined more exactly what is meant by "sensitive" and "insensitive" to rifampicin. Some people call an enzyme that is inhibited by 100 μg per ml sensitive, while an enzyme not inhibited by 10 μg is sometimes called "insensitive". However, the coli polymerase is inhibited by 0.001 μg/ml. I would like to ask Dr. KÜNTZEL at what concentration he got to the inhibition.

H. KÜNTZEL: We got inhibition with 4 μg/ml. I agree that this is much more than you need for the *E. coli* polymerase.

P. CHAMBON (Strasbourg): A question to both Dr. KÜNTZEL and Dr. WINTERSBERGER: Dr. RUTTER (San Francisco) has recently reported (Symposium on Gene Expression, Calcutta, India, February, 1973) that yeast mitochondria contain a RNA polymerase of 500000 MW with large subunits of 190000 and 150000 MW. Do you think that this finding could fit with the MW of your enzymes?

H. KÜNTZEL: The MW of 500000 fits well with our data. Highly purified mitochondrial RNA polymerase from *Neurospora* grown in the presence of ethidium bromide sediments in a high salt glycerol gradient as a 16 S octamer of the 64000 MW subunits. However, it is rather unlikely that mitochondrial RNA polymerases from yeast or other organisms have much larger subunits than those of the *Neurospora* enzyme (64000) or *Xenopus laevis* enzyme (46000). The large subunits reported by Rutter are probably either aggregates of smaller peptides or are derived from contaminating nuclear polymerases.

E. Wintersberger: This finding can still fit as far as the molecular weight of 500000 is concerned. As outlined in my talk, mitochondrial RNA polymerases very easily form aggregates with molecular weights in this range. The subunits 190000 and 150000, of course, do not fit with our enzyme. In fact these subunits are close to those of the yeast nuclear RNA polymerase A. It is possible that they are due to incomplete dissociation of the mitochondrial enzyme in SDS. We found that rather drastic conditions (1 h, 60°) are necessary to completely dissociate the mitochondrial RNA polymerase by SDS and mercaptoethanol for SDS-polyacrylamide gel electrophoresis. In the absence of more information on the experiments of Rutter's group, I am afraid I cannot really compare their results with ours.

E. Fuchs (Heidelberg): Did you compare the activity of yeast mitochondrial DNA with yeast, *Neurospora* or rat mitochondrial RNA polymerase? Are there differences in activity?

E. Wintersberger: I have not done this. I have compared yeast mitochondrial and nuclear RNA polymerases in their activity with native yeast mitochondrial DNA as template. In this case, the nuclear enzymes transcribe mitochondrial DNA much less effectively than the mitochondrial polymerase.

Ch. Scholtiszek: How far does the origin of your particles fit the oncogen hypothesis put foreward by Hübner? To me it looks as if your results-are the first evidence that such things can happen.

H. Küntzel: From the chemical composition and morphology of our virus-like particles and from the genetic behavior of the mutant producing these particles, we suspect that they play an active role by infecting or killing wild type mitochondria in heteroplasmons and that they may belong to the group of inheritable extranuclear elements like the kappa factor of *Paramecium*, the sigma factor of *Drosophila* and the "vegetative death" factor of *Aspergillus*. Although such an intracellular infectivity would differ profoundly from the transforming activity of tumour viruses, we cannot overlook some structural similarities between virus-like particles from "abn-1" and RNA tumor viruses. Our finding that cellular DNA contains genes for oncorna virus-like particles which are normally not expressed indeed suggests a strong parallel to the oncogene hypothesis by Hübner and Todaro. Of course fungal virus-like particles are not tumor viruses, but the study of their inheritance, induction and mode of action may help to understand the relation between chromosomal, extrachromosomal and viral information.

Grunicke (Freiburg): Does rifampicin effect RNA synthesis of intact mitochondria?

H. Küntzel: In some cases (*e.g.*, rat liver, *Physarum*) rifampicin is reported to inhibit RNA synthesis in intact mitochondria. In other cases (*e.g.*, yeast, *Neurospora*) no inhibition is seen.

E. Bautz (Heidelberg): How big are the particles that you find in relation to your mitochondria?

H. Küntzel: Our mitochondria are about 1 μ in diameter. The particles I mentioned are polymorphic; they tend to swell. Their core is about 0.12 μ, the outer diameter varies between 0.4 and 0.6 μ.

Translation I

Structure and Function of Bacterial Ribosomes

H. G. Wittmann

Max-Planck-Institut für Molekulare Genetik, Berlin-Dahlem, W. Germany

In recent years considerable progress has been made in the analysis of the structure of ribosomal components and their role in the various steps of protein biosynthesis. Numerous articles summarize various aspects of the structure, function and genetics of bacterial ribosomes, especially of *Escherichia coli*, and those which appeared since 1970 are listed here.

References

Beaudet, A. L., Caskey, C. T.: Polypeptide chain termination. Frontiers of Biology (North Holland) **27**, 133—172 (1972).

Crichton, R. R., Erdmann, V. A., Wittmann, H. G.: Bacterial ribosomal proteins. In: Neurath, H., Hill, R. L. (Eds.): The Proteins, 3rd Edn., Vol. IV. New York: Academic Press (in press) 1973.

Davies, J., Nomura, M.: The genetics of bacterial ribosomes. Ann. Rev. Genet. **6**, 203—234 (1972).

Fellner, P.: The primary structure of the 16 S and 23 S ribosomal RNAs. Biochimie **53**, 573—583 (1972).

Garrett, R. A., Wittmann, H. G.: Structure of bacterial ribosomes. Advan. Prot. Chem. **27**, in press (1973).

Garrett, R. A., Wittmann, H. G.: Structure and function of the ribosome. Endeavour **32**, 8—14 (1973).

Haenni, A. L.: Polypeptide chain elongation. Frontiers of Biology (North Holland) **27**, 33—54 (1972).

Kurland, C. G.: The proteins of the bacterial ribosome. Advan. Prot. Synth. **1**, 179—228 (1971).

Kurland, C. G.: Structure and function of the bacterial ribosome. Ann. Rev. Biochem. **41**, 377—408 (1972).

Kurland, C. G., Donner, D., van Duin, J., Green, M., Lutter, L., Randall-Hazelbauer, L., Schaup, H. W., Zeichhardt, H.: Structure and function of the ribosome. Febs Symp. **27**, 225—235 (1972).

Monier, R.: Structure and function of ribosomal RNA. Frontiers of Biology (North Holland) **27**, 353—394 (1972).

Nierhaus, K., Weber, J.: Struktur und Funktion der Ribosomen. Umschau **72**, 346—352 (1972).

Nomura, M.: Bacterial ribosome. Bact. Rev. **34**, 228—277 (1970).

Nomura, M.: Assembly of bacterial ribosomes. Science **179**, 864—873 (1973).

Osawa, S., Otaka, E., Takata, R., Dekio, S., Matsubara, M., Itoh, T., Muto, A., Tanaka, K., Teraoka, H., Tamaki, M.: Ribosomal protein genes in bacteria. Febs Symp. **23**, 313—336 (1972).

Revel, M.: Polypeptide chain initiation: the role of ribosomal protein factors and ribosomal subunits. Frontiers of Biology (North Holland) **27**, 87—131 (1972).

Rudland, P. S., Clark, B. F. C.: Polypeptide chain initiation and the role of a methionine tRNA. Frontiers of Biology (North Holland) **27**, 55—86 (1972).

Traub, P.: Structure, function and *in vitro* reconstitution of *Escherichia coli* ribosomes. Curr. Topics Microbiol. **52**, 1—93 (1970).

Wittmann, H. G.: A comparison of ribosomes from prokaryotes and eukaryotes. Symp. Soc. Gen. Microbiol. **20**, 55—76 (1970).

Wittmann, H. G.: Ribosomal proteins from prokaryotes. Febs Symp. **23**, 3—17 (1972).

Wittmann, H. G.: Ribosomal proteins of *Escherichia coli*: Their structure and functional role in protein biosynthesis. Febs Symp. **27**, 213—224 (1972).

Wittmann, H. G., Crichton, R. R.: Structure and function of ribosomal proteins of *Escherichia coli*. IV International Biophysics Congress (Moscow), in press (1972).

Wittmann, H. G., Crichton, R. R., Stöffler, G.: Structure and function of ribosomal proteins of *Escherichia coli*. Biochem. Soc. Symp. (Glasgow) in press (1973).

Wittmann, H. G., Stöffler, G.: Structure and function of bacterial ribosomal proteins. Frontiers of Biology (North Holland) **27**, 285—351 (1972).

Initiation of Protein Synthesis in Prokaryotes

M. Grunberg-Manago, Th. Godefroy-Colburn, A. D. Wolfe, P. Dessen*, D. Pantaloni*, M. Springer, M. Graffe, J. Dondon and A. Kay

Institut de Biologie Physico-chimique, Paris, France
**Laboratoire d'Enzymologie du C.N.R.S., Gif-sur-Yvette, France*

With 16 Figures

Initiation of protein synthesis appears to be a process of great complexity in both prokaryotic and eukaryotic systems. A possible reason for such complexity, particularly in comparison with the process of elongation of nascent proteins, may reside in gene regulation operating through the selection of mRNA or cistrons by ribosomes and protein factors. It is known that in prokaryotes the main control occurs at the transcriptional level. However, for polycistronic mRNA's there are a few facts consistent with a regulation at the translational level, and this is particularly shown in the case of RNA phage infection of *E. coli* (MS2, Qß, R17) where the three cistrons of the phage RNA are translated with different frequencies *in vivo* [1] although they are present in equal concentrations. Therefore, it is conceivable that a control exists at the translational level which might have a specific function in fine regulation of bacterial protein synthesis.

We would like to discuss our present knowledge of initiation with special emphasis on the role of initiation factors.

It has been known since the work of Marcker and Sanger [2] that in all organisms studied so far, there are at least two classes of methionine-accepting tRNA species, one of these methionyl-tRNA's being responsible for the initiation of protein synthesis.

In prokaryotes, the methionine of the initiator tRNA is formylated at the NH_2 position, after its attachment to the tRNA, by an enzyme called transformylase. This tRNA (designated as $tRNA^{fMet}$ or $tRNA_f^{Met}$) reacts differently than the other tRNA for methionine ($tRNA_m^{Met}$) in which the methionine cannot be formylated by the enzyme.

Soon after the discovery of fMet-tRNA, formylmethionine was shown to start two, and then all three, *in vitro* synthesized proteins from the cistrons of small RNA phages [3–5] (*i.e.*, maturation and coat protein, and synthetase). This indicates that fMet-tRNA is the chain initiator for each of the proteins designated by a polycistronic mRNA, such as phage mRNA. *In vivo* [6] a peptidase, the specificity of which is not well defined but which has a preference for methionine, removes this amino acid after it has been deformylated. A deformylase [6–8], isolated from *E. coli* extracts, deformylates formylmethionine after it has been incorporated into peptides; this enzyme is labile and inhibited by -SH present in *in vitro* mixtures for protein synthesis. In *E. coli* systems, an artifical N-acetylated tRNA (N-acetyl-phenylalanyl-tRNA) can be substituted for fMet-tRNA in the initiation of protein synthesis: at low Mg (5 mM) in the presence of initiation factors, it initiates the poly U-dependent synthesis of polyphenylalanine. No other blocked aminoacyl-tRNA can start protein synthesis.

In the eukaryote cytoplasmic machinery, two methionyl-tRNA's are also found, one of which can be formylated by the *E. coli* enzyme and is the initiator tRNA; formylation, however, does not occur during initiation of protein synthesis.

The nucleotide sequences of both *E. coli* $tRNA^{fMet}$ and $tRNA^{mMet}$ have been determined [9, 10], as well as those of yeast [11], rat liver [12], and myeloma initiator tRNA's [13]. Comparison of these sequences does not reveal any unusual characteristics which might account for the uniqueness of the role of initiator tRNA's. One of the important differences between initiator tRNA and Met-$tRNA^{mMet}$ is the former's coding properties. fMet-$tRNA^{fMet}$ binds to *E. coli* ribosomes in the presence of either the ApUpG or the GpUpG triplet, the binding to AUG being very efficient, even at low Mg (5–10 mM) [14, 15]; Met-$tRNA^{mMet}$, on the other hand, binds to *E. coli* ribosomes *only* in response to the AUG codon, and this binding requires high Mg (30 mM).

The existence of a large pool of 30 S and 50 S ribosomal subparticles, besides those ribosomes engaged in polysomes, suggested a dissociation of 70 S particles after messenger reading. There is good evidence that the two subunits have to separate at the beginning of each cycle of protein synthesis, and that only the 30 S subunit is able to bind mRNA and fMet-tRNA. The best evidence was obtained by a series of experiments [16] in which heavy 70 S ribosomes, labeled with N^{15} and deuterium, were mixed with light 50 S subunits, poly (A, U, G), fMet-tRNA, and protein factors involved in initiation; the experiments showed that fMet-tRNA was bound to hybrid 70 S ribosomes consisting of heavy 30 S and light 50 S. In addition, fMet-tRNA binding occurs quite well on 30 S ribosomes alone, whereas binding of all the other aminoacyl-tRNA's only occurs when the 50 S subunits are also present [17].

The ribosomes are in dynamic equilibrium with their subunits as shown by Spirin [18] and by Infante and Baierlein [19]; the equilibrium is extremely sensitive to the concentration of divalent cations. To determine the distribution of ribosomal species, investigators have mainly relied on ultracentrifugation. However, it has been shown that hydrostatic pressure [18, 19], produced by high centrifugal fields, affects the ribosome – subunits equilibrium. Attempts to overcome this difficulty have used a fixation procedure (with glutaraldehyde or formic acid) which "freezes" the particle distribution. This procedure, however, is only valid if one can demonstrate that the rate of fixation is much faster than the rate of association and dissociation. Recently, studies of the association equilibrium of the 50 S and 30 S subunits have been carried out with the light scattering technique [20]. This technique has the advantage of permitting observation of complexes which are not very stable and of eliminating influence of pressure; it has the disadvantage of not being very sensitive and does not permit characterization of changes in a small part of the ribosome population. Furthermore, some artifact could occur from aggregation, and the ribosomal distribution should therefore be checked by another technique.

We will compare, further on, the results obtained with this technique and those obtained with sucrose gradient centrifugation in relation to the role of initiation factors. First, we would like to give some of our data on the distribution of 70 S ribosomes and subunits as a function of Mg concentration. The percentage of 70 S ribosomes obtained by light scattering (Fig. 1) agrees with the results obtained by gradient centrifugations (within 10–15%), and stop-flow turbidimetric measurements show that association

(Fig. 2) and dissociation are very fast processes which markedly depend on the magnesium concentration, especially under 5 mM. The amount of 30 S and 50 S depends on the washing of the ribosomes (probably because of the extent of contamination by factors); it also depends on the concentration of monovalent cations, on the temperature, and even on the batch of ribosomes used.

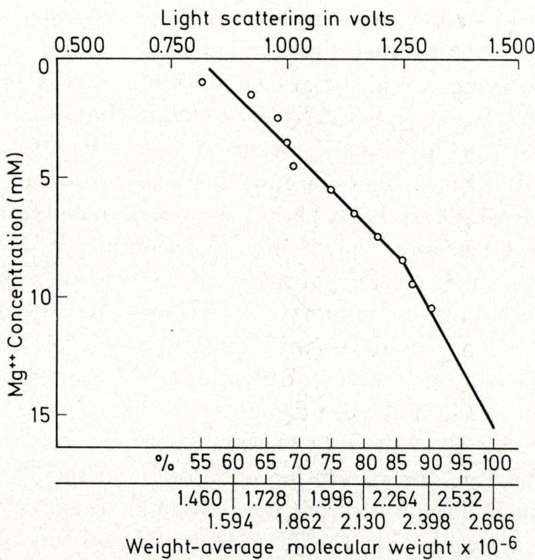

Fig. 1. Weight-average molecular weight of *E. coli* ribosomes as a function of magnesium concentration. The observed light scattering at 90° is depicted on the left, while the corresponding percentage change in total light scattering is depicted on the right. The ribosome concentration was 10.6 A_{260} units or a total of 254 pmoles/ml of 70 S ribosomes. The incubation mixture (total volume 2 ml) contained also: ammonium chloride, 50 mM; Tris (pH 7.5), 10 mM; β-mercaptoethanol, 6.6 mM; and varying concentrations of Mg-acetate. The light scattering was measured with a FICA instrument equipped with a Wang computer, series 700. The molecular weight-average was calculated from the light scattering values taken at different angles

All the experiments were performed with ribosomes that had been pre-activated by heating 10 or 20 min at 37 °C in 10 or 15 mM Mg^{2+}; when the ribosomes were assayed under optimal conditions for binding, it was found that 0.8 pmole of fMet-tRNA were bound per 1 pmole of ribosome.

The initiation process requires GTP and several protein factors which are loosely bound to the 30 S ribosome subunits but which can be extracted by washing the ribosomes with NH_4Cl (0.5–1.5 M) and are, therefore, well separated from elongation factors.

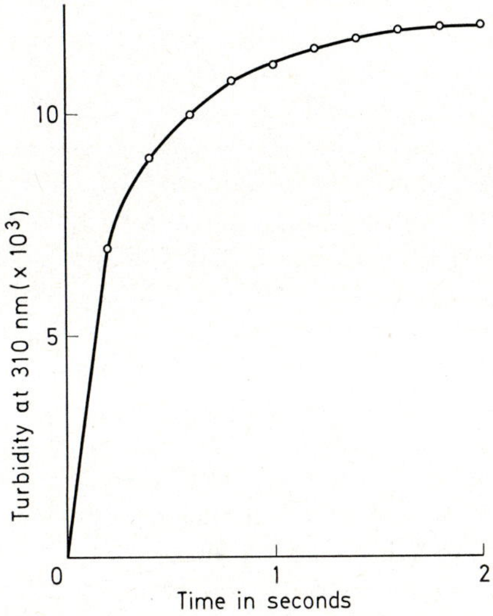

Fig. 2. Stop-flow turbimetric measurements of the kinetics of association of ribosomal subunits. The ribosomes, initially in a 15 mM Mg-acetate buffer, were dissociated by dilution in 1 mM Mg-acetate, 10 mM Tris (pH 7.5), 50 mM NH_4Cl, 6.6 mM β-mercaptoethanol, and kept for no longer than 15 minutes at room temperature. They were mixed in the stopflow apparatus with an equal volume of 16 mM Mg buffer (otherwise identical to the previous one), and the turbidity at 310 nm was recorded *vs* time, using a Durrum spectrofluorometer. The final concentration of ribosomes was 10.6 A_{260}/ml. The kinetics of association is biphasic (Wolfe, A. D., Dessen, P., Pantaloni, D., Febs Letters, in press (1973)); this curve shows the first phase of the process; the second one is completed within less than one minute

At first the nomenclature for the proteins required to translate natural and synthetic mRNA's was somewhat confused, but a uniform nomenclature was adopted in 1972. Three classes of factors have been isolated on DEAE-cellulose or -Sephadex columns by several laboratories: they are now called IF-1, IF-2, and IF-3. They are requisite for the formation of the initiation complex. In view of their cooperative effects, it is difficult to assign an unambiguous function to each individual factor, especially since they have a tendency to form complexes (particularly IF-1 and IF-2); this is probably related to their action under physiological conditions; IF-2·IF-3 complexes have also been observed [21], and it is therefore very difficult to obtain factors completely devoid of each other. 1 μg factor/100 μg ribosomes, or less, is enough to reach saturation, which emphasizes the difficulty of ascertaining whether they are contaminated or not by other proteins. The other problem is to obtain ribosomes completely devoid of factors. Usually, carefully salt-washed ribosomes do not contain appreciable amounts of IF-2, but have enough IF-1 and IF-3 to allow the positioning of fMet-tRNA on the ribosomes. This is why it was believed that IF-2 activity could be assayed without addition of the other factors and that it had an activity of its own for binding fMet-tRNA to ribosomes; with purified factors and ribosomes extensively washed on Diaflo XM 300 membranes, however, only a very slight activity could be detected in the absence of the two other factors. The fact that the ribosomes are very active in the presence of the three factors suggests that the extensive washing has not removed any other ribosomal protein necessary for activity.

Figure 3 shows a separation on DEAE-Sephadex of factors tested with ribosomes which had been washed once in 1.5 M NH_4Cl. The usual assay is that of fMet-tRNA binding activity (using Millipore filtration) or incorporation of fMet-tRNA into TCA precipitable material in the presence of synthetic polymers containing the initiation codon, or in the presence of phage RNA such as R17, Qβ, or MS2, or else "T4 mRNA" extracted from phage-infected cultures, either after 2–3 min (early mRNA) or after 13 min (late mRNA). The activity of the fractions from the DEAE-Sephadex column is first assayed without any addition of factors: the stimulating fraction in this assay corresponds to IF-2. By saturating the initiation assay mixture with IF-2, another factor, IF-1, is located which increases the stimulation of fMet-tRNA binding (particularly during the first three minutes). IF-3 activity is demonstrated by exploring the stimulating activity (using mostly late T4 mRNA or R 17 RNA) when IF-1 and IF-2 are both in saturating amounts.

As can be seen (Fig. 3), IF-3 can be separated by DEAE fractionation into different fractions, one of which stimulates more with T4 mRNA than with R17 RNA. Elution on DEAE-Sephadex is in the following order: IF-1, IF-3, IF-2. The pooled fractions can be purified to homogeneity by hydroxylapatite fractionation for IF-2, and carboxymethylcellulose and phosphocellulose for IF-1 and IF-3.

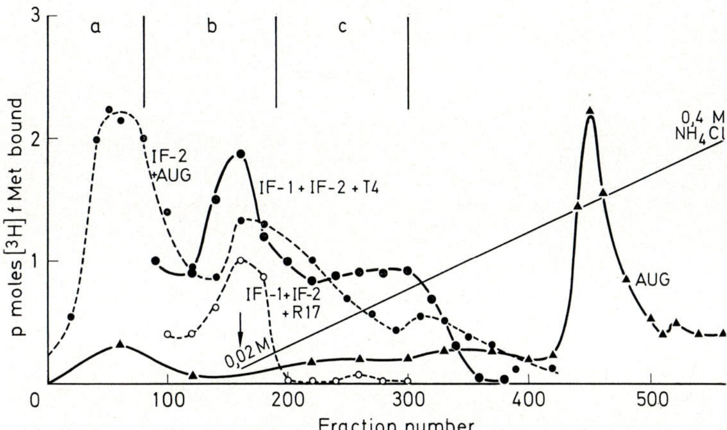

Fig. 3. Separation of initiation factors on DEAE-Sephadex. Crude initiation factors (1300 mg/270 ml) were dialyzed against: Tris (pH 7.5), 20 mM; PO_4 (pH 7.5), 5 mM; glycerol, 5%; β-mercaptoethanol, 7 mM, and layered onto a DEAE-Sephadex column (⌀ 6 cm, Ht 30 cm) previously treated twice with dimethyl dichlorosilane and equilibrated in the same buffer. The column was then washed with the same buffer to which had been added 0.02 mM NH_4Cl (1.5 liters). The elution was done with 6 liters of a 0.02—0.4 M NH_4Cl gradient in the same buffer. The fractions collected were 9 ml. Three IF-3 were separated and purified from the three fractions, *a*, *b*, and *c*. For the purification of the crude factors, see [22]

A fourth class of factors, called interference factors (*i*), which change the specificity of IF-3 for mRNA recognition has also been prepared by Revel [23] and by Ochoa [24].

Figure 4 shows the binding kinetics observed with the various factors in the presence of poly (A, U, G) at 5 mM Mg with ribosomes extensively washed by ultrafiltration on Diaflo membrane XM 300. It can be seen that with IF-2 alone very little binding occurs, addition of IF-1 stimulates the rate of binding, and IF-3 very markedly increases the total amount bound in the presence of a mixture of 70 S ribosomes, 30 S + 50 S subunits, and of GTP. Table 1 shows that all three factors are required in order to incorporate (^{14}C)valine in the presence of R17 phage RNA. In the presence of 30 S alone, we find that IF-1 stimulates the

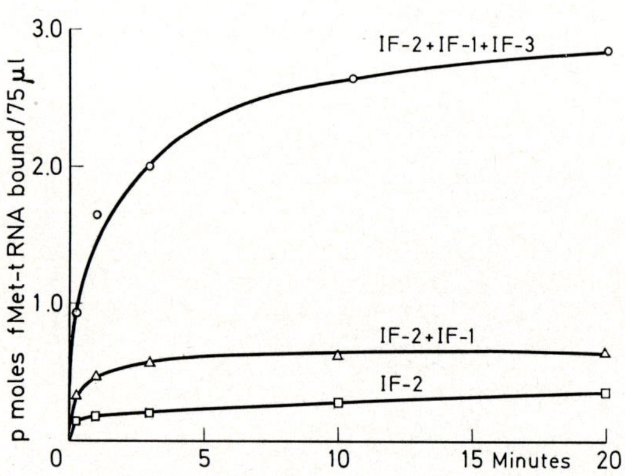

Fig. 4. Binding kinetics of fMet-tRNA in the presence of the initiation factors. The binding medium (400 µl) contained: 1.7 A_{260} units of ribosomes (preactivated for 20 min at 37 °C in Mg-acetate, 15 mM; NH_4Cl, 240 mM; Tris-HCl (pH 7.5), 150 mM); poly (A, U, G), 1.5 A_{260} units; (^{3}H)fMet-tRNA, 5.0 A_{260} units (90 pmoles, 1400 cpm/pmole); GTP, 1 mM; Mg-acetate, 5 mM; NH_4Cl, 80 mM; Tris-HCl (pH 7.5), 50 mM. Where indicated, highly purified initiation factors were added: IF-1, 0.46 µg; IF-2, 4 µg; IF-3, 2.3 µg. Binding was performed at 37 °C. 75 µl aliquots were removed at time intervals indicated on the figure and the binding measured by the Millipore filtration technique

binding of fMet-tRNA, but that IF-3 has very little stimulatory effect, and sometimes even an inhibitory effect which will be discussed later.

The initiation factors act as catalysts in the formation of the functional 70 S initiation complex, since none of them are found in this complex. They are thus recycled, as shown also by the kinetic experiments (the molar amount of complex obtained may by much greater than that of either IF-1, IF-2, or IF-3 added).

Table 2 summarizes the nature and activities of the respective factors: IF-2, the most acidic protein, does not only stimulate fMet-tRNA binding, but also hydrolyzes GTP in the presence of ribosomes, although washed ribosomes, or IF-2, singly hydrolyze very little GTP. IF-1, the most basic protein, increases the affinity of ribosomes for IF-2, and facilitates the recycling of IF-2;

Table 1. Incorporation of (^{14}C) valine with R 17 mRNA

	cpm [a]
IF	2110
IF-2	227
IF-2 + IF-3	641
IF-1 + IF-2 + IF-3	1358

[a] After subtraction of blank (500 cpm).

The incubation mixture (200 µl) contained: Tris (pH 7.5), 50 mM; NH_4Cl, 100 mM; β-mercaptoethanol, 7 mM; Mg-acetate, 7.5 mM; GTP, 0.9 mM; ATP, 0.44 mM; ribosomes, 1.8 A_{260} units; mixture of (^{12}C)aminoacyl-tRNA's and (^{14}C)valyl-tRNA (spec. act. 15132 cpm/A_{260}), 1.19 A_{260} units; 105000 g supernatant, 1.1 A_{260} units (0.54 A_{280}); crude initiation factors, 0.05 mg; IF-1, 0.2 µg; IF-2, 0.4 µg; IF-3, 0.05 µg.

Incubation, 30 min at 37 °C, was stopped by addition of 2 ml NaCl and 2 ml of 10% TCA. The mixture is then heated 10 min at 85 °C, filtered on Whatman GF/C glass filters, washed three times with 1% TCA and counted in an Intertechnique scintillator.

it also increases the dissociation activity of IF-3. IF-3 stimulates the binding of fMet-tRNA by increasing the affinity of ribosomes for IF-2; in its presence the 70 S ribosomes are dissociated into 30 S and 50 S subunits.

We would now like to present a more comprehensive and detailed description of the nature and activities of these factors:

IF-2: IF-2 is a slightly acidic protein which includes two subcomponents with a different chromatographic behavior on DEAE-cellulose, but an indentical behavior towards fMet-tRNA binding. The molecular weights of these subcomponents are 100,000–91,000 and 80,000; it is possible that one derives from the other by proteolysis, the difference in their molecular weight corresponding to a peptide which would have been removed. Results of

TRAUT and HERSHEY[1] indicate that the fingerprints of these two IF-2 are quite similar. IF-2 carries a sulfhydryl group necessary for its activity.

Using (^{32}P)-labeled IF-2 (obtained by phosphorylating the factor with rabbit muscle protein kinase [25]) FAKUNDING and

Table 2. Characteristics of initiation factors

Factors	Molecular weight	Characteristics
IF-2, IF-1, IF-3		Catalyze the binding of fMet-tRNA$_f$ to ribosomes at the messenger initiation site; found only on 30 S subunits in normal conditions.
IF-2	100000—91000 and 80000	Absolutely required for fMet-tRNA binding; GTPase activity; the only initiation factor found in a 70 S complex in the presence of a non-hydrolyzable analogue of GTP.
IF-3	21500 and 23500	Favors dissociation of 70 S ribosomes; increases the affinity of IF-2 and IF-1 for the 30 S subunits; required for initiation on natural messengers and synthetic ones at low concentration.
IF-1	9400	Increases the affinity of IF-2 and IF-3 for the 30 S subunits; increases the dissociating activity of IF-3; helps recycle IF-2.
(*i*)	Several species 100000—30000	Changes the relative efficiency of translation of different cistrons.

HERSHEY [26] investigated the binding and release of IF-2 by sucrose density centrifugation analysis. They showed that it binds to 30 S ribosomal subunits in the absence of the other components of the initiation complex (Fig. 5); this binding was only

[1] Personal communication.

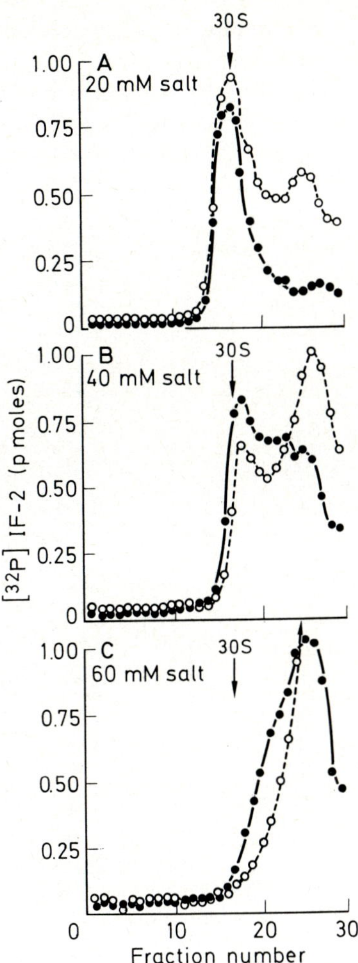

Fig. 5. Effect of GTP and salt concentration on IF-2 binding to 30 S ribosomes. Reaction mixtures (0.1 ml) contained: Tris-HCl (pH 7.4), 50 mM; NH_4Cl, 100 mM; Mg-acetate, 5 mM; β-mercaptoethanol, 7 mM; GTP (closed circles only), 0.2 mM; 30 S ribosomes, 60 pmoles; (^{32}P) IF-2 (1800 cpm/pmole), 22 pmoles. After 10 min incubation at 30 °C, the mixtures were chilled and layered onto 4.6 ml linear 7—25% sucrose gradients containing: Tris (pH 7.4), 50 mM; Mg-acetate, 5 mM; β-mercaptoethanol, 7 mM; GTP (closed circles only), 0.04; and NH_4Cl, 20, 40 or 60 mM (as indicated). The gradients were centrifuged at 40000 rpm for 180 min at 4 °C in a Beckman SW-56 rotor (From [26])

observed with the 30 S, and no interaction with 50 S or 70 S free of bound mRNA could be observed.

This binding is stabilized by the addition of GTP, both to the reaction mixture for binding and to the gradients. This effect does not occur with GDP nor with GMP-PCP (a non-hydrolyzable analogue of GTP where a methylene bridge is substituted for oxygen between the β and the γ phosphorus atoms). The addition of IF-1 and IF-3 greatly enhances the stability of IF-2 binding to the 30 S (Fig. 6).

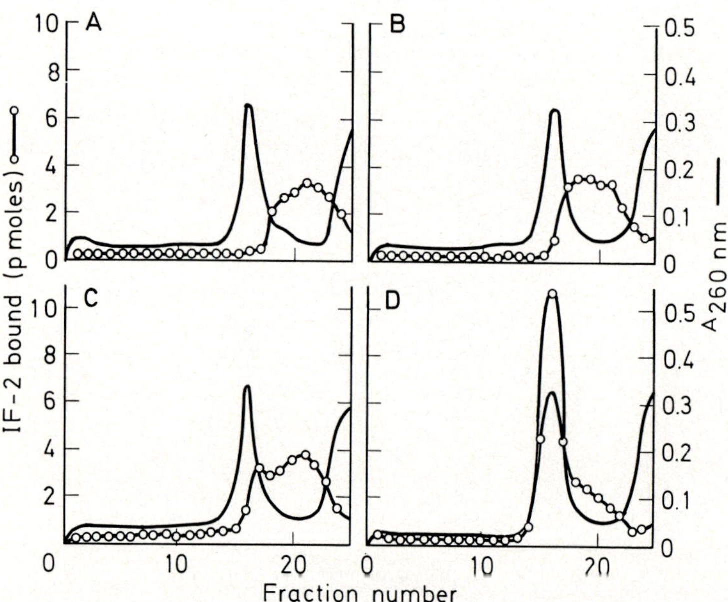

Fig. 6. Effect of IF-1 and IF-3 on IF-2 binding to 30 S ribosomes. Reaction mixtures (0.1 ml) contained: Tris-HCl (pH 7.4), 10 mM; NH_4Cl, 100 mM; Mg-acetate, 6 mM; dithiothreitol, 1 mM; GTP, 1 mM; 30 S ribosomes, 125 pmoles; (^{32}P)IF-2 (100 cpm/pmole), 66 pmoles; and where indicated, IF-1, 75 pmoles; and IF-3, 80 pmoles. The mixtures were incubated for 5 min at 30 °C, diluted with 0.3 ml buffer (Tris-HCl (pH 7.4), 10 mM; Mg-acetate, 10 mM; β-mercaptoethanol, 7 mM), and layered onto 13 ml linear 7—20% sucrose gradients containing the above buffer, plus 20 mM NH_4Cl. The gradients were centrifuged at 39,000 rpm for 180 min at 4 °C in a Beckman SW-40 rotor (From [26]). *A* IF-2; *B* IF-1 + IF-2; *C* IF-3 + IF-2; *D* IF-1 + IF-2 + IF-3

IF-2 has yet another activity: it hydrolyzes GTP, but only in the presence of 70 S or 30 S + 50 S ribosomes; this hydrolysis, however, occurs in the absence of the other components of the initiation complex [27, 28]. Such activity, uncoupled from the initiation of protein synthesis, resembles that of the polypeptide chain elongation factor, EF-G, but can be distinguished from this factor by its lack of susceptibility to fusidic acid, an antibiotic which strongly inhibits the EF-G-dependent GTPase. In the presence of all the initiation complex components, still more GTP is hydrolyzed; the stimulation of hydrolysis over that of the uncoupled activity is called "coupled GTPase". The results of Fakunding and Hershey [26] demonstrate that in the presence of the AUG triplet and fMet-tRNA, IF-2 remains bound to the 70 S ribosomes if GTP hydrolysis is prevented (by the presence of GMP-PCP, for example), but that this factor is released in the presence of GTP (Fig. 7). This suggests that GTP hydrolysis is necessary for the recycling of IF-2. It has been suggested that initiation complexes, analogous to the elongation complexes ($EF\text{-}T_u \cdot GTP \cdot$ aminoacyl-tRNA, and $GTP \cdot EF\text{-}T_u$) are formed, consisting of IF-2 · GTP [28] and IF-2 · fMet-tRNA · GTP [29 – 31]. Although a stable IF-2 · GTP or IF-2 · GDP complex is difficult to demonstrate, GTP – and to a lesser extent GDP – was reported to protect IF-2 against heat inactivation or inactivation by N-ethyl-maleimide [28, 32], and evidence has been presented for an IF-2 · fMet-tRNA · GTP complex [29–31]. However, the affinity of IF-2 for GTP, or for the GTP · fMet-tRNA association, is probably at least two orders of magnitude lower [28] than the affinity of $EF\text{-}T_u$ for GTP or for the GTP · amino-acyl-tRNA association ($10^4\ M^{-1}$). Furthermore, in contrast to $EF\text{-}T_u$, IF-2 has less affinity for GDP than for GTP [28]; in any event, the affinity is too low for the complex to be stable, unless some other protein is involved; it is therefore doubtful that this complex is an intermediate in the formation of the initiation complex. Moreover, with highly purified factor, no GTP · IF-2 nor GTP · fMet-tRNA · IF-2 complex could be observed [33], and the latter complex is therefore questionable.

Both ribosomal subunits, 50 S and 30 S, are necessary for GTPase activity, whether coupled or uncoupled, which implies that an interaction between the two subunits is necessary for GTP hydrolysis to occur. Experiments with thiostrepton [34] suggest that the site of IF-2-dependent GTP hydrolysis is close to the elongation factor site on the 50 S subunit since this antibiotic which inactivates the A site (by binding 1/1 mole to the 50 S subunit) and inhibits EF-G-dependent GTPase activity inhibits the IF-2-dependent GTPase to approximately the same extent (albeit

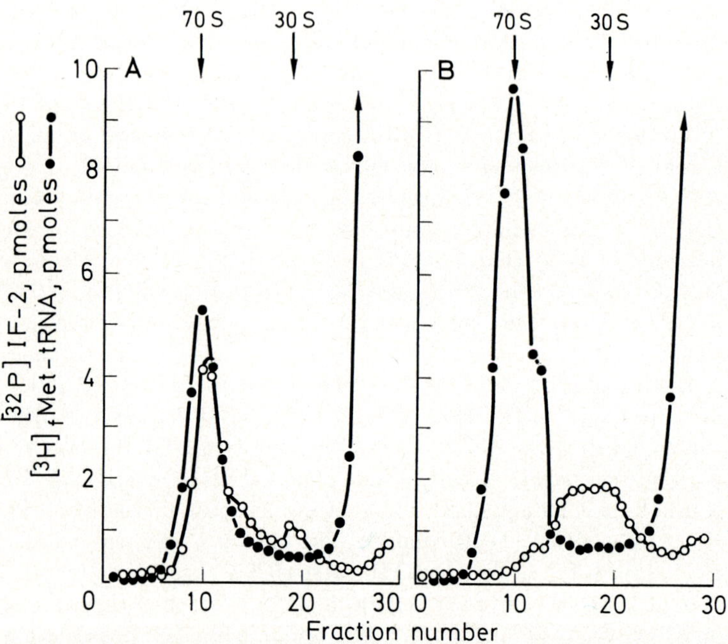

Fig. 7. Interaction of IF-2 with the 70 S initiation complex. Complete 30 S initiation complex was formed as follows: reaction mixtures (0.1 ml) contained: Tris-HCl (pH 7.4), 10 mM; NH_4Cl, 100 mM; Mg-acetate, 5 mM; dithiothreitol, 1 mM; 30 S ribosomes, 60 pmoles (0.8 A_{260} units); GMP-PCP (panel A) or GTP (panel B), 0.2 mM; ApUpG, 25 μM; (^{32}P)IF-2 (320 cpm/pmole), 4 μg; IF-1, 0.4 μg; and (^{3}H)fMet-tRNA (spec. act. 2.8 Ci/mmole), 40 pmoles. The mixtures were incubated for 10 min at 30 °C, 66 pmoles of 50 S ribosomes were then added, and the incubation was continued for an additional 5 min. The mixtures were then chilled and layered onto 4.6 ml linear 7—25% sucrose gradients containing: Tris-HCl (pH 7.4), 10 mM; NH_4Cl, 40 mM; Mg-acetate, 5 mM; dithiothreitol, 1 mM; and either GMP-PCP (panel A) or GTP (panel B), 0.04 mM. The gradients were centrifuged at 40000 rpm for 160 min at 4 °C in a Beckman SW-56 rotor. (From [26])

slightly less) (Fig. 8). Furthermore, it has been shown that treatment of the 50 S ribosomal subunit with 50% ethanol and 1.0 M NH_4Cl results in the loss of several ribosomal proteins, especially the acidic proteins L 7 and L 12 [35, 36]; this loss is accompanied by a sharp reduction in the ribosomal-dependent GTPase activities

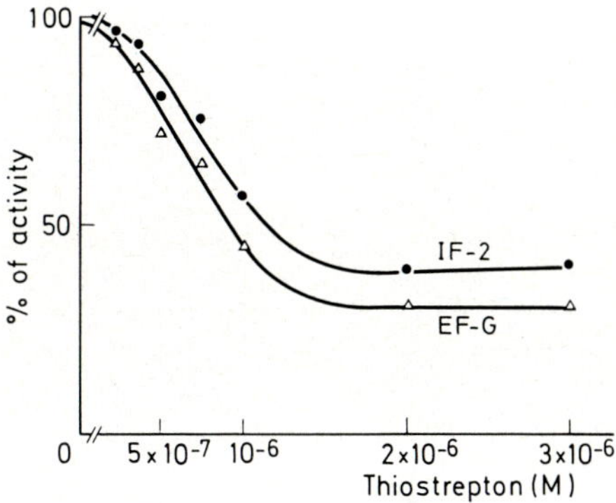

Fig. 8. Thiostrepton inhibition of EF-G- and IF-2-mediated GTPase. The incubation mixture (25 µl) contained: Tris (pH 7.4), 50 mM; NH_4Cl, 80 mM; Mg-acetate, 10 mM; β-mercaptoethanol, 7 mM; 70 S ribosomes, 1 A_{260} unit; (^{32}P) GTP (spec. act., 2.14—0.88 Ci/mmole), 720 pmoles; IF-2, 2.45 µg; EF-G, 0.3 µg; and thiostrepton in Me_2SO solution, as indicated. The final Me_2SO solution was always kept less than 2%; under these conditions no appreciable GTPase stimulation occurs. Inhibition was calculated as compared to the sample without Me_2SO. The molecular weight of thiostrepton was taken as 1616, according to ANDERSON et al. Nature (London) **225** (1970) 233. (From [34])

of the elongation factors (EF-G and EF-T); this reduction is reversed by adding back either protein. The removal of L12 also has an inhibitory effect on the ribosomal-dependent GTPase and on the fMet-tRNA binding activities of IF-2, but to a lesser extent than for the EF-G-dependent GTPase [37]. The addition of L12 to ethanol-treated 50 S particles restores the level of ribosomal-dependent GTPase activities (both coupled and uncoupled) of IF-2 to that found when untreated 50 S particles are used (Table 3).

The effect of L12 upon IF-2-dependent attachment of fMet-tRNA to ethanol-treated ribosomes was also investigated. Addition of 50 S subunits increases the factor-dependent fMet-tRNA binding that occurs with the 30 S subunits alone. The stimulation obtained by the addition of ethanol-treated 50 S particles is only 30–40 % that obtained when 50 S subunits, reconstituted with L12, are added.

Table 3. Effect of L 12 on IF-2-and EF-G-dependent GTPase activities

Additions	GTP-γ-^{32}P hydrolyzed (pmoles)		
	30 S + 50 S + IF-2	30 S + 50 S + IF-2 + poly (A, U, G) + fMet-tRNA	30 S + 50 S + EF-G
Untreated 50 S particles			
Expt. 1	287.8	697.3	—
Expt. 2	—	—	—
NH_4Cl-ethanol treated 50 S particles + L 12			
Expt. 1	336.1 (100)	680.0 (100)	—
Expt. 2	78.2 (100)	284.9 (100)	134.5 (100)
NH_4Cl-ethanol treated 50 S particles			
Expt. 1	111.1 (33.0)	233.0 (34.3)	—
Expt. 2	32.7 (41.8)	95.5 (33.7)	24.1 (17.9)

50 S and 30 S subunits (Expt. 1, 23.5 pmoles each; Expt. 2, 11.7 pmoles each) were pre-incubated with or without L 12 (Expt. 1, 5 µg; Expt. 2, 1.5 µg) for 15 min at 37 °C in final conditions (in 50 µl) of Tris-HCl, pH 7.5, 60 mM; $MgCl_2$, 10.5 mM; NH_4Cl, 100 mM; β-mercaptoethanol, 7 mM. To this were added: IF-2, 1.5 µg (in both Expts.) or EF-G, 0.18 µg; GTP-γ-^{32}P, 2000 pmoles (1 pmole gives about 100 cpm in both Expts.); poly (A, U, G), 0.15 A_{260} units; and fMet-tRNA, 0.5 A_{260} units (where indicated), together with Tris-HCl, NH_4Cl, and $MgCl_2$ to maintain the pre-incubation salt conditions (final volume 100 µl). Incubation was for 20 min at 37°. Activity was measured as the liberation of $^{32}P_i$ [1]. The control values obtained with factors alone and/or 30 S + 50 S alone (10—20 pmoles) have been subtracted. Figures in bracket are percentage of activity (reconstituted 50 S being taken as 100%). (From [37]).

Our results show that the lack of L12 impairs all the IF-2 activities associated with the 50 S subunits, and that these activities can be restored by addition of L12. However, the GTPase activity of IF-2 is less affected by the loss of L12 than is the GTPase activity of elongation factor EF-G. L7 and L12 can therefore be said to play a role, though probably indirectly, in the binding of fMet-tRNA, as well as in the coupled and uncoupled GTP hydrolyses.

GTP hydrolysis, concomitant with IF-2 removal, is required before fMet-tRNA can react with puromycin, or with the next amino acid. In the presence of GMP-PCP no reaction occurs because, as was shown by BENNE and VOORMA [38], the IF-2 remaining on the ribosome prevents puromycin attachment. This is consistent with the results of FAKUNDING and HERSHEY, discussed above, on the difference in binding of IF-2 to 70 S in the presence of GTP or of GMP-PCP. However, it is not the hydrolysis of GTP *per se* which is required for the release of IF-2, since DUBNOFF et al. [39] have shown that the 30 S initiation complex, depleted of GTP, will nonetheless react in the presence of 50 S to form the fMet-tRNA · puromycin complex. It could mean that in the complete absence of GTP, IF-2 leaves the 70 S complex. This is also consistent with the observation of FAKUNDING and HERSHEY [26] on the binding of IF-2 to ribosomes as analyzed by centrifugation: when GMP-PCP was included in the binding reaction mixture but was omitted from the sucrose gradient, IF-2 was not found in the 70 S region as was the case when GMP-PCP had been added to the gradient (Fig. 7).

The role of GTP in initiation (which can, though with less efficiency, be replaced in that case by GMP-PCP) would therefore be to increase the affinity of IF-2 for the initiation complex, and the role of GTP hydrolysis would be to increase the rate at which IF-2 leaves the 70 S complex. The fact that a GTPase activity is observed with IF-2 alone, in the absence of fMet-tRNA, indicates that in the presence of GTP and IF-2, junction of the two ribosomal subunits occurs. This is consistent with the light scattering experiments showing an increase of light scattering upon addition of IF-2 (Fig. 9a). IF-2 could therefore be considered as an association factor interacting with both the 30 S and 70 S ribosomes at a specific time during protein synthesis. But in the absence of all the other components of protein synthesis, the complex is not stable enough to withstand high centrifugal fields [26].

IF-1: IF-1 is a small protein, the most basic of the three, with a molecular weight of 9,400; it is an integral part of the initiation complex containing 30 S, AUG, fMet-tRNA, IF-2, and GTP, but it is not bound to the 30 S in the absence of the other initiation components, as shown by the experiments of HERSHEY et al. [40]. They studied the binding of (^{3}H)-labeled IF-1 to 30 S ribosomes by sucrose gradient centrifugation: in the absence of IF-2 no binding was detectable; in the presence of IF-2, a small amount of radioactivity could be detected in the 30 S region and there was considerable trailing of counts from this region to the top of the gradient. That IF-1 binding occurs in the presence of IF-2 is also demonstrated by the increase of affinity of 30 S for IF-2, as discussed above. We also observed this by light scattering: when IF-1 is added to ribosomes in the absence of the other components of initiation complex, there is an increase in light scattering which is consistent with a slight association of the ribosomal subunits to form the 70 S ribosomes; addition of IF-1 in the presence of IF-2 also tends to increase the subunit association due to IF-2 (Fig. 9a). This effect is consistent with our observations showing that the uncoupled 70 S-dependent IF-2 GTPase activity is stimulated by IF-1 (provided it is devoid of IF-3). The previously observed inhibitory effect of IF-1 on IF-2 uncoupled GTPase [28] could be explained by a small contamination of IF-1 by IF-3.

When all the initiation complex components are included (IF-2, AUG, fMet-tRNA), there is a considerable stimulation of IF-1 binding to the 30 S ribosomal subunits [40]. After addition of the 50 S subunits, IF-1 is released from the initiation complex; this adjunction is sufficient, and release does not require GTP hydrolysis since it occurs in the presence of GMP-PCP [40]. For a long time, it was believed that the role of IF-1 consisted solely in stabilizing IF-2-dependent fMet-tRNA · ribosome binding [41] since its stimulatory action is less pronounced at low temperature. It has also been reported [42] that IF-1 lowers the K_m for the initiation triplet ApUpG. Recently it was suggested that the action of IF-1 was very similar to that of elongation factor EF-T_s insofar as it appeared to change the stoichiometric IF-2-dependent fMet-tRNA · ribosome binding into a catalytic one [43].

In the absence of IF-1, or in its presence but with GMP-PCP replacing GTP, the IF-2-promoted binding of fMet-tRNA to the 70 S initiation complex is stoichiometric; when GTP hydrolysis

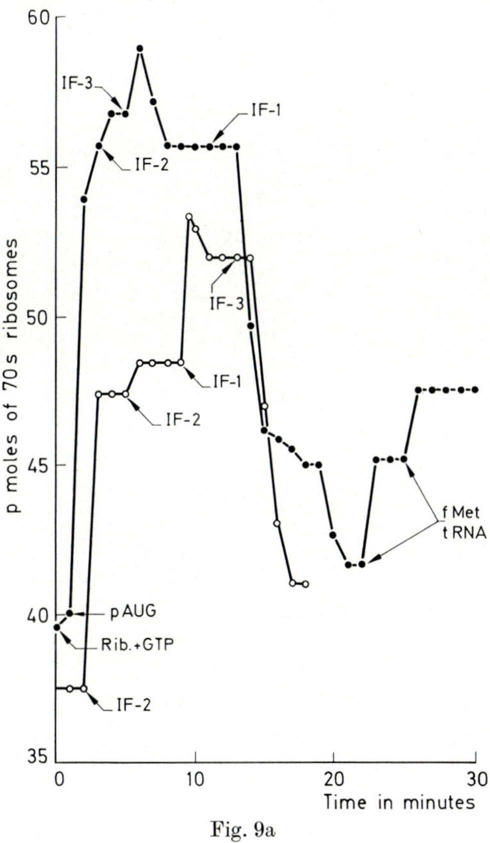

Fig. 9a

Fig. 9 a and b. Effect of selected addition of the components of initiation complex upon the apparent concentration of 70 S ribosomes. (a) The scattered light was measured at 90° angle in a solution containing (per ml.): ribosomes, 3.0 A_{260} units; GTP, 1 mM; Mg-acetate, 5 mM; NH_4Cl, 50 mM; Tris-HCl (pH 7.5), 10 mM; β-mercaptoethanol, 6.6 mM. When indicated, the following components were added: poly (A, U, G), 0.05 A_{260} units; IF-1, 0.8 μg; IF-2, 1 μg; IF-3, 1 μg; fMet-tRNA, 0.865 A_{260} units (8 pmoles of fMet). (b) Same system as for (A), except for the concentrations of: ribosomes (5 A_{260} units/ml); IF-3 (2 μg/ml); and poly (A, U, G) (0.2 A_{260} units/ml). No IF-2 was used

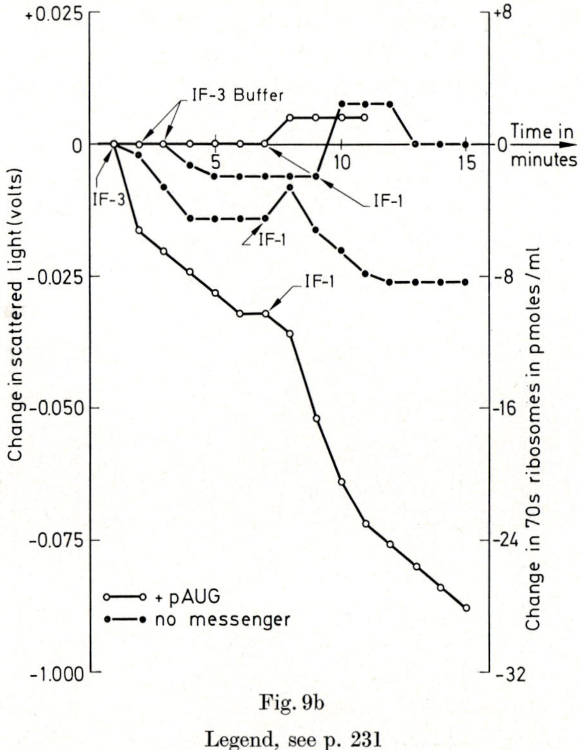

Fig. 9b

Legend, see p. 231

is possible and IF-1 is present, the binding of fMet-tRNA is catalytic [43, 44] (Fig. 10). However, the analogy with EF-T_s could not be drawn: the stimulatory effect is more pronounced during the first three minutes of incubation, and also, there is stimulation in the presence of GMP-PCP. The affinity of IF-2 for GDP being low, a protein factor analogous to EF-T_s is not needed in the initiation step.

Finally, IF-1 also markedly stimulates the dissociation activity exhibited by IF-3, as we shall discuss shortly.

IF-3: IF-3 is a protein (or a group of proteins) with a molecular weight of 21,500–23,500. Enhancement of fMet-tRNA binding to ribosomes by IF-3 is more striking with natural mRNA, and it was suggested that this factor recognizes a longer sequence of the message than the initiation triplet ApUpG or GpUpG. However, even with synthetic polymers, when the concentration of poly AUG is decreased, the percentage of stimulation by IF-3 is increased over the background without IF-3 and comes closer to that observed with natural messenger (Table 4). The effect of synthetic polymer concentration has been reported by WAHBA et al. [45].

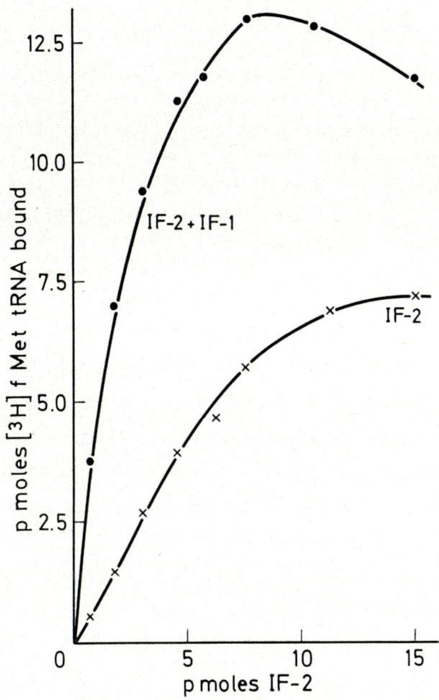

Fig. 10. Influence of saturating quantities of IF-1 upon IF-2-dependent f(^{3}H)Met-tRNA ribosomal binding. Samples (0.1 ml) contained: Tris-HCl (pH 7.5), 50 mM; NH_4Cl, 80 mM; Mg-acetate, 5 mM; GTP, 1 mM; high salt washed *E. coli* ribosomes, 2.0 A_{260} units; poly (A, U, G), 0.3 A_{260} units; f(^{3}H)-Met-tRNA (spec. act. 4.5 Ci/mM), 1.0 A_{260} units. IF-1 (3.7 μg/ml added where indicated) was purified by chromatography on DEAE-Sephadex and phosphocellulose. IF-2 (amounts added, calculated by assuming a molecular weight of 80,000, as indicated in the Fig.) was purified by chromatography on DEAE-Sephadex and hydroxylapatite. Incubation, 20 min at 37 °C, was stopped by addition of 3 ml chilled incubation buffer. Samples were then filtered over nitrocellulose filters (Millipore), washed 3 times with buffer, dried, and counted in toluene, POPOP and PPO with 38% efficiency. The amounts of fMet-tRNA bound in the absence of factors has been subtracted

Table 4. Effect of mRNA concentration on the stimulation by factors (pmoles fMet-tRNA bound)

Messenger	IF-1 + IF-2	IF-1 + IF-2 + IF-3
Poly (A, U, G) ($0.15\ A_{260}$ units)	3.03	10.1
Poly (A, U, G) ($0.015\ A_{260}$ units)	0.48	4.06
R 17 RNA ($0.77\ A_{260}$ units)	0	3.35

The incubation mixture (100 μl) contained: Tris (pH 7.5), 50 mM; NH_4Cl, 50 mM; Mg-acetate, 5 mM; β-mercaptoethanol, 7 mM; GTP, 1 mM; ribosomes, 1.8 A_{260} units; messenger as indicated; (^{3}H) fMet-tRNA, 1.58 A_{260} units (S.A. 2219 cpm/pmole); IF-1, 0.28 μg; IF-2, 0.61 μg; IF-3, 0.019 for poly (A, U, G) and 0.028 μg for R 17 RNA.

Incubation 30 min at 30 °C; filtration on Millipore nitrocellulose filters which were then washed three times with the same buffer, dried and counted.

IF-3 readily binds to the 30 S particles, whether the two other initiation factors are present or not [46, 47]. In the presence of viral mRNA, it also binds to the 30 S; moreover, the association between 30 S and viral mRNA requires the presence of IF-3, but IF-1 and IF-2 can be omitted; these last two factors enhance the binding of MS2 RNA provided, however, IF-3 is present [48].

It is also known that IF-3 is released when the 70 S ribosome complex is formed [46, 47], and it was reported by the group of Bosch [49] that IF-3 is released from the 30 S ribosomal subunits before their junction with the 50 S, when fMet-tRNA is bound. This could explain the fact that fMet-tRNA binding is inhibited by an excess of IF-3 (and vice-versa) in the presence of 70 S ribosomes; this inhibition is even stronger with the 30 S subunits, which explains Dubnoff's [39] and our observation that IF-3 does not stimulate the binding to a great extent in the presence of this ribosomal subunit, and is even often inhibitory.

IF-3 also exhibits a dissociating activity: when ribosomes are incubated with this factor (at 5 mM Mg) and centrifuged on a sucrose gradient, the amount of 30 S and 50 S found is higher than in the absence of IF-3 [22, 39, 46, 48, 50, 51]; this activity is markedly stimulated by the presence of IF-1, as has also been shown by

MIALL and TAMAOKI [52], but it is not observed in the presence of IF-1 alone (Figs. 9b and 11). The factor-dependent dissociation of 70 S ribosomes is markedly increased by the presence of mRNA (Figs. 9b and 12). The experiments made use of light scattering as well as centrifugation techniques in order to insure results which

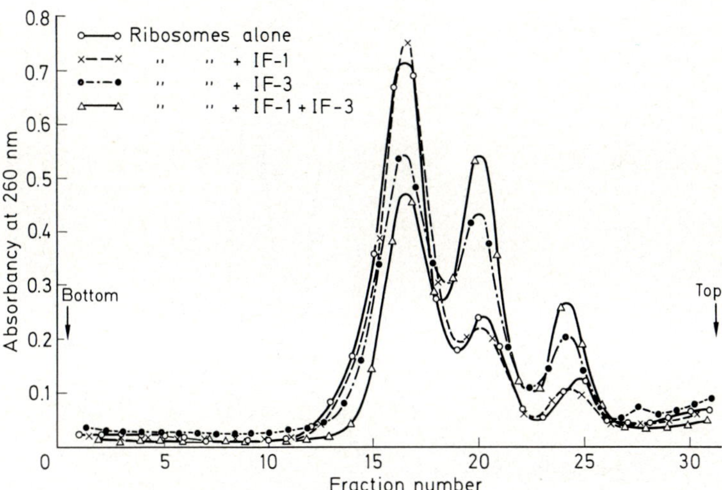

Fig. 11. Factor-mediated dissociation of 70 S ribosomes. The 70 S ribosomes (12 A_{260} units) were first pre-activated for 20 min at 37 °C in 200 µl of: Tris-HCl (pH 7.5), 170 mM; NH_4Cl, 170 mM; and Mg-acetate, 17 mM. 30 µl aliquots (0.8 A_{260} units) were diluted to 120 µl with water in the presence of initiation factors (IF-1, 0.4 µg; IF-3, 0.8 µg), as indicated. The factors were pure, as judged from polyacrylamide-SDS gel electrophoresis, and the amount added was calculated to give roughly a 1/1 molar ratio to ribosomes. After 15 min at 37 °C, the ribosomes were analyzed by centrifugation in 10—30% (w/w) sucrose gradients containing: Tris-HCl (pH 7.5), 10 mM; NH_4Cl, 50 mM; Mg-acetate, 5 mM; the samples were centrifuged for 90 minutes at 50.000 rpm in the SW 56 Spinco rotor

would not be influenced by hydrodynamic pressure. With light scattering, only a slight dissociating activity could be observed for IF-3 unless IF-1 was also present (Fig. 9b); moreover, even this activity was weak compared to that found by sucrose gradient centrifugation. The picture is quite different in the presence of mRNA. In the absence of any factors, the mixture of mRNA and ribosomes resulted in an increase in light scattering (suggesting formation of polysomes). Addition of IF-3 and IF-1 resulted in an immediate dissociation of polysomes and ribosomes (Fig. 9b). The effect of mRNA is probably not specific to the initiation triplet:

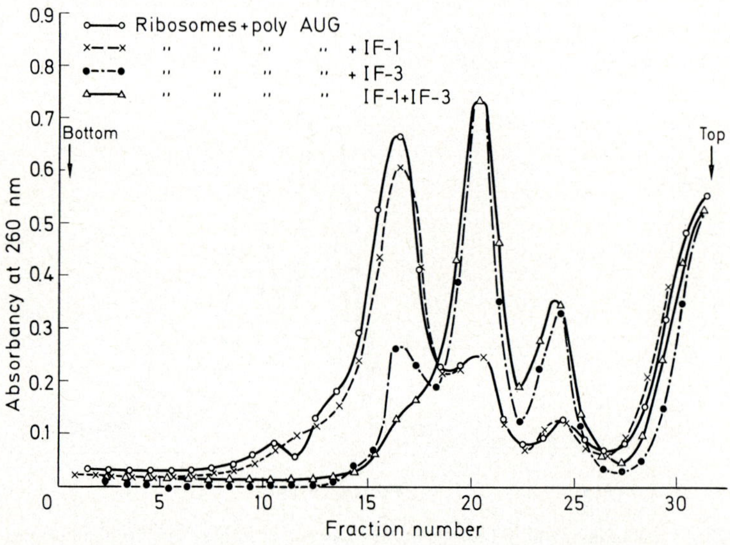

Fig. 12. Influence of poly (A, U, G) on the factor-mediated dissociation of ribosomes. The experiment was performed as in Fig. 11, except that poly (A, U, G) (0.65 A_{260} units) was added in each sample at the same time as the initiation factors

different synthetic polymers, especially ones containing G (AUG, UG, AG), stimulated the dissociation of ribosomes, but poly (AU) and poly (U) were inefficient. IF-3 showed a dissociating effect in the presence of mRNA, but no such dissociation was observed with mRNA and IF-1 alone.

There is good evidence that IF-3 is an anti-association factor which acts by binding to the free 30 S subunit, thereby pushing the equilibrium towards release of more 30 S ribosomes [53, 54]. What is the part played by IF-1 in the stimulation of the dissociating activity? This factor does not appear to be indispensable, but it does stimulate IF-3 activity: we can observe the effect of IF-3 alone in the formation of 30 S + 50 S from the 70 S ribosomes; but, using both sucrose gradient centrifugation and light scattering techniques, we have no evidence that IF-1 promotes dissociation of 70 S as suggested by NOLL and NOLL [55]. Since, when alone, this factor is not bound to either 70 S or 30 S, we would like to believe that it works in conjunction with IF-3 by increasing the rate of interaction with the latter, which results in pushing the equilibrium still further.

Before giving a tentative scheme for the initiation process, we would like to discuss the effect of antibiotics on the formation of the initiation complex, and particularly some of our results with streptomycin.

Tetracyclin [56] and fusidic acid [28] have little effect on initiation, but with streptomycin a strong inhibition of fMet-tRNA binding is observed *in vitro*, regardless of the mRNA used [57, 58]. This can be attributed to an interaction between the drug and the ribosomes since when streptomycin-sensitive ribosomes are replaced by ribosomes from streptomycin-resistant strains, no inhibition can be observed. It has also been shown that the antibiotic induces fMet-tRNA release from the initiation complex and does not cause dissociation of the total complex, which might explain the accumulation of monosomes observed by LUZZATO et al. *in vivo* [59]. Since the fMet-tRNA released was located in a puromycin donor site, it can be concluded that, in some way, streptomycin destroys the P site causing the release of initiator tRNA. When GMP-PCP replaces GTP, or when 30 S ribosomal subunits are used, no release of fMet-tRNA occurs.

An interesting effect was observed during that study: if streptomycin is added at intervals following the onset of fMet-tRNA binding, the antibiotic causes no release during the first minute (when binding is only 20 % of saturation level), but as the binding proceeds, the streptomycin dissociating effect appears [56, 57] (Fig. 13).

At low temperature (13 °C) no fMet-tRNA displacement was observed upon addition of streptomycin, whereas at that temperature fMet-tRNA reacts with puromycin, and dihydrostreptomycin is also bound (saturation level is reached in two minutes at this temperature [56]). It therefore appears that the initiation complex can adopt two possible conformations: one which would be susceptible to streptomycin and another which would not. With the 30 S ribosomes we never observed any effect of streptomycin.

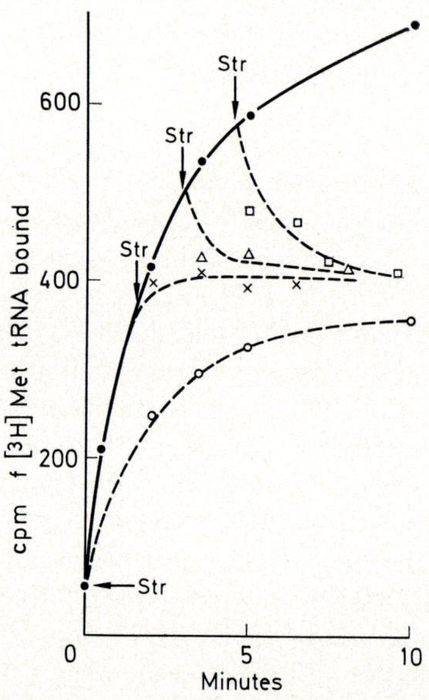

Fig. 13. Effect of streptomycin added in the course of complex formation. Streptomycin (4 μg/ml) was added in identical samples at different time intervals indicated by the arrows, and the fMet-(^{3}H)tRNA bound was measured by the Millipore filtration technique. O—O control without streptomycin. (From [56])

Figure 14 gives a tentative scheme for initiation complex formation. After dissociation of the 70 S has taken place (whether induced by factors or spontaneously), all three factors bind cooperatively to the 30 S. Binding of IF-3 or of IF-2 can occur singly on the 30 S ribosomes, but there is no evidence of such a binding with IF-1 alone. When IF-2 is present, 30 S and 50 S have a tendency to associate and form the 70 S. This is shown by light scattering and GTPase ac-

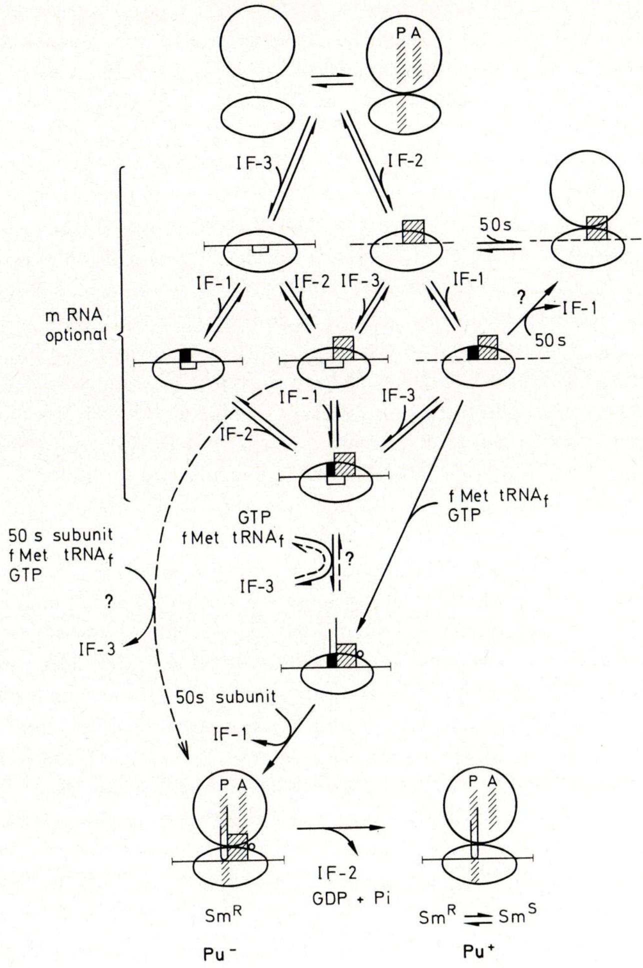

Fig. 14. Scheme for polypeptide chain initiation in prokaryotes

tivity, but this 70 S is not stable in sucrose gradient centrifugation. IF-1 stimulates the binding of IF-2 and the subsequent formation of the ribosomal duplex, and it also stimulates the dissociating activity of IF-3; moreover, the binding of IF-2 to 30 S ribosomes is also stimulated by IF-3. In short, each factor increases the affinity of the others, and eventually a 30 S complex is formed including all three factors. GTP and mRNA do not appear indispensable at this point: GTP increases the affinity of IF-2 and mRNA facilitates the binding. When fMet-tRNA is then added, IF-3 is released and a complex is formed: IF-2 · IF-1 · GTP · mRNA · fMet-tRNA · 30S. When the 50 S joins the 30 S, IF-1 is released while IF-2 stays on until GTP is hydrolyzed. Spontaneous release of IF-2 occurs concomitantly with GTP hydrolysis. When the hydrolysis is prevented (by use of GMP-PCP), IF-2 stays on the 70 S initiation complex. Thus in this scheme the 70 S initiation complex would have at least two conformations: one, containing IF-2 and GTP, which does not react with puromycin and another which does not contain IF-2 and GTP and reacts with puromycin; the latter is the functional form of the initiation complex in which fMet-tRNA is presumed to be in the P site. It was argued that in the complex containing IF-2 and GTP, the fMet-tRNA was in the A site [27], but this has been ruled out by the fact that tetracyclin does not inhibit formation of the initiation complex [56] and by experiments by THACH and THACH [60] showing that the messenger does not move with respect to the 30 S subunit during release of IF-2; this is in contrast to what occurs during EF-G-dependent translocation. These authors called this step "accommodation", and, assuming a rigid structure for the 70 S ribosome, gave a model in which fMet-tRNA was bound directly on the P site; the "accomodation" involved mere bending (or rotation) of the 3′ end of the tRNA in order to reach the catalytic site of the peptidyl transferase. This, however, does not readily explain the experiments with thiostrepton, nor the necessity of the L12 protein which show that the IF-2-dependent and EF-G-dependent GTPases overlap, both including elements of the A site. But one could assume that IF-2 is large enough (being a protein with a molecular weight of 80,000–100,000) to cover part of the A site as well as part of the P site. The presence of IF-2 would thus block the binding of puromycin (or of amino-acyl-tRNA), explaining the lack of reaction with the antibiotic.

Streptomycin sensitivity does not exactly parallel puromycin reactivity. The complex which does not react with puromycin is also resistant to streptomycin, but the complex which reacts with puromycin may be either partly sensitive to streptomycin (complex formed at 25°) or completely insensitive to this antibiotic (complex formed at 13°). These data suggest that after removal of IF-2 and GTP, fMet-tRNA can be bound to the ribosomes in at least two states, and initially at a site that shares some properties of the A site (streptomycin resistance) and of the P site (puromycin reactivity) [61].

One very important question is whether there is a control at the translational level in prokaryotes and the best chance is that it does exist for the translation of polycistronic mRNA. The data presented in Fig. 3 suggest that the stimulating activity of the different IF-3 fractions vary in the presence of different mRNA's. Revel et al. [62], Lee-Huang and Ochoa [63], as well as ourselves [22] have some evidence for different classes of IF-3, each stimulating differently in the presence of various mRNA's. However, although IF-3 binding to the 30 S ribosomes (as estimated by the dissociating activity) is increased by addition of some G-containing polynucleotides, there does not seem to be a specificity for any given sequence. Lee-Huang and Ochoa reported the existence of two IF-3 [63], one specific for late T4 mRNA and the other for the translation of early T4 mRNA and MS2 mRNA. Our own experiments indicate that when IF-3 is purified from different factions eluted from DEAE-Sephadex (Fig. 3, pools a, b, c), it stimulates fMet-tRNA incorporation or binding in the presence of synthetic polymers as well as R17 or T4 RNA, whatever the fraction used, but with some difference in efficiency for IF-3b and IF-3c (Table 5). Using R17 RNA as messenger, the ratio of translation of coat protein cistron as compared to the other cistrons (mainly synthetase) does not appear to change whatever IF-3 is used: it could be estimated from the ratio of (^{14}C)valine to (^{3}H)histidine incorporated into TCA precipitable material (since coat protein does not contain histidine) and from the direct determination of the amount of fMet-alanine dipeptide (which starts the coat protein) and of fMet-arginine (which starts the maturation protein) (Table 6).

As mentioned earlier, Revel et al. [23] found an interfering protein [called factor (*i*)] which could associate very tightly with the different IF-3 and would regulate the relative translation of T4 and MS2 mRNA's. With MS2 RNA (treated with formaldehyde to disrupt the secondary structure in order to make initi-

Table 5. Specificity of various fractions of IF-3 towards different mRNA's

	Binding of (^{3}H)fMet-tRNA			Incorporation of (^{14}C) valine		
	(in pmoles)			(in cpm)		
	R17	T4	T4/R17	R17	T4	T4/R17
IF-1 + IF-2	0	0		1000	1241	
IF-1 + IF-2 + IF-3a				2518	2810	1.1
IF-1 + IF-2 + IF-3b	2.09	2.34	1.11	2349	2142	0.96
IF-1 + IF-2 + IF-3c	0.67	1.38	2.06	829	1458	1.76

For binding, the incubation mixture and experimental procedure were the same as in Table 4, except for the concentrations of initiation factors and mRNA: IF-1, 0.4 μg; IF-2, 0.84 μg; IF-3 where indicated (IF-3a, 0.043 μg; IF-3b, 0.035 μg; IF-3c, 0.065 μg); and either R 17 RNA, 0.77 A_{260} units, or T 4 mRNA, 0.99 A_{260} units.

Incubation 20 min at 37 °C.

For incorporation see conditions in Table 6 (except for the amount of IF-1 and IF-2 which were the same as above).

Table 6. Incorporation of (^{3}H)histidine and (^{14}C)valine in the presence of different mRNA's

	R 17		T 4	
	Valine (pmoles)	Hist/Val ratio	Valine (pmoles)	Hist/Val ratio
IF-1 + IF-2	1.37	1.65	1.73	3.1
IF-1 + IF-2 + IF-3a (0.12 μg)	5.40	0.904	5.45	2.76
IF-1 + IF-2 + IF-3b (0.23 μg)	8.53	0.91	7.00	2.95
IF-1 + IF-2 + IF-3b pur. (0.24 μg)	8.51	1.08	6.15	2.83
IF-1 + IF-2 + IF-3c (0.25 μg)	4.23	0.8	4.63	2.26

The incubation mixture (200 μl) contained: Tris, pH 7.5, 50 mM; NH_4Cl, 60 mM; β-mercaptoethanol, 7 mM; ATP, 2 mM; GTP, 0.5 mM; Mg-acetate, 12.5 mM; pyruvate kinase, 40 μg; PEP, 5 mM; tRNA 1.25 A_{260} units; mixture of the common amino acids (minus histidine and valine), 20 nmoles of each one; (^{3}H) histidine; 7.4 nmole; (^{14}C) valine, 20 nmoles (spec. act. 4000 cpm/pmole of either one); ribosomes, 1.8 A_{260} units; 105000 g supernatant, 320 μg; formylfolate (previously reduced by H_2) in the presence T 4 mRNA, 0.99 A_{260} units; IF-1, 0.28 μg; IF-2, 0.61 μg; IF-3, as indicated.

Incubation (30 min at 37 °C) is stopped by addition of 2 ml NaCl and 2 ml 10% trichloracetic acid. After heating 10 min between 80 and 90 °C, the mixture is filtered on glass filters (GF/c Whatman), the filters are washed three times with 1% TCA, dried and counted.

ation available to all three cistrons), it inhibits initiation at the coat protein cistron (the main translated codon) but stimulates initiation at the synthetase cistron. In the case of late T4 mRNA, the overall translation is changed, but, with this template also, factor (*i*) stimulates the translation of certain cistrons while inhibiting that of others. This protein has been identified with the largest of the three host proteins present in the RNA phage replicase. REVEL et al. propose that different interfering factors exist and that they have an important function in *E. coli* at the translation level of the gene expression regulation (recently they isolated several of these factors[2]). They think factors (*i*) could act in a repressor-like fashion towards ribosome attachment to specific initiation sites and that this could, in an indirect way, increase binding at other sites when they are available. LEE-HUANG and OCHOA [23] have isolated two factors (*i*), one of which, ($i\alpha$), is identical with the first factor (*i*) isolated by REVEL's group. They have shown that factor ($i\alpha$) inhibits the translation of MS2 RNA, but they have no evidence of cistron-specificity; their second factor, ($i\beta$), inhibits the translation of late T4 mRNA; ($i\alpha$) and ($i\beta$) each work only in conjunction with the IF-3 specific for the messenger RNA involved. Our own data is not clear on these points, and we believe factors (*i*) are better demonstrated in eukaryotic systems.

We would finally like to report some of our results [64] showing that protein factors play an active role in the ribosomal site selection process of both messenger and transfer RNA's, and that they do not merely act by increasing the affinity of aminoacyl-tRNA and mRNA for their pre-existing attachment regions on the ribosome.

The evidence is based on the fact that the analog of fMet-tRNA (N-Ac-Phe-tRNA) does not correctly recognize its decoding site on the ribosome unless initiation factors are present (even at high Mg, under conditions where it is normally bound). The saturation curves of this tRNA, when added singly (at 10 mM Mg) or in the presence of IF-1 and IF-2 (at 5 mM Mg), show different characteristics (Fig. 15). Only in the presence of factors could ribosomes be saturated (approximately 1 pmole N-Ac-Phe-tRNA/1 pmole ribosome). Furthermore, only the factor-mediated reaction placed this

[2] Personal communication.

tRNA in the P site, as evidenced by the fast reaction with puromycin, the chasing with streptomycin and the lack of sensitivity towards tetracyclin (Table 7). 70 % of N-Ac-Phe-tRNA bound in the presence of factors (at a temperature of 30 °C) is released by streptomycin, as compared to 30 % release when non-enzymatically

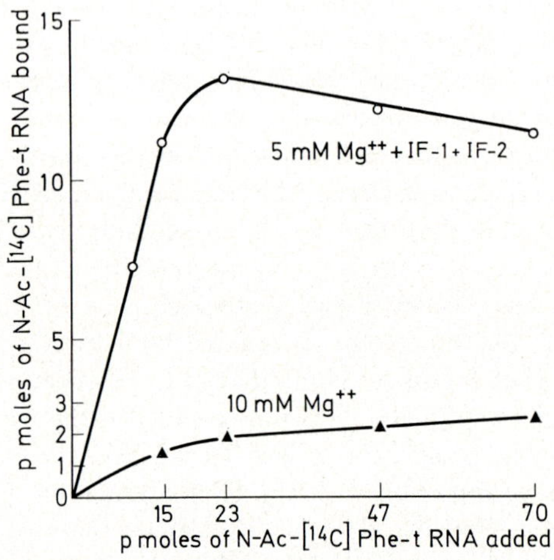

Fig. 15. Binding of N-Ac-(^{14}C)Phe-tRNA in the presence and absence of initiation factors, at 5 mM and 10 mM Mg^{2+}. The incubation mixture (50 μl) contained: Tris-HCl (pH 7.5), 50 mM; NH_4Cl, 80 mM; Mg-acetate, as indicated; GTP, 1 mM; poly (U), 0.2 A_{260} units; ribosomes, 0.8 A_{260} units; IF-1, 1.35 μg, and IF-2, 2.25 μg (where indicated); N-Ac-(^{14}C)Phe-tRNA, as indicated. Incubation 15 min at 25 °C. (From [63])

bound. As already noted with fMet-tRNA, temperature is important for the release by streptomycin; at temperatures lower than 30 °C, no release occurred.

During initiation there is another step which is quite important, the binding of the aminoacyl-tRNA which immediately follows the initiation codon. This step has seldom been separated from the elongation step, although it may be different from the other steps of the elongation process which follow. Notwithstanding, the fact is that a specific opening of the A site only occurs when initiator tRNA has been correctly bound to ribosomes, *i.e.*, in the presence of initiation factors. Accordingly, it is only when N-Ac-Phe-tRNA

Table 7. Effect of streptomycin and puromycin on the N-Ac-Phe-tRNA · poly(U) · ribosome complex

Expt. No.	Conditions	N-Ac-Phe-tRNA bound (pmoles)		Puromycin reaction binding (%)
		— Sm	+ Sm	
1	5 mM Mg (25 °C)[a]	1.8	—	15
2	10 mM Mg (30 °C)[a]	2	2.8	34
3	5 mM Mg + (IF-1 + IF-2) (25 °C)[a]	13.5	10.7	83
4	5 mM Mg + (IF-1 + IF-2) (30 °C)[a]	13.5	4.14	—

[a] Temperature of second incubation.

Same incubation mixture as in Fig. 15, except for N-Ac-(^{14}C)Phe-tRNA, 23.5 pmoles, and streptomycin sulfate: Expts. 2 and 4, 0.75 μg; Expt. 3, 1.5 μg.

For binding determination: in Expts. 2 and 4, after 15 min incubation at 25 °C, the temperature was shifted to 30 °C and incubation resumed for another 30 min; in Expts. 1 and 3, incubation was 30 min at 25 °C.

For puromycin reaction: after 15 min incubation at 25 °C, 100 μl of 0.7 mg/ml puromycin dihydrochloride solution were added to the samples prior to a further incubation of 5 min at 25 °C.

Amount of N-Ac-(^{14}C)Phe · puromycin formed was determined by the method of Leder and Byrsztyn (Biochem. Biophys. Res. Comm. **25**, 233, 1966). Amount of N-Ac-(^{14}C)Phe · puromycin formed without poly (U) (always <0.3 pmoles) was subtracted. (From [64]).

is prebound to the ribosomes in the presence of initiation factors that a stronger EF-T_u-dependent binding of Phe-tRNA occurs at the A site (Fig. 16); this requires the continuity of the mRNA; no such effect is observed in the presence of $(pU)_3$. This suggests that opening of the A site demands the correct adjustment of initiator tRNA (or of its analog) at the P site, in addition to the presence of factor EF-T_u and GTP. Thus, when present alone, aminoacyl-tRNA's do not recognize their respective decoding sites, and their interaction with ribosomes does not, apparently, suffice to promote the conformational changes that appear adequate for peptide chain growth.

In conclusion, the particular combination of ribosome activities expressed at any one time is controlled by the character of protein factors associated with ribosomes. We may thus infer that the structure of the ribosome is cyclically altered, either subtly or dramatically in response to the different sets of factors. Each initiation factor seems to modify the structure of the ribosome so as to allow a succession of steps (which are still not well known) monitored by the binding, the functioning, or the release of each factor. The last step results in an initiation complex having the exactly required conformation to allow elongation to start.

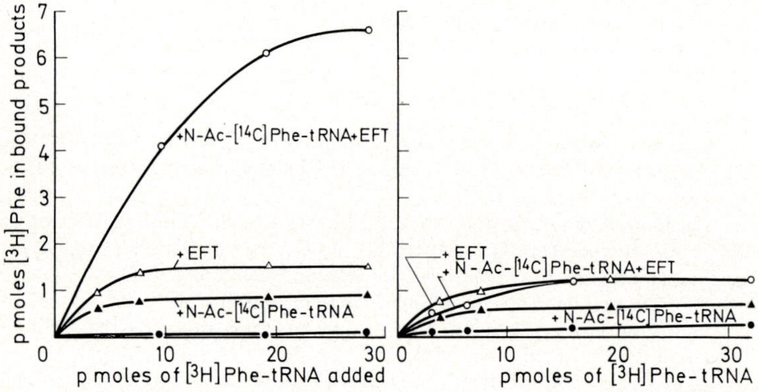

Fig. 16. (^{3}H) Phe in ribosome-bound products in the presence and absence of N-Ac-(^{14}C) Phe-tRNA. A first incubation was carried out under the same conditions as in Fig. 15 with 23.5 pmoles of N-Ac-(^{14}C) Phe-tRNA, where indicated, and Mg-acetate, 5 mM for the left hand panel, and 10 mM for the right. The samples were then chilled to 0 °C and (^{3}H) Phe-tRNA was added in the indicated amounts, and, where indicated, 8 μg of factor EF-T. A second incubation (final volume 55 μl) was carried out for 15 min at 0 °C. Binding without poly (U) was not subtracted

Acknowledgements

This work was supported by the following grants to Dr. Grunberg-Manago: Centre National de la Recherche Scientifique (G.R.N°18); Délégation Générale à la Recherche Scientifique et Technique (Convention 72.7.0388 Interactions Moléculaires en Biologie); and the Ligue Nationale Francaise contre le Cancer (Comité de Paris).

We wish to thank Dr. Fakunding and Dr. Hershey for having communicated their results to us before publication.

References

1. KOZAK, M., NATHANS, D.: Bact. Rev. **33**, 264 (1972).
2. MARCKER, K. A., SANGER, F.: J. molec. Biol. **8**, 835 (1964).

3. ADAMS, J.M., CAPECCHI, M.R.: Proc. nat. Acad. Sci. (Wash.) **55**, 147 (1966).
4. WEBSTER, R.E., ENGELHARDT, D.L., ZINDER, N.D.: Proc. nat. Acad. Sci. (Wash.) **55**, 155 (1966).
5. LODISH, H.F.: Nature (Lond.) **220**, 345 (1969).
6. TAKEDA, M., WEBSTER, R.E.: Proc. nat. Acad. Sci. (Wash.) **60**, 1487 (1968).
7. LIVINGSTON, D.M., LEDER, P.: Biochemistry **8**, 435 (1969).
8. ADAMS, J.M.: J. molec. Biol. **33**, 571 (1968).
9. DUBE, S.K., MARCKER, K.A., CLARK, B.F.C., CORY, S.: Nature (Lond.) **218**, 232 (1968).
10. CORY, S., MARCKER, K.A.: Europ. J. Biochem. **12**, 177 (1970).
11. SIMSEK, M., ZIEGENMEYER, P.J., HECKMAN, J., RAJBANDHARY, U.L.: Proc. nat. Acad. Sci. (Wash.) **70**, 1041 (1973).
12. PETRISSANT, G.: Proc. nat. Acad. Sci. (Wash.) **70**, 1046 (1973).
13. PIPER, P.W., CLARK, B.F.C.: FEBS Letters **30**, 265 (1973).
14. CLARK, B.F.C., MARCKER, K.A.: J. molec. Biol. **17**, 394 (1966).
15. GOSH, H.P., SOLL, D., KHORANA, H.G.: J. molec. Biol. **25**, 275 (1967).
16. GUTHRIE, C., NOMURA, M.: Nature (Lond.) **219**, 232 (1968).
17. GRUNBERG-MANAGO, M., CLARK, B.F.C., REVEL, M., RUDLAND, P., DONDON, J.: J. molec. Biol. **40**, 33 (1969).
18. SPIRIN, A.S.: FEBS Letters **14**, 349 (1971).
19. INFANTE, A.A., BAIERLEIN, R.: Proc. nat. Acad. Sci. (Wash.) **68**, 1780 (1971).
20. ZITOMER, R.S., FLAKS, J.G.: J. molec. Biol. **71**, 263 (1972).
21. GRONER, Y., REVEL, M.: Europ. J. Biochem. **22**, 144 (1971).
22. GRUNBERG-MANAGO, M., RABINOWITZ, J.C., DONDON, J., LELONG, J.C., GROS, F.: FEBS Letters **19**, 193 (1971).
23. GRONER, Y., POLLACK, Y., BERISSI, H., REVEL, M.: Nature (Lond.) New Biol. **239**, 16 (1972).
24. LEE-HUANG, S., OCHOA, S.: Biochem. Biophys. Res. Commun. **49**, 371 (1972).
25. FAKUNDING, J.L., TRAUGH, J.A., TRAUT, R.R., HERSHEY, J.W.B.: J. biol. Chem. **247**, 6365 (1972).
26. FAKUNDING, J.L., HERSHEY, J.W.B.: J. biol. Chem. **248**, 4206 (1973).
27. KOLAKOFSKY, D., DEWEY, K.F., HERSHEY, J.W.B., THACH, R.E.: Proc. nat. Acad. Sci. (Wash.) **61**, 1066 (1968).
28. LELONG, J.C., GRUNBERG-MANAGO, M., DONDON, J., GROS, D., GROS, F.: Nature (Lond.) **226**, 505 (1970).
29. RUDLAND, P., WHYBROW, W.A., CLARK, B.F.C.: Nature (Lond.) New Biol. **231**, 76 (1971).
30. LOCKWOOD, A.H., CHAKRABORTY, P.R., MAITRA, U.: Proc. nat. Acad. Sci. (Wash.) **68**, 3122 (1971).
31. GRONER, Y., REVEL, M.: J. molec. Biol. **74**, 407 (1973).

32. Mazumder, R., Chae, Y. B., Ochoa, S.: Federation Proc. **28**, 597 (1969).
33. Springer, M., Dondon, J., Grunberg-Manago, M. (unpublished experiments), Clark, B. F. C. (personal communication).
34. Grunberg-Manago, M., Dondon, J., Graffe, M.: FEBS Letters **22**, 217 (1972).
35. Hamel, E., Koka, M., Nakamoto, T.: J. biol. Chem. **247**, 805 (1972).
36. Sander, G., Marsh, R. C., Parmeggiani, A.: Biochem. Biophys. Res. Commun. **47**, 866 (1972).
37. Kay, A., Sander, G., Grunberg-Manago, M.: Biochem. Biophys. Res. Commun. **51**, 979 (1973).
38. Benne, R., Voorma, H. O.: FEBS Letters **20**, 347 (1972).
39. Dubnoff, J. J., Lockwood, A. M., Maitra, U.: J. biol. Chem. **247**, 2884 (1972).
40. Hershey, J. W. B., Dewey, K. F., Thach, R. E.: Nature (Lond.) **222**, 944 (1969).
41. Chae, Y. B., Mazumder, M., Ochoa, S.: Proc. nat. Acad. Sci. (Wash.) **63**, 828 (1969).
42. Mazumder, R.: FEBS Letters **18**, 64 (1971).
43. Benne, R., Arentzen, R., Voorma, H. O.: Biochim. biophys. Acta, **269**, 304 (1972).
44. Kay, A. C., Grunberg-Manago, M.: Biochim. biophys. Acta, **277**, 225 (1972).
45. Wahba, A. J., Iwasaki, K., Miller, M. J., Sabol, S., Sillero, M. A. G., Vasquez, C.: Cold Spr. Harb. Symp. quant. Biol. **34**, 291 (1969).
46. Sabol, S., Ochoa, S.: Nature (Lond.) New Biol. **234**, 233 (1971).
47. Thibault, J., Chestier, A., Vidal, D., Gros, F.: Biochimie **54**, 829 (1972).
48. Sabol, S., Sillero, M. A. G., Iwasaki, K., Ochoa, S.: Nature (Lond.) **228**, 1269 (1970).
49. Vermeer, C., de Kievit, R. J., van Alphen, W. J., Bosch, L.: FEBS Letters **31**, 273 (1973).
50. Subramanian, A. R., Davis, B. D.: Nature (Lond.) New Biol. **228**, 1254 (1971).
51. Subramanian, A. R., Davis, B. D., Beller, R. J.: Cold Spr. Harb. Symp. quant. Biol. **34**, 223 (1969).
52. Miall, S. H., Tamaoki, T.: Biochemistry **11**, 4826 (1972).
53. Kaempfer, R.: J. molec. Biol. **71**, 583 (1972).
54. Sabol, S., Meier, D., Ochoa, S.: Europ. J. Biochem. **33**, 332 (1973).
55. Noll, M., Noll, H.: Nature (Lond.) New Biol. **238**, 225 (1972).
56. Zagorska, L., Dondon, J., Lelong, J. C., Gros, F., Grunberg-Manago, M.: Biochimie **53**, 63 (1971).
57. Lelong, J. C., Cousin, M. A., Gros, D., Grunberg-Manago, M., Gros, F.: Biochem. Biophys. Res. Commun. **42**, 530 (1971).
58. Modolell, J., Davis, B. D.: Proc. nat. Acad. Sci. (Wash.) **67**, 1148 (1970).

59. LUZZATO, L., APIRION, D., SCHLESSINGER, D.: Proc. nat. Acad. Sci. (Wash.) **60**, 873 (1968).
60. THACH, S. S., THACH, R. E.: Proc. nat. Acad. Sci. (Wash.) **68**, 1791 (1971).
61. SPRINGER, M., DONDON, J., GRAFFE, M., GRUNBERG-MANAGO, M., LELONG, J. C., GROS, F.: Biochimie, **53**, 1047 (1971).
62. REVEL, M., AVIV, H., GRONER, Y., POLLACK, Y.: FEBS Letters **9**, 213 (1970).
63. LEE-HUANG, S., OCHOA, S.: Nature (Lond.) New Biol. **234**, 236 (1971).
64. SPRINGER, M., GRUNBERG-MANAGO, M.: Biochem. Biophys. Res. Commun. **47**, 477 (1972).

Discussion

H. NOLL (Evanston): We have claimed that the factor F1 facilitates the dissociation of 70 S ribosomes by increasing the rate constant k_1 of the dissociation reaction at Mg^{2+} concentrations that keep the equilibrium very far toward association. Under these conditions the effect of IF-1 in the absence of IF-2 and IF-3 can only be shown in exchange experiments, not in, *e.g.*, light scattering studies. I also want to point out that it is very important what kind of ribosomes you use. Marianne has already mentioned that there are tight couples and loose couples. In our experiments, we have used only tight couples, and they can be converted 100% into initiation complexes.

E. FUCHS (Heidelberg): How specific is the initiation at high Mg^{2+} concentrations with met-tRNA?

M. GRUNBERG-MANAGO (Paris): The initiation is specific until 10—12 mM Mg^{2+}; at higher concentrations the initiation step no longer depends upon factors.

E. FUCHS: Can Mg^{2+} be substituted by polyamines?

M. GRUNBERG-MANAGO: At low concentration polyamine can replace most of the Mg^{2+}; small amounts of Mg^{2+} are however still necessary. At high concentrations of polyamines the binding of initiator tRNA is non-specific and no longer depends upon messenger triplet.

Translational Control in Eukaryotic Organisms

William A. Strycharz, Marjut Ranki and Kjeld A. Marcker

Department of Molecular Biology, Aarhus University, Aarhus, Denmark

The identification of bacterial initiation factors has been pursued for several years; the elucidation of the precise roles of these factors in protein biosynthesis is now reasonably well established. During the past few years, considerable effort has been directed toward the search for similar factors from mammalian systems. These investigations have further supported the conclusion that protein synthesis in prokaryotic and eukaryotic organisms involves essentially similar mechanisms.

Specific initiation factors have been shown to bind the initiator tRNA to ribosomes; these are referred to as IF_1 and IF_2 in *Escherichia coli* [3] and IF-M_1 and IF-$M_{2\,A+B}$ in rabbit reticulocytes [9]. IF_3 has been shown, in the case of *E. coli*, to be required for the translation of natural mRNA [7]. A similar factor, IF-M_3, has been isolated from rabbit reticulocyte [6, 8] and rabbit liver ribosomes [5] and appears to be required, in addition to IF-M_1 and IF-$M_{2\,A+B}$, for the translation of globin mRNA in the reticulocyte cell-free system [1].

Recently, Groner et al. have reported the existence of messenger-discriminating factors in *E. coli* [2]. Evidence has been presented that indicates that these protein factors affect the recognition of mRNA initiation sites by the ribosome. Since these factors have been shown to be associated with the initiation factor IF_3, thereby altering its specificity, they have been termed "interference factors" or "i-factors." The intriguing question arises, therefore, as to whether the expression of genetic information in eukaryotes might be regulated in a similar manner.

Recent investigations in this laboratory have led to the identification in Krebs II ascites cells of a supernatant factor, IF_{EMC}, which is absolutely required for the initiation of translation of encephalomyocarditis viral RNA (EMC-RNA) [12]. IF_{EMC} was purified by means of ammonium sulfate fractionation followed by chromatography on hydroxyapatite, DEAE-Sephadex A-50, Sephadex G-100 superfine, and phosphocellulose. The resultant preparation was shown, by means of SDS polyacrylamide gel electrophoresis, to contain a single major component with a molecular weight of approximately 53000 daltons [11]. IF_{EMC} is not required for the translation of synthetic polynucleotides (*e.g.*, poly $(AUG)_n$ or $(AUG)U_n$), hemoglobin mRNA, or MS2 RNA. It is concluded, therefore, that this factor represents a messenger-specific initiation factor.

Examination of initiation peptides in studies using the antibiotic sparsomycin suggested the presence in ascites cells of other messenger-specific factors required for the initiation of translation of globin α-chain mRNA, globin β-chain mRNA, MS2 RNA, and perhaps other natural mRNA's [10]. Although phosphocellulose-purified IF_{EMC} was not required for the initiation of globin α-chains and was not sufficient for the initiation of globin β-chains in the fractionated system, the addition of crude ascites supernatant to the system did result in the initiation of β-chains. These findings suggested that an additional factor, present in crude ascites extracts, was required for β-chain initiation. Preliminary results suggest that a hemoglobin mRNA-specific factor is eluted from phosphocellulose by means of a KCl gradient slightly earlier than IF_{EMC}. This component apparently controls the initiation of β-globin mRNA translation.

Nudel et al. have recently isolated a similar type of protein factor from the rabbit reticulocyte ribosome. This factor has a molecular weight of approximately 65,000 daltons and stimulates the translation of α-globin mRNA and TMV RNA in an ascites cell-free system [4]. It has little effect on β-globin mRNA translation and no effect with mengoviral RNA. The data from our laboratory, together with that from Nudel and coworkers, suggest, therefore, that the synthesis of hemoglobin α- and β-chains is independently controlled by two distinct messenger-discriminating factors.

In addition to studies with EMC-RNA and hemoglobin mRNA, we have also investigated the interaction of IF_{EMC} with other natural mRNA's. In view of the lability of the phosphocellulose-purified IF_{EMC}, the Sephadex G-100 superfine-purified factor was used in these investigations. Our data suggest that both Rous Sarcoma viral RNA and mengoviral RNA translation are stimulated to a significant extent. It is not yet certain whether these latter two activities represent functions of IF_{EMC} or whether they are associated with factors distinct from IF_{EMC}.

A protein component of high molecular weight is also required for the translation of EMC-RNA, as well as other natural mRNAs, in the ascites cell-free system. The results of our investigations suggest that this factor is closely related to the IF-M_3 preparations from rabbit reticulocyte and rabbit liver [5].

The translation of EMC-RNA in the basic fractionated cell-free system requires, in addition to IF_{EMC}, elongation factor I (EF-I), partially purified from ascites supernatant by ammonium sulfate fractionation and Sephadex G-200 chromatography. The ascites ribosomes used in these studies were approximately 60% saturated with EF-I; however, addition of the Sephadex G-200-purified EF-I beyond saturating levels resulted in the significant further stimulation of EMC-RNA translation. This observation suggested the presence in the EF-I preparation of another essential factor; in view of its exclusion on Sephadex G-200, the activity was ascribed to the putative IF-M_3.

Further fractionation of the Sephadex G-200-purified preparation was achieved by KCl gradient elution on DEAE-Sephadex A-50. EF-I activity, monitored by poly U-directed polyphenylalanine synthesis, was eluted over the relatively broad range of 0.15–0.28 M KCl. IF-M_3-type activity, monitored by EMC-RNA translation, was eluted slightly later; maximum activity appeared at approximately 0.22 M KCl.

Appropriate column fractions were pooled and the IF-M_3-type factor was further purified by stepwise-KCl gradient elution on phosphocellulose. Maximum EF-I activity was localized in the 0.25 M KCl-containing fraction; although EF-I activity was also present in the 0.20 and 0.30 M KCl fractions, no activity was detected in the 0.50 M KCl-containing fraction. IF-M_3-type activity was eluted at higher concentrations; maximum activity was obtained with the 0.30 M KCl fraction while considerable activity was also eluted with 0.50 M KCl.

The availability of a Sephadex G-200-purified EF-I fraction deficient in IF-M_3-type activity has made it possible to further define the nature of the latter factor. A comparison of a normal with an IF-M_3-deficient EF-I preparation in EMC-RNA translation is given in Table 1. Although the two preparations were

Table 1. Stimulation of EMC-RNA translation in an IF-M_3-deficient ascites system by addition of rabbit liver IF-M_3

IF-M_3-containing EF-I	IF-M_3-deficient EF-I	IF_{EMC}	rabbit liver IF-M_3	^{35}S-methionine incorporated (pmol × 10^{-2})
+	—	+	—	3.22
—	+	+	—	0.34
—	+	+	+	2.39
—	+	—	+	0.35

The measurement of EMC-RNA-directed ^{35}S-methionine incorporation in the basic fractionated ascites cell-free system has been described elsewhere [12]. A standard 50 μl reaction contained the following: 0.75 A_{260} ribosomes; 15—50 μg Sephadex G-200-purified EF-I; 3 μg Sephadex G-100 superfine-purified IF_{EMC}; 0.10—0.20 A_{260} ^{35}S-Met-tRNA; 0.10 A_{260} EMC-RNA; 20 mM Tris-HCl, pH 7.5; 3 mM $MgCl_2$; 80 mM KCl; 6 mM 2-mercaptoethanol; 1 mM ATP; 0.1 mM GTP; 100 μg creatine phosphate; 10μg creatine phosphokinase. Approximately 30 μg rabbit liver IF-M_3 saturated the system. Incorporation in the presence of EMC-RNA was corrected for background incorporation in the absence of added mRNA; the latter varied from 0.0017 to 0.0037 pmoles ^{35}S-methionine.

shown to have equal activity in polyphenylalanine synthesis, the IF-M_3-deficient EF-I fraction did not significantly support the translation of EMC-RNA. Addition of rabbit liver IF-M_3 (kindly provided by W. F. ANDERSON) to the system, however, restored more than 70% of the ^{35}S-methionine incorporation obtained in the complete system. (Similar results were obtained with rabbit reticulocyte IF-M_3.) In the absence of IF_{EMC}, no ^{35}S-methionine incorporation occurred in the rabbit liver IF-M_3-supplemented system. It is concluded that IF-M_3 and IF_{EMC}, two distinct protein factors, are both required for the optimal translation of EMC-RNA in the ascites cell-free system.

The availability of the IF-M_3-deficient EF-I preparation has also made it possible to further monitor the activity of the phosphocellulose-purified ascites IF-M_3. Addition of the previously mentioned 0.50 M KCl-containing phosphocellulose fraction to the IF-M_3-deficient system resulted in the partial restoration of EMC-RNA-directed ^{35}S-methionine incorporation.

The observations presented in this report are consistent with the hypothesis that translational control of protein synthesis in eukaryotic organisms exists at the level of chain initiation. Two different types of protein factors are involved. In the fractionated ascites cell-free system, EMC-RNA translation requires the presence of a high molecular weight factor, IF-M_3, as well as a lower molecular weight component, IF_{EMC} · IF-M_3 is neither tissue-specific nor messenger-specific. Rabbit liver IF-M_3 can support globin synthesis in the reticulocyte cell-free system [5]; both rabbit liver and rabbit reticulocyte IF-M_3 can substitute for ascites IF-M_3. A common IF-M_3 is therefore utilized for both globin mRNA and EMC-RNA translation. These observations imply that control of the template specificity of ribosomes in higher organisms is associated with messenger-discriminating factors such as IF_{EMC}. Evidence is accumulating to suggest the existence of other similar factors, each presumably facilitating the preferential translation of certain classes of mRNA.

Two mechanisms for the control of EMC-RNA translation by IF_{EMC} may be considered. IF_{EMC} might interact stoichiometrically with IF-M_3, the resultant complex thereby increasing the affinity of the ribosome for the initiation site of EMC-RNA. This model is similar to that proposed for the *E. coli* "*i*-factors" [2]. Alternatively, IF_{EMC} might interact directly with the initiation site of EMC-RNA, perhaps by "melting out" the initiation site, thereby facilitating the interaction of EMC-RNA with IF-M_3. Although no stoichiometric relationship between IF_{EMC} and IF-M_3 is implicit in this latter model, such a mechanism could control the specificity of the mRNA-ribosome interaction. Either of these two mechanisms might be extended to include translational control of protein synthesis in other systems.

References

1. Crystal, R. G., Nienhuis, A. W., Prichard, P. M., Picciano, D., Elson, N. A., Merrick, W. G., Graf, H., Shafritz, D. A., Laycock, D. G., Last, J. A., Anderson, W. F.: FEBS Letters **24**, 310 (1972).

2. Groner, Y., Pollack, Y., Berissi, H., Revel, M.: Nature (Lond.) New Biol. **239**, 16 (1972).
3. Lucas-Lenard, J., Lipmann, F.: Ann. Rev. Biochem. **40**, 409 (1971).
4. Nudel, U., Lebleu, B., Revel, M.: Proc. nat. Acad. Sci. (Wash.) **70**, 2139 (1973).
5. Picciano, D. J., Prichard, P. M., Merrick, W. C., Shafritz, D. A., Graf, H., Crystal, R. G., Anderson, W. F.: J. biol. Chem. **248**, 204 (1973).
6. Prichard, P. M., Picciano, D. J., Laycock, D. G., Anderson, W. F.: Proc. nat. Acad. Sci. (Wash.) **68**, 2752 (1971).
7. Revel, M., Lelong, J. C., Brawerman, G., Gros, F.: Nature (Lond.) **219**, 1016 (1968).
8. Schreier, M. H., Staehelin, T.: Nature (Lond.) New Biol. **242**, 35 (1973).
9. Shafritz, D. A., Anderson, W. F.: Nature (Lond.) **227**, 918 (1970).
10. Smith, A. E., Wigle, D. T.: Europ. J. Biochem. **35**, 566 (1973).
11. Wigle, D. T.: Europ. J. Biochem. **35**, 11 (1973).
12. Wigle, D. T., Smith, A. E.: Nature (Lond.) **242**, 136 (1973).

Discussion

Ch. Jungwirth (Würzburg): Is the concentration of the EMC RNA-specific factor altered in EMC-infected cells compared to uninfected cells?

K. A. Marcker (Aarhus): No. We have never detected any difference between infected and uninfected cells.

Th. Staehelin (Basel): I would like to say a word of caution to the question of how washing of ribosomes results in supernatant proteins. We know that M3 has the highest affinity to ribosomes. However, when you wash ribosomes with 0.1 M KCl, you will elute a certain amount of this factor. An even larger amount is eluted with 0.3 M KCl. Thus, the distinction between bound and supernatant factors depends very much on the treatment of the ribosomes.

G. Schütz (New York): 1. I assume that you have checked the translation of EMC RNA and globin mRNA at the same $MgCl_2$ concentration. It is however known that these two mRNA's require different $MgCl_2$ concentrations for optimal translation. Did you check the stimulation of translation of these two RNA's by the factor at different $MgCl_2$ concentrations?

K. A. Marcker: This is true, but you cannot change the rate of the translation of the different messengers by changing the $MgCl_2$ concentration.

G. Schütz: 2. Since the proportion of α and β chains synthesized is dependent on the mRNA concentration used, I wonder whether you have used different mRNA concentrations.

K. A. Marcker: We have not tried different mRNA concentrations.

J. Drews: Can you use your small factor to form an initiation complex with EMC RNA *in vitro?*

K. A. Marcker: Yes, I think we could, but we have not tried. We have been more concerned to find other factors before we went to functional studies.

The Mechanism of Subunit Interaction as a Key to the Understanding of Ribosome Function

HANS NOLL[1], MARKUS NOLL[2], BERN HAPKE[3] AND GERBRAND VAN DIEIJEN

Department of Biological Sciences, Northwestern University, Evanston, USA

With 34 Figures

1. Introduction

The most intriguing feature of the ribosome is its highly asymmetrical construction of two unequal subunits that, in the case of prokaryotic organisms, are described by their 30 S and 50 S sedimentation values. Although the pioneering work of TISSIÈRES and WATSON [1] established that Mg^{2+} ions were essential to maintain the associated state and hinted at the existence of a dynamic equilibrium, a rather static view of ribosome structure dominated the thinking for nearly a decade. Perhaps consonant with the prevailing mores of the time, it was taken for granted that the union between the two subunits was a marriage for life. This vision seemed to satisfy an inner need, for how else could it be explained that the scientists would suppress their natural curiosity and refrain from a closer scrutiny of the relationship between the subunits? However, the less reverent and more open climate of the sexual revolution and women's lib was not going to bypass the ribosome, and soon representatives of the younger generation [2] boldly proclaimed that ribosomes not engaged in protein synthesis normally existed as free subunits that would enter a temporary union only

---

[1] Lifetime Career Professor of the American Cancer Society.

[2] Present address: MRC Laboratory of Molecular Biology, Hills Rd. CB2 2QH, England

[3] Present address: Nuclear Products Div., 3 M Center, St. Paul, MN 55101, USA.

for the purpose of creating a new polypeptide chain. Although this radical view subsequently turned out to be untenable, by removing an old taboo it led to fresh experiments and marked the beginning of a more dynamic view of the ribosome.

After the elegant experiments of MESELSON and co-workers [3], it could no longer be denied that ribosomes in the cell did indulge in partner swapping, although their experiments could not tell the extent or frequency of this practice. It is becoming clear, however, that healthy ribosomal subunits prefer togetherness and occur as couples under physiological conditions even when not engaged in translation [4, 5]. The present debate centers mostly around the question of the stability of their relations, *i.e.*, whether these couples exchange partners only occasionally or whether they are naturally promiscuous. We are interested in the intimate relations between the subunits because we would like to know how the stability of their union affects their performance in protein synthesis and their fate throughout their life cycle. If we disregard ribosome synthesis and degradation, the events commonly summarized as ribosome cycle consist of initiation, chain elongation, termination and release [6]. As we shall see, the stability of the ribosome couples profoundly influences these events.

It has been known for a long time that ribosomes engaged in protein synthesis and carrying a nascent polypeptide chain are much more stable with regard to dissociation than vacant ribosomes. We have found that vacant ribosomes, as commonly prepared, normally consist of a mixture of what I shall call "tight" and "loose" couples as well as varying amounts of free subunits. In the tight couples the 30 S and 50 S partners are strongly attached to each other even at relatively low Mg^{2+} concentrations (3–4 mM) and under physiological conditions ($\cong$ 5 mM Mg^{2+}) exchange partners only occasionally and for special reasons, such as initiation and release [7]. The loose couples, on the other hand, are very busy exchanging partners all the time under physiological conditions and appear to be dissociated significantly at 5 mM Mg^{2+} [8, 9]. I am obliged to report, without doubt to the displeasure of some of my friends, that the tight couples are the most active ribosomes and represent the physiologically normal situation, whereas the loose couples, even though they may have more fun, are somehow defective.

Our results imply that loose couples were originally tight couples that, as the result of some as yet unknown traumatic experience of the 50 S subunit, have lost a good deal of attraction for each other. Nevertheless, when tested for translation of R17 RNA, loose couples are nearly as active in protein synthesis as tight couples, and at higher Mg^{2+} concentrations behave much like tight couples. Finally, to conclude this preliminary survey of the social life of ribosomes, the so-called "native" subunits should be mentioned. Our experiments revealed that native subunits consist largely of active 30 S and dead 50 S particles [10, 11]. Apparently, the 50 S partner is very vulnerable and rapidly goes to pieces when not protected by association with its 30 S counterpart. We have shown that ribosomal subunits that have lost the ability to unite spontaneously are invariably inactive in protein synthesis with natural messengers and so the association test is both the simplest and the most stringent structural criterion for ribosome activity.

2. Active Ribosomes

Nearly all studies of ribosome function in the past have been carried out with preparations in which not more than a small fraction of the ribosomes were active. It became increasingly clear that many of the important problems of the detailed mechanisms of protein synthesis and translational control, of the meaning of the fractional proteins, etc., could only be resolved with better defined ribosome preparations in which all particles were active. It seemed likely that many of the conflicting results published in the literature and obtained with the same highly purified factor preparations were attributable to the use of different and poorly characterized ribosomes. In our efforts to remedy the situation, we learned to prepare bacterial ribosomes that could all be converted to an initiation complex using the RNA of the small coli phage R17 as messenger [5, 7, 10–13].

It turned out that most ribosomes ($\cong 80\%$) in a freshly prepared crude S-30 extract are vacant couples at 10 mM Mg^{2+}, and that all vacant couples are active. The large scale preparation of such ribosomes in a B-29 zonal rotor is illustrated in Fig. 1. If the fractions corresponding to the 70 S peak are collected and recentrifuged in the same manner, pure tight couples free of detectable contamination by subunits (< 1 %) are obtained, as evident from the sedimentation pattern of an analytical sucrose gradient (Fig. 2).

The first clue that the 70 S ribosome couples rather than the free subunits were the active particles came from the type of ex-

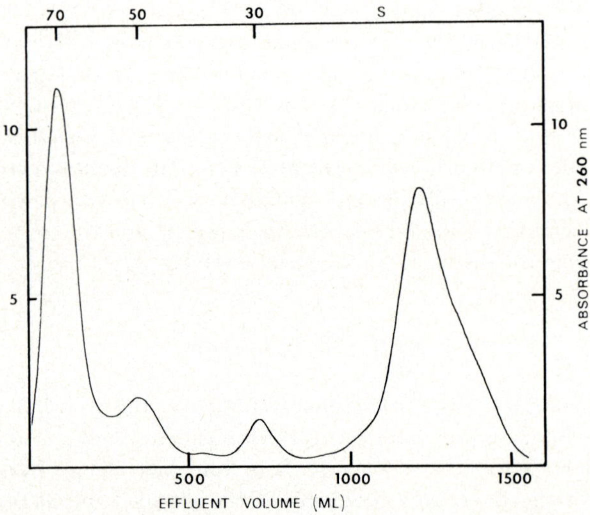

Fig. 1. Sedimentation pattern of a fresh S-30 extract in a zonal rotor. The extract was prepared from *E. coli* MRE 600 cells by grinding with alumina and centrifuged in a B29 IEC zonal rotor as described previously [5, 11]

periment illustrated in Fig. 3. If ribosomes from an S-30 extract are allowed to stand at 0 °C for prolonged periods, the subunits increase at the expense of the 70 S couples (a–c) and the ability to form an initiation complex declines to about the same extent (d–f). By testing each kind of subunit with an excess of active particles of the complementary type both for the ability to form couples at 10 mM Mg^{2+} (association test) and for initiation activity (activity test), we found that a lesion in the 50 S subunit was responsible for the loss of activity because only the 50 S but not the 30 S subunits were unable to reform active couples [11].

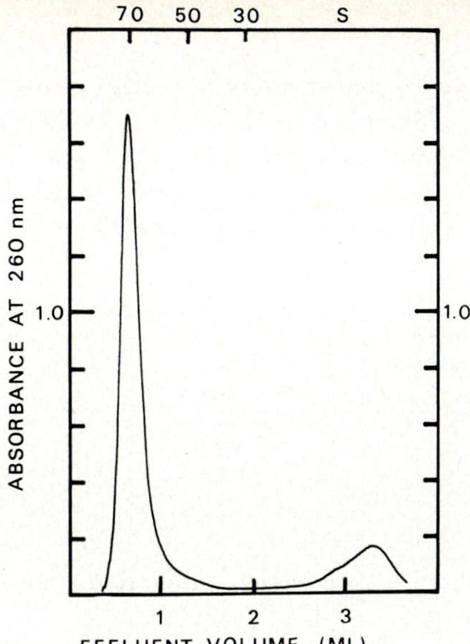

Fig. 2. Sedimentation pattern of vacant ribosome couples free of contaminating subunits. For experimental details see [10]

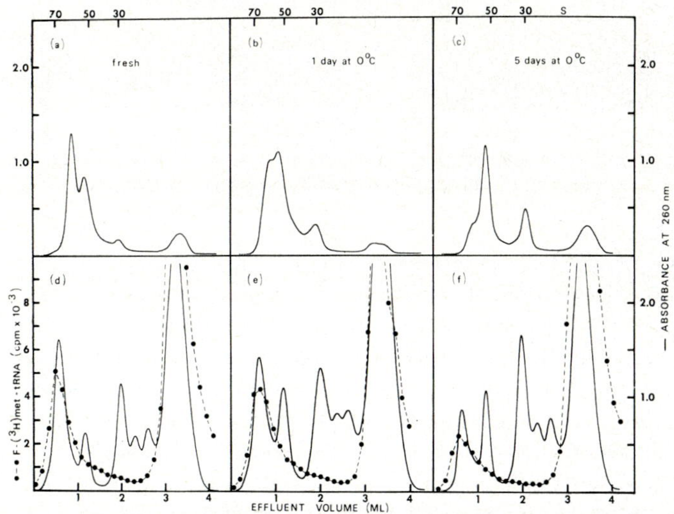

Fig. 3a—f. Irreversible dissociation and inactivation of ribosomes in solutions kept at 0 °C for prolonged periods. The ribosome preparations shown in (a—c) were tested for their ability to form an initiation complex (d—f). For details see [10, 11]

The quantitative conversion of pure tight couples into an initiation complex is illustrated in Fig. 4. In this experiment initiation mixtures containing a limiting amount of pure tight ribosome couples were analyzed on sucrose gradients after incubation with increasing concentrations of radioactive fMet-tRNA. The Mg^{2+}

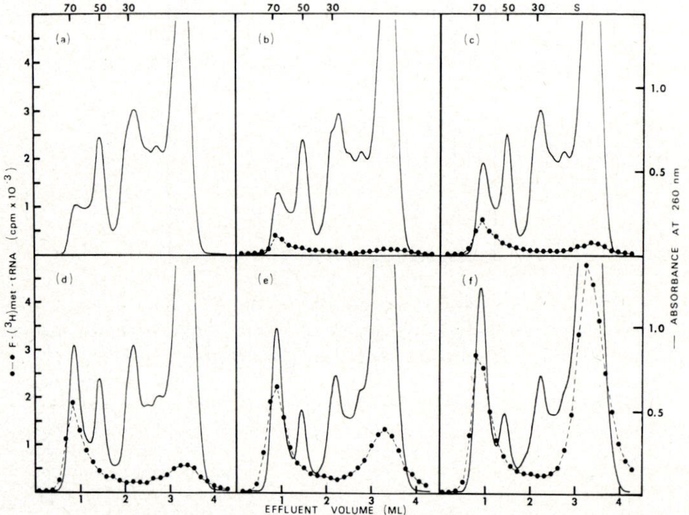

Fig. 4a—f. Conversion of pure tight couples into an initiation complex as a function of limiting fMet-tRNA concentrations [10, 12]

concentration in the gradients was reduced to 3 mM in order to cause dissociation of most vacant couples while preserving the majority of the initiation complexes. The sedimentation patterns in Fig. 4 illustrate strikingly the progressive formation of the initiation complex by the increase in the 70 S absorbance peak and by the proportional binding of initiator tRNA. Thus, if we plot the 70 S peak area as a function of the radioactive fMet-tRNA bound (Fig. 5), a straight line is obtained with a slope corresponding to one molecule of initiation complex per molecule of fMet-tRNA bound and with an intercept reflecting the residual 70 S particles that resisted dissociation in the absence of added fMet-tRNA. Since we started with pure couples, the extent of conversion (ca.

65 %) can be estimated directly from the fraction of 50 S subunits remaining at saturation. The value of 65 % is an underestimate because at 3 mM Mg^{2+} in the gradient the initiation complex begins to break down. I have chosen this example because it illustrates the high initiation efficiency of these ribosomes in a direct way without the complications and uncertainties inherent in specific

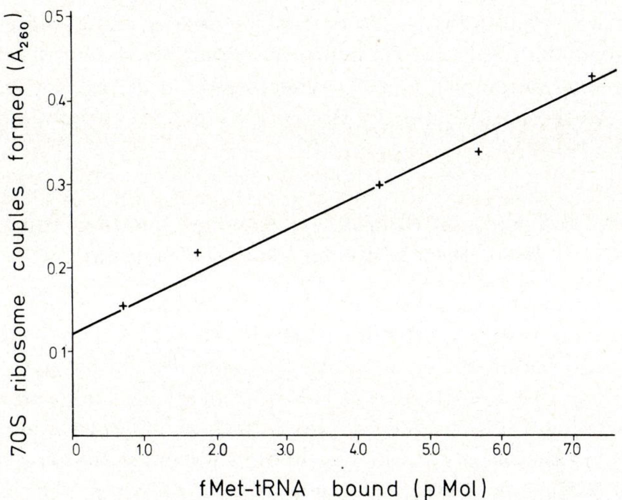

Fig. 5. Plot of 70 S peak area in Fig. 4 as a function of f-^{3}H-Met-tRNA bound to 70 S initiation complex. The linear regression was determined by a *Hewlett-Packard* 9020 calculator and gave a slope of 0.041 A_{260} units/pMol fMet or 24.1 pMol fMet/A_{260} unit of initiation complex and an intercept of 0.121 A_{260} units. The intercept represents couples stable at 3 mM Mg^{2+} in the absence of fMet-tRNA. Since these couples are active, they would all have been converted into initiation complexes at saturation (Fig. 4, gradient f) and the intercept would be zero. Assuming saturation at 7.2 pMol fMet-tRNA bound, the corrected slope would be 0.058 A_{260} units/pMol fMet or 17.2 pMol fMet/A_{260} unit of initiation complex. If each ribosome (0.24 A_{260}/pMol) binds one fMet-tRNA, 17.2 pMol ribosomes would contribute 0.72 A_{260} units and the remaining .28 A_{260} units would correspond to ribosome-bound R 17 RNA. If each initiation complex contained an intact messenger, R 17 RNA should acount for 40% of the A_{260} of the complex. Determinations with labeled mRNA showed, however, that the average messenger size of the complex was only 70% of the intact molecule [12], in agreement with the value obtained from Fig. 5 if the intercept is zero at 7.2 pMol fMet-tRNA bound. Data from [10, 12]

activity data, recoveries, molecular weights of ribosomes and messengers and other sources of error. Nevertheless, we also have determined the efficiency of conversion by a careful analysis of the sedimentation patterns at 6 mM Mg^{2+} in the gradients. By comparing the recovery of fMet-tRNA radioactivity and 70 S absorbance with the known inputs after correcting for the losses at the walls of the tube, we found that as much as 85–90% of vacant ribosome couples could be converted into an initiation complex [12].

It was important to show that all of these initiation complexes are functional. This was examined by testing the extent of release of fMet by puromycin and by extending the nascent coat protein chain to the hexapeptide. Both reactions were essentially quantitative [10, 12].

3. Active Site of Ribosomes: Topology and Structural Dynamics of Interface between Subunits

Extensive studies confirmed the validity of the concept that interaction between subunits is the most sensitive and critical structural manifestation of active ribosomes. So far we have seen no exceptions to the general rule that may be summarized as follows: vacant ribosome couples regardless of preparation and origin are always active, whereas subunits present in the dissociated state at Mg^{2+} concentrations above 10 mM are active only to the extent that they can be converted to couples by addition of an excess of active subunits of the complementary type.

The more profound implication of this rule is that initiation and, as we shall see, all known biochemical or biophysical functions of the ribosome require the integrity of critical sites located at the interface between subunits and necessary for their interaction. Indeed, if we look at the ribosome as a giant enzyme, we could say that the interface between the subunits is the active site. This is the essence of Spirin's semi-mechanistic model in which the subunits are joined by some sort of hinge [14]. In confirmation of his model, we have shown that translocation and amino-acyl-tRNA binding involve the alternating weakening and strengthening of the interaction between subunits [15]. Other evidence has been summarized by Spirin. Some of the more striking examples are the protection against RNase attack of a 30 nucleotide sequence containing the initiation site [16] and the protection from proteolytic attack of a 30 amino acid sequence at the growing point of the nascent polypeptide chain [17].

Apparently in the vacant couple these delicate structures are protected from enzymatic attack and thermal denaturation by the formation of a fairly stable complex between the complementary surfaces, especially at higher Mg^{2+} concentrations. During initiation a partial or complete separation of this complex is necessary for the insertion of messenger RNA and two transfer RNA molecules [7]. It is difficult to imagine how this could take place without extensive reorientation of the interface topology. It appears that in the course of evolution nature staged the elaborate ballet of allosteric sequences performed by the multitude of protein factors involved in chain initiation, elongation, termination and release in order to ensure in the most efficient manner the required topological adjustments at the interface as messenger RNA and transfer RNA molecules are inserted and removed in precise order. According to this concept, each step in the assembly and breakdown of the active complex corresponds to a different and precisely defined state of subunit interaction. It follows that it should be possible to identify these states by quantitative measurements of the interaction between subunits.

4. Quantitative Measurements of Subunit Interaction

It is convenient to define as the ground state the interaction between subunits in vacant couples as described by the equilibrium

$$70\,\mathrm{S} \underset{k_2}{\overset{k_1}{\rightleftharpoons}} 50\,\mathrm{S} + 30\,\mathrm{S} \tag{1}$$

Quantitative assays of the interaction have involved:

a) evaluation of the dissociation constant $K_{diss} = k_1/k_2$ by determinations of the position of the equilibrium at various Mg^{2+} concentrations;

b) evaluation of the rate k_1 of dissociation; and

c) evaluation of the hydrostatic pressure required to bring about a certain degree of dissociation at a given Mg^{2+} concentration.

For all three measurements ultracentrifugation has been the major tool although, more recently, measurements of light scattering have proved both convenient and useful [8]. The advantages of measuring the extent of association or dissociation by the intensity of the scattered light are obvious, because it makes it possible to follow the kinetics of the reaction in solution without influencing the equilibrium. Yet, since the method, unlike centrifugation, does not look at the individual particle species, it depends for calibration on verification of the distribution by ultracentrifugation. Ultracentrifugation, on the other hand, is capable of telling how many particles of each kind are present in a mixture, but it has to contend with complications arising from shifts in the equilibrium caused by changes in concentration or pressure during centrifugation. Unfortunately, these complications have led some investigators to dismiss summarily the validity of results obtained by sedimentation analysis. As I shall show, it is possible by judicious design of the experiment and critical evaluation of the results not only to avoid the pitfalls mentioned but to turn these seemingly undesirable complications to advantage. Finally, electrophoresis in gels with molecular sieving action would seem to hold considerable promise as an analytical separation method for ribosomes because it avoids the complications of centrifugation [18, 19].

5. Position of Dissociation/Association Equilibrium as a Function of the Mg^{2+} Concentration

The preparation of pure vacant couples not contaminated with subunits and of uniform biological activity made it meaningful to measure the dissociation $K = k_1/k_2$ under equilibrium conditions at 0° and 37 °C and at various Mg^{2+} concentrations. Control experiments established that at 15 mM Mg^{2+} and 4 °C no dissociation takes place on the gradient. It was found, however, that in solution the association reaction was fast even at 0° while dissociation was completely inhibited. Consequently, the dissociation seen on cold sucrose gradients is much less than that in the incubation mixture at 37° because cooling to 4° produces a considerable shift toward association. To measure the true dissociation at 37 °C, we resorted to a displacement labeling technique: a large excess of radioactive 30 S subunits was added to the equilibration mixture before cooling in order to shift the equilibrium completely toward association [10]. The true dissociation may then be calculated from the number of radioactive subunits incorporated into the 70 S peak. A further correction has to be made for the unequal losses of 70 S, 50 S and 30 S particles during centrifugation because of collisions with the wall of the tube ("wall effect") [10, 12].

The gradients in Fig. 6 show the expected increase in the transfer of radioactivity from the 30 S to the 70 S peak as the Mg^{2+}

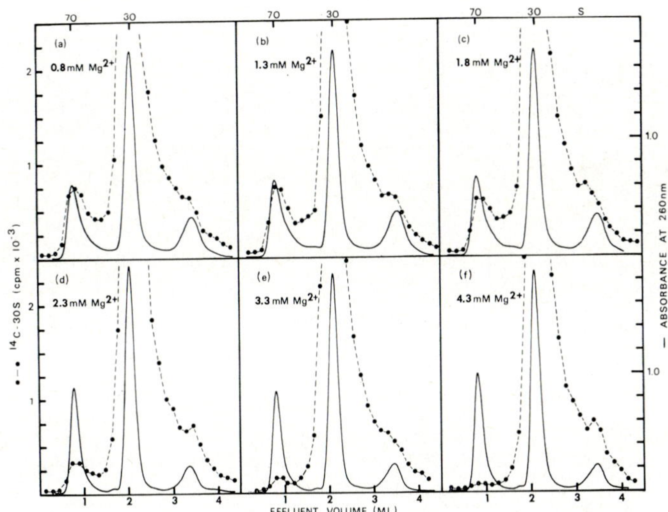

Fig. 6a—f. Dissociation of pure tight couples at 37 °C determined at various Mg^{2+} concentrations by displacement of equilibrium with excess of radioactive 30 S subunits. Before addition of the ^{14}C-labeled 30 S subunits, the mixtures were equilibrated for 10 min at 37 °C in buffer solutions (50 mM HEPES, pH 7.6; 50 mM NH_4Ac) containing the Mg^{2+} concentrations indicated. Data from [10, 12]

concentration is lowered from 4.3–0.8 mM. The absence of a peak corresponding to free 50 S subunits implies that we are dealing with a true equilibrium, the position of which depends on the concentration of each component. Similar experiments measured the dissociation at 0°; we computed the dissociation

$$D_e = (c_{50})_e / (c_{70})_0 = (c_{30})_e / (c_{70})_0, \qquad (2)$$

in which $(c_{70})_0$ = initial conc. of 70 S particles

$(c_{50})_e$ = equilibrium conc. of 50 S particles, etc., and plotted it as a function of the Mg^{2+} concentration (Fig. 7). The curves show that below 3 mM Mg^{2+} the equilibrium is strongly dependent on the temperature and very sensitive to small changes in the Mg^{2+} concentration. The narrow range of dissociation (60 % of the particles dissociating between 1.5 and 2.5 mM Mg^{2+} at 37°) implies that the population is fairly homogeneous.

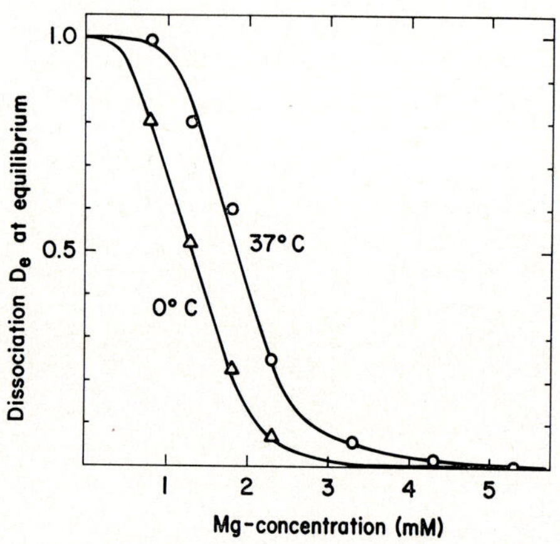

Fig. 7. Dissociation of pure tight ribosome couples as a function of Mg^{2+} concentration at 0 °C and 37 °C. Each point was determined from a gradient as in Fig. 6. Data from [10, 12]

Zitomer and Flaks have published similar measurements obtained by light scattering methods [8]. Their results are strikingly different in that their curves are much flatter with a much higher 50 % dissociation point (9 mM Mg^{2+} at 37°) suggesting a more heterogeneous and much less stable population of ribosome couples. Indeed, their ribosomes, prepared according to Kurland [20], had been treated with high concentrations of $(NH_4)_2$ SO_4 and incubated at 37°. These procedures convert nearly all tight couples into loose couples and subunits [9].

6. Determination of Dissociation Rate k_1 by Subunit Exchange

For a complete description of the equilibrium, we need to know one of the rate constants in addition to the dissociation constant. The rate of spontaneous dissociation, k_1, was determined at 37°

and 5 mM Mg^{2+} from the initial rate at which pure labeled 50 S particles appeared in the 70 S ribosomes after they were mixed with pure non-radioactive vacant couples. In all of these exchange experiments the purity of the particles is essential for contamination with the complementary subunit would lead to the immediate

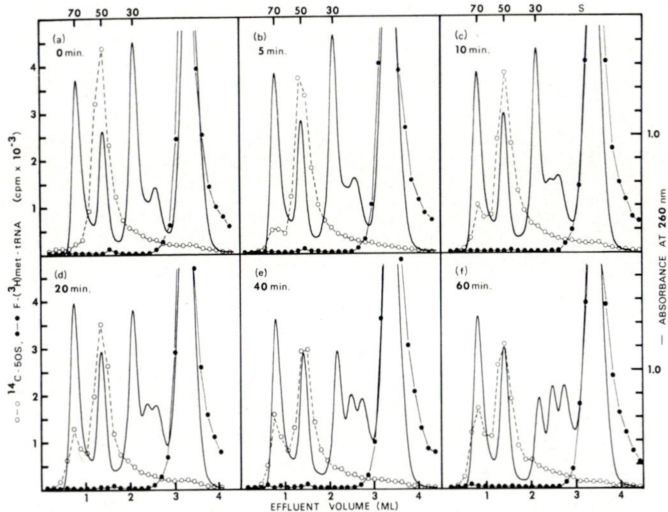

Fig. 8a—f. Rate of subunit exchange of tight couples in the absence of initiation factors. The incubation mixtures contained non-radioactive pure tight couples, ^{14}C-50 S subunits and all the components for chain initiation except initiation factors. The peak near 30 S corresponds to intact R17 RNA; during incubation it is slowly degraded into 21 S and 14 S fragments. No binding of fMet-tRNA to the 70 S peak is observed in the absence of initiation factors. The exchange rate was the same when mRNA and fMet-tRNA were omitted. The Mg^{2+} concentration in the gradient was 15 mM. Data from [10, 12]

formation of radioactive couples and hence to a large 0-time background and corresponding error. The progress of exchange over a 60 min period is illustrated in the series of gradients in Fig. 8. The purity of the particles, *i.e.*, the absence of contaminating 30 S subunits, is borne out by the fact that at 0-time no radioactive 50 S subunits are present in the 70 S peak.

The extent of exchange plotted in Fig. 9 follows closely the theoretical curve given by the solution of the differential equation describing the exchange process [10]:

$$\frac{(c_{70})_0 + (^*c_{50})_0}{(^*c_{50})_0} \frac{(^*c_{70})_t}{(c_{70})_0} = 1 - \exp\left(- \frac{(c_{70})_0 + (^*c_{50})_0}{(^*c_{50})_0} k_1 t\right). \tag{3}$$

The rate constants of dissociation

$$k_1 = \frac{1}{(c_{70})_0} \left(\frac{dc_{50}}{dt}\right)_{t=0} \tag{4}$$

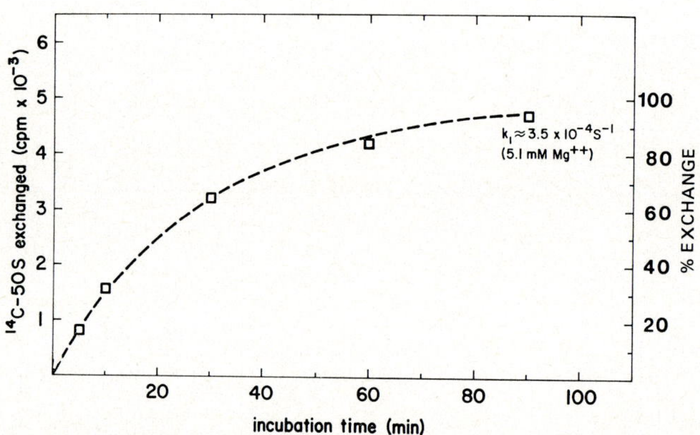

Fig. 9. Extent of exchange as a function of incubation time. The experimental points were determined from gradients similar to those in Fig. 8, except that in this experiment mRNA and fMet-tRNA were omitted. The k_1-value was determined from the slope of the initial rate or computed from the theoretical curve (dotted line) selected according to the best fit. Data from [10]

have been determined from the slopes of the initial rate of incorporation of ^{14}C–50 S subunits into ribosome couples, $(dc_{50}/dt)_{t=0}$, and the initial concentrations of vacant couples, $(c_{70})_0$.

The important point of Fig. 9 is that in the case of tight couples the rate of dissociation at 5 mM Mg^{2+} is rather slow, corresponding to a 50 % exchange time of 20 min. By contrast, loose couples exchange an order of magnitude faster.

As expected, the rate of spontaneous dissociation is strongly dependent on the Mg^{2+} concentration and increases exponentially as the Mg^{2+} concentration is lowered (Fig. 10). Thus, an about 30-fold rise in k_1 is observed as the Mg^{2+} concentration is reduced from 5 to 2 mM. It follows that information on the *effective* Mg^{2+} concentration present in a reaction mixture is essential if exchange rates measured under different conditions are to be compared.

The steep change in subunit interaction produced by small shifts in the Mg^{2+} concentration is a powerful device for control by amplification for, by covering or uncovering in this critical con-

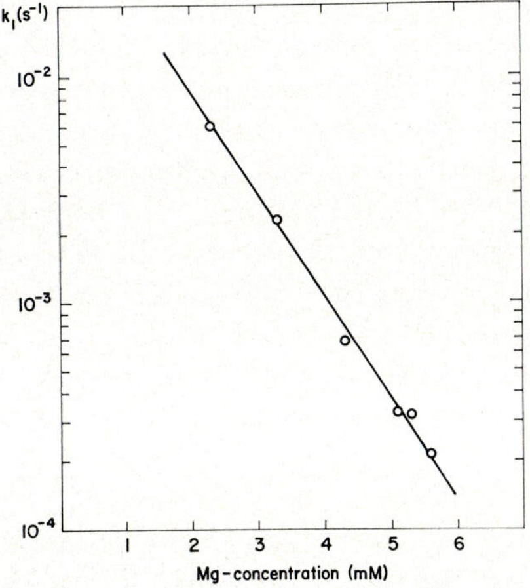

Fig. 10. Mg^{2+}-dependence of rate constant k_1 of dissociation. Data from [10]

centration range some of the Mg^{2+} binding sites with a suitable protein, a dramatic shift in the rate of dissociation or association could be accomplished. I shall return to this idea later in connection with our model of initiation.

7. Ribosome Heterogeneity: Tight and Loose Couples

Attempts to characterize ribosomes structurally without reference to function led to the concept of ribosome heterogeneity with all its tantalizing paradoxes as stated most forcefully by KURLAND [21]. The most trivial, and in a way most pleasing, resolution of this paradox would be the demonstration that active ribosomes that are all in the same functional state are homogeneous structurally. Our isolation of ribosomes that can all be converted into an initiation complex has made it possible to test this question.

By a reversal of what I have called CRICK's second dogma, viz. "If you don't understand function, study structure" [22], we have tried to understand ribosome structure from our knowledge of function.

Our observations that the ability of ribosome subunits to unite spontaneously is a stringent requirement for biological activity suggested that important insight into the mechanism of ribosome function might be gained by examining the structural features of both subunits involved in this interaction. Of most immediate interest was obviously the question of what determined the quality of this interaction. The answer to this question required a reasonably rapid quantitative or at least semi-quantitative test for scanning ribosomes with regard to the strength of this interaction. For this purpose we took advantage of the observation that "vacant" ribosome couples from bacteria, *i.e.*, ribosomes not engaged in translation and free of mRNA and peptidyl-tRNA, dissociate during centrifugation in gradients when the hydrostatic pressure exceeds a certain critical value [5, 13, 19, 23–25]. In particular, we exploited the finding that pressure-induced dissociation depends in opposite ways on the rotor speed and the Mg^{2+} concentration in the gradient. Using this technique in combination with our high-resolution density gradient analysis, we found that gentle inactivation of either subunit abolishes the ability for interaction in a stepwise rather than a gradual process.

Our analyses revealed that all ribosomes can be classified according to three major structural modifications of their subunits:

$$\text{30 S-a} \rightleftharpoons \text{30 S-b}, \quad \text{30 S-a} \rightarrow \text{30 S-c} \leftarrow \text{30 S-b} \quad (\text{30 S-c inactive})$$

$$\text{50 S-a} \rightarrow \text{50 S-b}, \quad \text{50 S-a} \rightarrow \text{50 S-c} \leftarrow \text{50 S-b} \quad (\text{50 S-c inactive})$$

Both subunits may occur in two modifications *a* or *b* which are either active or can be activated. Particles that have lost the ability to join and thus are irreversibly inactivated are defined as being in the c-state. Dissociation by exposure to low Mg^{2+} converts the 30 S particle into the b form that is inactive in translation because it cannot reassociate with any 50 S particles. Thermal activation according to ZAMIR et al. [26] restores both biological activity and the capacity to combine with 50 S subunits at the same rate. Typically, activation is accomplished by heating at 37 °C and

10 mM Mg^{2+} for 60 min. Thus, activation reconverts 30 S-b into 30 S-a subunits.

If 30 S-a subunits are mixed at 10 mM Mg^{2+} with 50 S subunits prepared by dissociation at low Mg^{2+} concentrations, two types of association products result: tight and loose couples. The

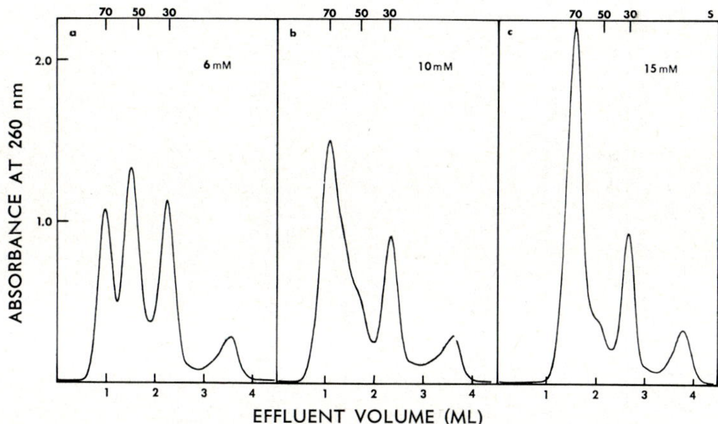

Fig. 11 a—c. Different stabilities of tight and loose couples revealed by centrifugation at increasing Mg^{2+} concentrations. Subunits prepared by dissociation of 70 S ribosomes at low Mg^{2+} concentrations were mixed in TMND buffer (10 mM Mg^{2+}, 160 mM NH_4^+, 50 mM Tris, pH 7.5, 10 mM 2-mercaptoethanol) at 0 °C for 15 min and the mixtures analyzed on gradients containing 100 mM NH_4^+ and 6 (a), 10 (b) and 15 mM Mg^{2+} (c). (The 30 S subunits had been heat-activated before storage at —60 °C.) The same results were obtained when the association buffer contained the same Mg^{2+} concentration as the gradient used for analysis. Centrifugation was at 30,000 rpm. Data from [9]

two populations of particles are revealed by low speed gradients (30,000 rpm) containing different Mg^{2+} concentrations (Fig. 11). At 6 mM Mg^{2+} we see, in addition to subunits, a sharp 70 S peak of the tight couples; at 10 mM Mg^{2+} the loose couples are just beginning to dissociate on the gradient, while at 15 mM Mg^{2+} the association products are nearly all sedimenting as a stable 70 S peak. The incipient dissociation of the loose couples at 10 mM Mg^{2+} and 30,000 rpm is nearly completed at the hydrostatic pressure generated at 60,000 rpm and the two populations of particles can now be identified by two peaks (Fig. 12) corresponding to about 68 S (tight couples) and 57 S (loose couples). Evidently, the 30 S-a subunits combine with two populations of 50 S subunits: 50 S-a particles that interact strongly to form tight aA couples or 50 S-b particles that engage in a much weaker union with their 30 S counterparts to yield loose aB couples[4]. If what appears as two

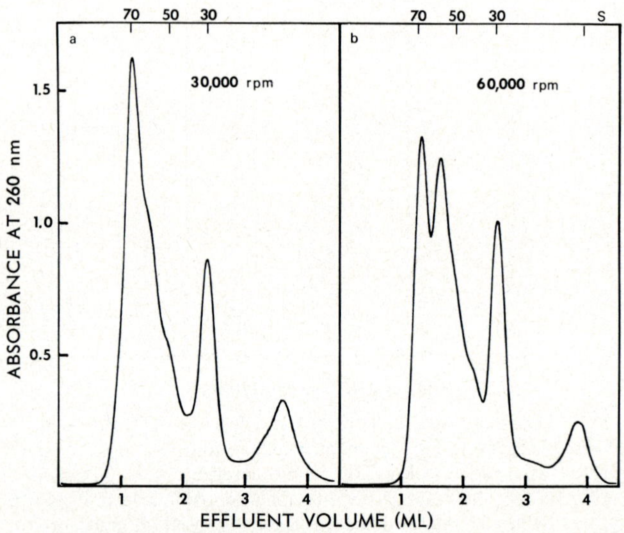

Fig. 12a and b. Different stabilities of tight and loose couples to pressure-dependent dissociation. Association mixtures prepared as indicated in legend to Fig. 11 were analyzed on gradients containing 10 mM Mg^{2+} and centrifuged at (a) 30,000 or (b) 60,000 rpm. Data from [9]

distinct types of particles by the stability criteria chosen is not simply the manifestation of an equilibrium mixture, reisolation should produce only one species. Indeed, 50 S-b particles isolated as free subunits from a 6 mM Mg^{2+} gradient will reassociate with 30 S-a subunits to give only the loose aB couples from which they were derived (Fig. 13b). The complete absence of tight couples is strikingly illustrated by superposition of the sedimentation pattern

[4] For the sake of convenience, we denote the states of the small subunit with a or b, those of the large subunit with A or B.

in Fig. 13b with that produced by reassociation of 30 S-a subunits with a mixture of 50 S-a and 50 S-b particles (Fig. 13a).

The observation that fresh extracts that had never been exposed to low Mg^{2+} concentrations contain only aA and no aB couples implies that loose couples are derived from the native tight couples as a result of damage suffered in the course of preparative proce-

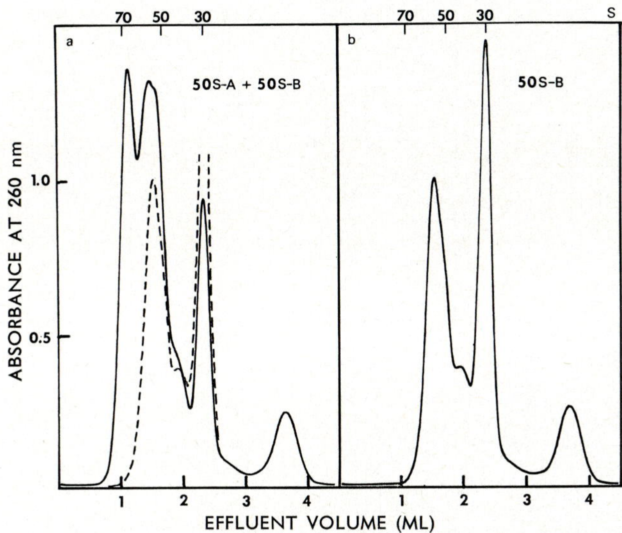

Fig. 13a and b. Homogeneity of loose couples after preparative separation from tight couples. (a) Association product of pre-activated 30 S subunits with a mixture of 50 S-a and 50 S-b particles. (b) Association product of 30 S subunits with 50 S-b subunits. The 50 S-b particles were prepared by collecting from a zonal rotor the 50 S particles that failed to associate with active 30 S subunits at 6 mM Mg^{2+}. The sedimentation pattern of loose couples (b) has been plotted as dotted line in (a) to show the absence of tight couples. Data from [9]

dures. Particularly harmful appears to be dissociation at low Mg^{2+} concentrations. However, the reformation of significant amounts of tight couples from subunits suggests that dissociation *per se* (or the structural changes in the 50 S subunit resulting from Mg^{2+} removal) is not responsible for the change to the b-form but rather facilitates conversion by exposing critical sites at the interface to irreversible thermal denaturation or to enzymatic damage by nucleases or proteases, or by facilitating the loss from the ribosome of a critical protein. Thus, free 50 S particles could be converted into the b-form with a half-time of about 60 min during incubation in buffer at 10 mM Mg^{2+} and 37 °C. This is apparent in Fig. 14a–d from the increase in loose couples at the expense of tight couples

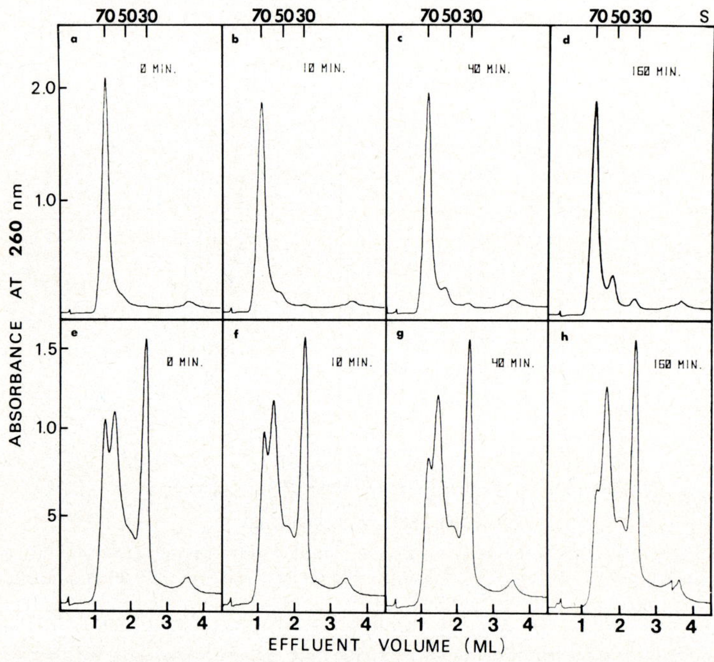

Fig. 14a—h. Conversion of 50 S-a to 50 S-b in free and associated state. Tight couples (a—d) or 50 S subunits (e—h) were incubated in TMND buffer at 37 °C for the indicated periods. The couples were then analyzed on sucrose gradients (a—d) containing 6 mM Mg^{2+} (5 h at 30,000 rpm), whereas the 50 S subunits were allowed to reassociate with 30 S-a particles at 0 °C and analyzed on sucrose gradients (e—h) containing 10 mM Mg^{2+} (1.2 h at 60,000 rpm). Data from [9]

observed in the association test with 30 S-a subunits. Similar treatment of vacant couples that had never been dissociated, on the other hand, produces loose couples at a much slower rate, as evident from the small proportion of subunits produced in a gradient containing 6 mM Mg^{2+} (Fig. 14a–d).

According to our scheme, the following four combinations of couples may exist: aA, aB, bA, bB. As already pointed out, the last two couples are not observable under the conditions of our studies because the b-form of the 30 S subunit is unable to interact with either form of the 50 S subunit. However, it has been reported previously [27] that so-called "interaction-free" subunits could be prepared that, although unable to interact with each other alone, could form couples upon incubation at 30 °C with both mRNA (poly U) and unacylated tRNA (HO-$tRNA_{phe}$). These couples, which contained one molecule of HO-$tRNA_{phe}$ and sedimented with a rate corresponding to about 60 S, could be converted quantitatively into a more stable 70 S complex by the EF-T-dependent binding of one molecule of phe-tRNA. Although it was clear that these interaction-free subunits were some sort of artifact, we exploited them to prove the 2-site model by rigorous stoichiometry. We further showed that this "primitive" 70 S initiation complex could be reconverted into a less stable 60 S post-translocation complex by incubation with EF-G and GTP [15].

We now find that these complexes formed from interaction-free subunits are bB couples. As shown in Fig. 15a–f, pure 50 S-b (a) and 30 S-b (b) subunits fail to join on incubation without (c) or with poly U (d), apparently because the thermal energy provided at 30 °C is not sufficient for the 30 S-b → 30 S-a conversion. However, with both poly U and HO-$tRNA_{phe}$ present, a 60 S complex is formed (e) which is converted into a 70 S complex by the enzymatic binding of phe-tRNA (f). The b-state of the 30S subunit in these complexes is deduced from the absence of a 60-70S peak in the sedimentation patterns (Fig. 15 c, d) after incubation and centrifugation of the subunits at 10 mM Mg^{2+}.

8. Functional Properties of Tight and Loose Couples

In order to examine the performance of tight and loose couples, we have tested them in five assays: (i) binding of tRNA-OH, (ii) rate of phenylalanine polymerization at limiting ribosome concentration, (iii) extent of initiation complex formation with R17 RNA, (iv) stability of 70 S phe-tRNA initiation complex during translocation, and (v) oligophenylalanine synthesis by spontaneous translocation.

The results of these experiments convey the general impression that with the conversion from tight to loose couples and finally to interaction-free subunits, the performance degenerates in the sense that the more exacting and sophisticated functions disappear first while the most primitive activities survive to various degrees. The

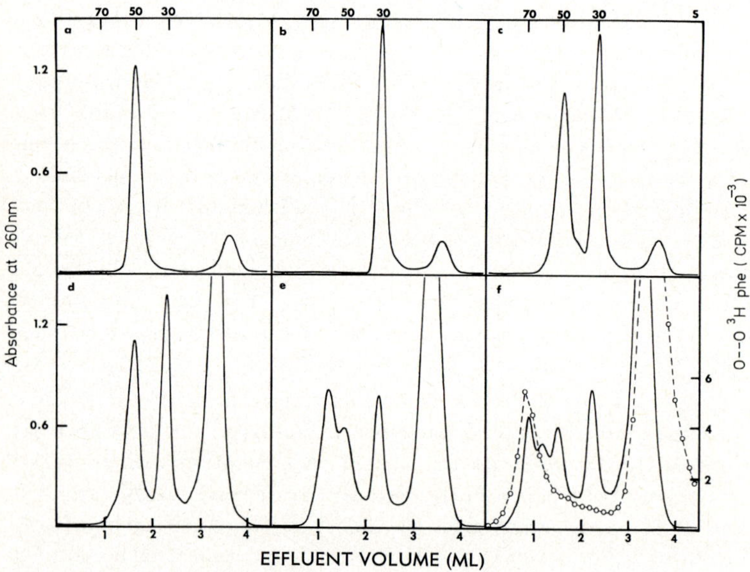

Fig. 15a—f. Failure of 50 S-b and 30 S-b subunits to form couples unless supplied with both messenger (poly U) and codon-specific uncharged tRNA ($tRNA_{phe}$). Gradient (a): 50 S-b subunits, (b) 30 S-b subunits, (c) 50 S-b plus 30 S-b subunits, (d) same as (c) plus $tRNA_{phe}$ or poly U, (e) same as (c) plus both $tRNA_{phe}$ and poly U, (f) same as (e) plus GTP, ^{3}H-phe-tRNA and EF-T. Experimental conditions as described by SCHREIER and NOLL [15]. Data from [9]

most striking example is the finding that interaction-free subunits (bB) are incapable of initiation with R17 even though they are still able to form an initiation complex with poly U and HO-$tRNA_{phe}$ (Fig. 16) and to synthesize polyphenylalanine (Table 1). Since, in contrast to primitive chain initiation, recognition of a natural messenger in its native state requires both IF-3 [7] and certain features

Table 1. Functional properties of tight and loose couples[a]

Type of particle or mixture of subunits	Binding of ^{32}P-tRNA$_{phe}$[b] (relative amounts bound to or trailing (t) behind peak in ± to ++++ or molecules / ribosome		Rate of poly U-directed phe polymerization[c] ^{3}H-phe incorp.	**EF-G-promoted Translocation** Stability of couples[d] before / after translocation		Binding of ^{3}H-phe-tRNA and tRNA$_{phe}$[e] before / after translocation		**Spontaneous Translocation**[f] mono-phe	oligo-phe
	— poly U	+ poly U	cpm	before	after	before	after	(%)	(%)
30 S-a	$+^t$	+++							
30 S-b	$\pm^t$	$\pm^t$							
aA	0.5	1.0	9000	stable	stable	tight	loose	82	18
aB	0.25	1.0	3500	stable	dissoc.	tight	loose	61	39
bB	$\pm^t$	$\pm^t$	1300	stable	dissoc.	tight	loose	61	39

[a] Data from [9].

[b] The components were incubated at 0 °C for 30 min and the mixture analyzed on sucrose gradients (10 mM Mg^{2+}, 60000 rpm).

[c] Polymerization was carried out with EF-T and EF-G according to SCHREIER and NOLL [15], ribosomes being limiting.

[d] The ribosome couples analyzed here are in the form of what we have called a primitive initiation complex containing phe-tRNA in the A-site and tRNA$_{phe}$ in the P-site [15, 27]. This complex is stable to dissociation under the conditions of centrifugation used. After treatment with EF-G and GTP only the aB and bB complexes become susceptible to dissociation by pressure.

[e] In what we define as tight binding the radioactive tRNA molecules cosediment with the A_{260} peak of the ribosome, whereas in loose binding the radioactivity trails behind.

[f] If reisolated initiation complexes containing phe-tRNA in the A-site and tRNA$_{phe}$ in the P-site are incubated at 37 °C without elongation factors and phe-tRNA, oligophenylalanine chains are formed slowly because of exchange and spontaneous translocation (Ref. [15] and unpublished results of SCHREIER and NOLL). Initiation complexes from aB and bB subunits form much more oligopeptides under the same conditions than those from aA couples.

of the 30 S subunit [28], it would be plausible if the impairment affects mostly some of these more subtle structures that have evolved later. The change of the interface structure reflected in the weakening of subunit interaction would likewise tend to make

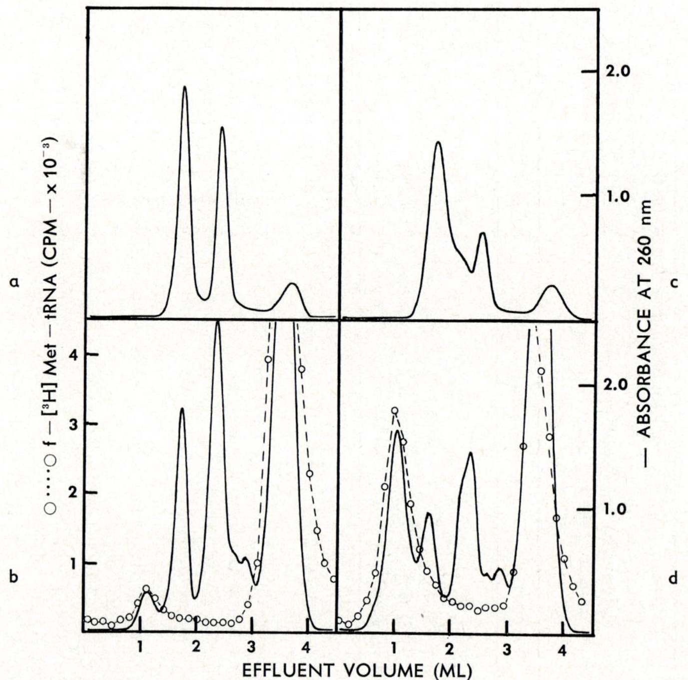

Fig. 16. Failure of 30 S-b subunits to initiate with natural mRNA (R 17 RNA). Mixtures of 30 S-b and 50 S-b particles (a) fail to associate upon incubation with the components of the R17 initiation system (b). After converting 30 S-b into 30 S-a by thermal activation, aB couples are formed (c) that are active in initiation (d). The residual initiation complex formed in (b) is explained by a 5—10% contamination of the 30 S subunits with 30 S-a particles. Incubation was at 30 °C to prevent conversion of 30 S-b to 30 S-a. Data from [65]

interaction with the other components of the active complex more sloppy. Indeed, the results in Table 1 show that both the binding of $tRNA_{phe}$ and the enzymatic control of translocation deteriorate. Apparently, the greater play of the subunits in their relative motions that results from the aA → aB conversion favors unspecific translocation [15]. In line with this interpretation is the finding that the increased susceptibility to pressure-induced dissociation associated with translocation (seen as a shift in the 70 S peak to 60 S in the case of bB and aB couples) is not observed with the tight aA couples although they exhibit the same weakening of tRNA binding upon incubation with EF-G and GTP (Table 1). The smaller shift observed by CHUANG and SIMPSON (29) suggests that their ribosomes consisted of a mixture of aA and aB couples.

The operationally defined terms "tight" and "loose" couples imply that susceptibility to pressure is a valid measure of the strength of interaction between subunits. I am afraid that physical chemists might consider this to be rather loose talk. Although our results imply that it is a useful reflection of ribosome stability, the dissociation on the gradient is obviously a rather complicated function of the affinity between the subunits. However, the direct and precise measures of affinity such as the dissociation constant at equilibrium or the rate of exchange require much more cumbersome experimental procedures. In any case, if dissociation by pressure is a meaningful gauge, we would expect that loose couples show a much greater rate of exchange than tight couples. In particular, we would predict that k_1 for loose couples in contrast to tight couples is still appreciable at 10 mM Mg^{2+}. That this is the case is suggested by an experiment in which aA and aB couples were compared at low (3.5 mM) and high (9.5 mM) Mg^{2+} concentration in regard to initiation with R17 RNA. We have shown previously that (i) dissociation into subunits is required for initiation with native R17 RNA and (ii) no measurable dissociation (or exchange) and hence no initiation is observed at 9.5 mM Mg^{2+} and 50 mM NH_4^+ with tight aA couples [7]. The results in Fig. 17 indeed show that while in the case of the aA couples initiaton is almost completely suppressed at high Mg^{2+}, the aB couples are able to maintain nearly half of their original initiation efficiency. The sedimentation pattern in Fig. 17c clearly shows that tight couples in contrast to loose couples fail to bind the R17 messenger at 9.5 mM Mg^{2+}. This example illustrates rather strikingly how drastically the quality of the ribosomes may alter their functional parameters.

Although the initiation experiment in Fig. 17 implies that loose couples exchange subunits much more rapidly than tight couples, we have so far not verified this by direct measurements. However, in other experiments we have found that the rate of subunit exchange is extremely sensitive to various treatments of ribosomes. Thus, washing tight couples with 0.5–1.0 M NH_4^+ more than doubles their rate of exchange without converting them into loose

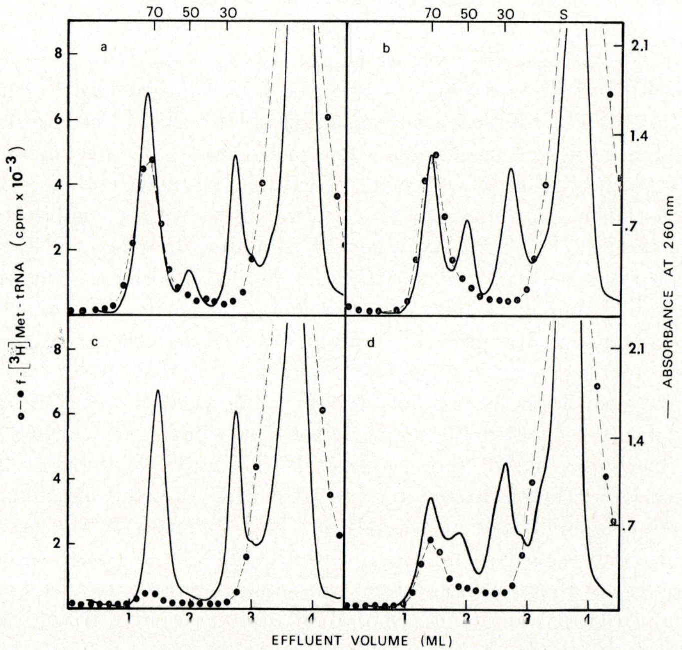

Fig. 17a—d. Different Mg^{2+} requirements of tight and loose couples in initiation with R17 RNA. Initiation complex formed with tight couples (a, c) or loose couples (b, d) at 3.5 (a, b) and 9.5 mM Mg^{2+} (c, d). Data from [65]

couples (HAPKE and NOLL, unpublished results). This is not in conflict with the apparent homogeneity of aA couples; it simply means that the rate of exchange is a more sensitive measure of subunit interaction than susceptibility to pressure-induced dissociation as defined in our assay and classification system.

9. Modification of the Subunit Interface by RNase and Proteases: Production of "Strange Particles"

So far we have not been able to pinpoint the molecular nature of the lesion responsible for the 50 S-a → 50 S-b conversion. It seemed conceivable that it was caused by lytic enzymes, since both RNase [10, 13, 30] and proteases [10] have been reported to bring about dissociation. However, treatment of mixtures of aA and a*B* couples with either free or fiber-bound pancreatic RNase failed to produce the changes characteristic of a 50 S-a → 50 S-b conversion.

If ribosomes are incubated with sufficiently low concentrations of pancreatic RNase or chymotrypsin to prevent extensive disintegration, a time-dependent irreversible dissociation into subunits is observed. The rate of this dissociation increases rapidly as the Mg^{2+} concentration is reduced (Fig. 18, 19) and corresponds

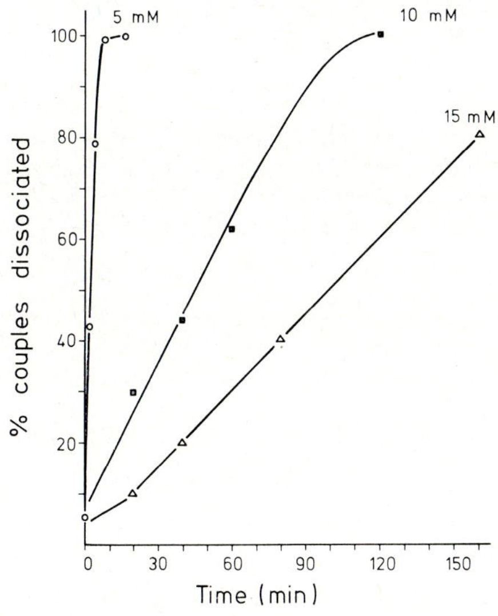

Fig. 18. Rates of RNase-induced irreversible dissociation of tight couples at various Mg^{2+} concentrations. Each point was determined from the relative peak areas in a sucrose gradient. aA couples that had been washed with 1 M NH_4Cl were incubated for the indicated periods at 37 °C in 50 mM Tris, pH 7.5, 160 mM NH_4^+, 10 mM 2-mercaptoethanol and the Mg^{2+} concentrations shown in the graph. The RNase concentration was 0.01 μg/ml. Data from [9]

to the rate of spontaneous dissociation. This suggests that the enzyme attacks the interface as it becomes exposed. When subunits were treated separately with fiber-bound RNase and then tested for reassociation with their untreated counterpart, the 50 S subunits were found to be much more susceptible to enzymatic

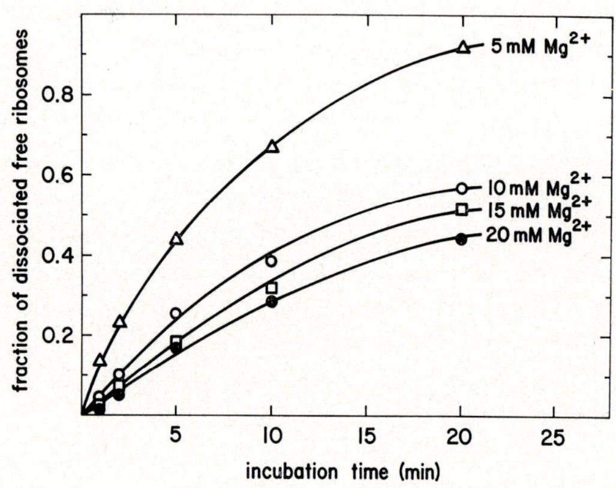

Fig. 19. Rates of chymotrypsin-induced irreversible dissociation of tight couples at various Mg^{2+} concentrations. 1 M NH_4Cl-washed aA couples from MRE 600 cells were incubated at 37 °C in 50 mM HEPES pH 7.6, 50 mM NH_4^+, 10 mM 2-mercaptoethanol, 0.2 μg/ml of chymotrypsin. Data from [10]

inactivation than the 30 S subunits. If a mixture of 50 S-a and 50 S-b particles are used, there seems to be a gradual disappearance of both tight and loose couples in the association product (Fig. 20a–d) and a corresponding decline of poly U-dependent polyphenylalanine synthesis (Fig. 20e–h).

There seems to be no correlation between the a and b state of the 50 S particle and changes in its 23 S rRNA detectable in sucrose gradients. Analysis of the RNA released with SDS reveals that most preparations of 50 S subunits contain variable proportions of the 17 S/15 S pair of fragments resulting from a break in the parent 23 S molecule. We have seen tight couples with nearly all the 23 S RNA broken and loose couples with nearly all the 23 S RNA intact. To give an example, a mixture of about equal parts 50 S-a and 50 S-b contained about 65% intact 23 S rRNA and 35% fragments (Fig. 21e). Very short exposure to fiber-bound RNase in the cold nearly completed the scission of 23 S RNA (Fig. 21f) without a corresponding reduction of tight couples in the association product (Fig. 21a, b). Continued incubation with RNase led to the progressive disappearance of tight

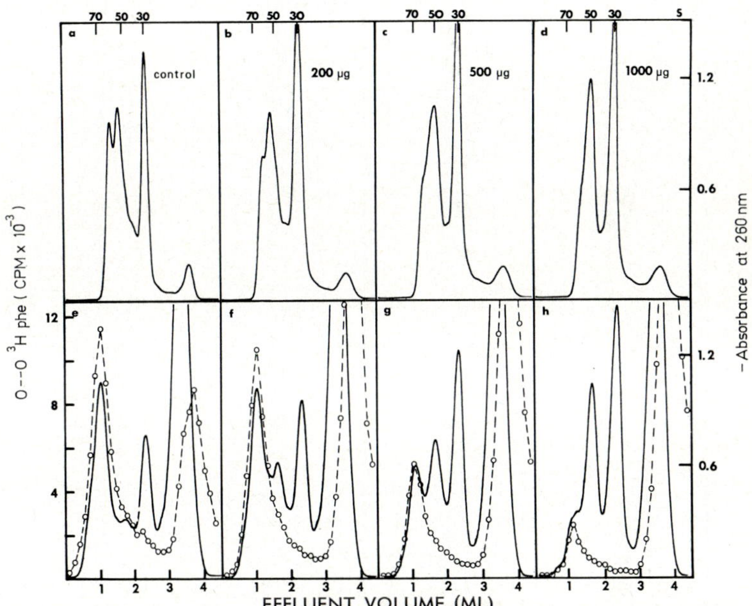

Fig. 20a—h. Inactivation of 50 S subunits with respect to couple formation (a—d) and poly U-dependent incorporation (e—h). The 50 S subunits were treated with 200—1000 μg/ml of fiber-bound RNase at 30 °C for 10 min. After removing the enzyme, samples were taken for association tests with 30 S-a subunits or for initiation assays with $tRNA_{phe}$, phe-tRNA, poly U and GTP. The gradients contained 10 mM Mg^{2+}. Data from [9]

couples and a broadening of the peak corresponding to loose couples (Fig. 21c, d) while there is no further apparent change in the sedimentation pattern of the rRNA (Fig. 21g, h). In contrast to the highly specific event responsible for the 50 S-a → 50 S-b conversion, mild treatment with RNase gradually reduces the affinity of the 50 S subunits for their 30 S partners until interaction and incorporation ability in response to poly U are abolished. Evidently, even the most gentle treatment with RNase reduces the affinity of the 50 S subunits for their 30 S particles to such an extent that they will dissociate over a broad range of conditions at which aB couples are still perfectly stable. We have called the broad

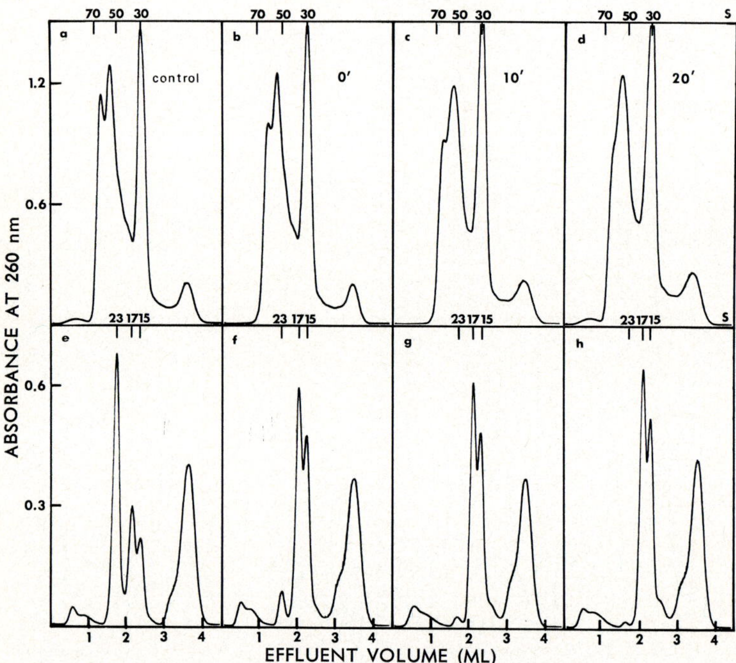

Fig. 21a—h. Formation of "strange" particles upon association of RNase-treated 50 S subunits with 30 S-a subunits. RNase treatment was as indicated in Fig. 20 except that the RNase concentration was constant (200 μg/ml) and the time varied as shown. Samples were treated with SDS to release ribosomal RNA and analyzed on sucrose gradients (e—h), other samples were tested for reassociation with 30 S-a subunits (a—d). Data from [9]

spectrum of loose and extra loose couples generated by RNase "strange particles" because they are not normally encountered in ribosome preparations.

It is rather intriguing that the drastic changes in subunit interaction resulting from mild endonuclease treatment of the 50 S subunit are not associated with breaks leading to major fragments.

It is equally interesting that the one major scission promoted by this treatment, *i.e.*, the break into the 17 S/15 S fragments, appears to have no influence on subunit interaction. Since fragments >50 nucleotides are detectable by our high resolution sucrose gradient analysis, it follows that the RNase-sensitive regions must be near the end of the 23 S or 5 S RNA. Preliminary experiments have not revealed any major breaks in 5 S RNA. At the moment our findings are best explained by postulating that the gradual weakening of subunit interaction produced by mild RNase treatment is the result of an accumulation of breaks in or the progressive removal of nucleotides from the ends of either 23 S or 5 S RNA. By contrast, the aA → aB conversion appears not to involve any nuclease action at all. Rather, it seems to be the result of a specific event at a structurally strategic site.

10. Analysis of Active Ribosomes and Subunit Interface by 2-D Polyacrylamide Gel Electrophoresis

The highly specific nature of the trauma responsible for the change from tight to loose couples pointed to the loss or damage of a specific ribosomal protein of the 50 S subunit. Several clues led us to take a closer look at the ribosomal proteins as the primary agents in subunit interaction.

(i) We consistently observed that crude preparations of initiation factors were capable of converting at least 85% of ribosomes into an initiation complex whereas pure factors gave conversion yields of at most 60%. To explain this difference we postulated that the population of vacant ribosomes was probably not homogeneous and some of them ($\cong$ 30%) might lack one or several structural proteins present in crude factors without which they are not competent for initiation [10].

(ii) Chromatography of crude initiation factors yielded a heat stable, trypsin-sensitive fraction that was highly active in stabilizing loose couples [10]. Incubation with this factor, which seems to resemble a preparation described by Garcia-Patrone et al. [31], results in an extensive conversion of loose couples into faster sedimenting particles resembling tight couples (Fig. 22).

Examination of the crude factors by 2-D PGE [32] revealed the presence of 22 well separated spots, 14 of which could be identified with known ribosomal proteins (L1, L6/11, L8/9, L13, L24, L25, L29, L30, L34 – S5, S8, S15, S16, S17) (Fig. 23). The purified chromatographic fraction containing the stabilizing or cou-

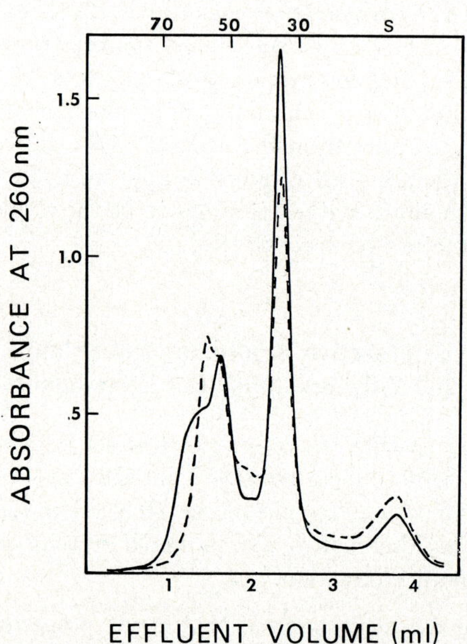

Fig. 22. Effect of coupling factor on sedimentation behavior of loose couples. Sedimentation pattern of aB couples incubated in buffer (dotted line) or buffer plus coupling factor (solid line). Data from [56]

pling factor produced 9 spots (Fig. 24), of which 4 were identified as L1, L8/9, L25, S15/16. Two additional spots correspond to either of the following 3 pairs: L6 – L11, L6 – S5, L11 – S5. The remaining three spots have not as yet been identified with any known components. Since L1 was by far the most prominent spot, its identity with the stabilizing activity is an intriguing possibility that we are now testing.

The observation that practically all ribosomes are active when fortified with ribosomal wash rich in ribosomal proteins suggests that some or most of the fractional proteins listed by KURLAND are not present in whole copy amounts because of losses during the isolation procedure. We sought further information on this point by comparing the quantitative composition of inactive 30 S-c particles with that of 30 S-a particles that are part of n60 S couples.

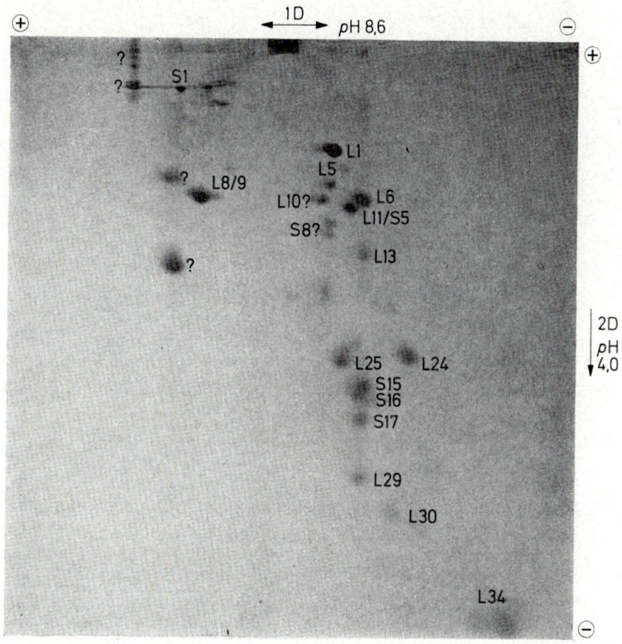

Fig. 23. 2-D polyacrylamide gel electropherogram of crude initiation factors [10]. The spots corresponding to known ribosomal proteins are identified by their numbers according to WITTMANN [35]. Data from [32]

The results in Table 2 reveal significant increases in the following proteins: S5, S6, S9/11, S10, S18 and S20. In view of the fact that 30 S-c particles have lost their ability to interact with 50 S subunits, it would not be surprising if some of these proteins are located at the interface. Indeed, independent evidence points to the location of S9, S11, and S20 at the interface.

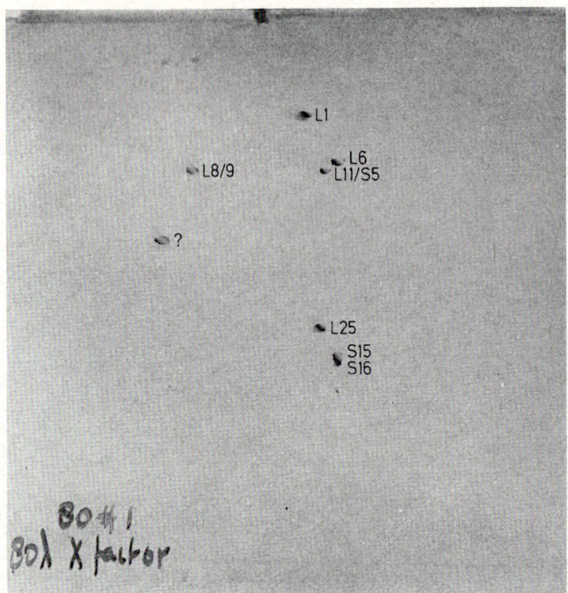

Fig. 24. 2-D PGE of purified ribosomal wash containing coupling factor activity. Data from [56]

Another protein, S12, is more than twice as abundant in our preparations in comparison to published values. Several laboratories exploited the finding that the uncoupled EF-G-dependent GTPase requires both subunits for activity [33, 34]. Thus, HIGHLAND et al. [34] compiled a list of interface proteins required for coupling by using specific antibodies directed against pure ribosomal proteins to test which monovalent Fab fragments would inhibit both GTPase activity and the ability to form couples (Table 3). In their stringent test both S9 and S11 were found to be necessary for subunit interaction. On the other hand, S11 and S12 have been shown to interact with the 23 S rRNA of the 50 S subunit [35]. Since S12 has been shown to be responsible for the binding of streptomycin, the postulated role of this protein in subunit interaction would also explain the increased resistance to pressure-dependent dissociation observed after exposing vacant couples to streptomycin (M. NOLL and H. NOLL, unpublished observations). Finally, the two pro-

Table 2. Average copy numbers of proteins in active and inactive preparations of 30 S ribosomes[a]

Type of particle	Kurland[b] K	W	D and N	30 S-c[c]	30 S-a[d]
Protein					
S 1	0.14	0.10	0.37	0.44	0.24
S 2	0.55	0.40	0.68	0.56	0.66
S 3	0.71	0.75	0.64	0.58	0.69
S 4	0.89	1.20	0.89	0.88	0.84
S 5	0.80	1.20	0.89	0.86	*1.24*
S 6	0.80	0.10	0.67	0.56	*0.89*
S 7	0.89	1.00	0.77	0.75	0.71
S 8	0.90	1.00	0.94	0.77	0.92
S 9+11	1.46	1.55	1.41	1.16	*1.68*
S 10	0.79	0.60	0.85	0.51	*1.05*
S 12	—	0.20	0.48	0.44	0.47
S 13	—	0.70	0.74	0.60	0.81
S 14	0.52	0.40	0.46	0.52	0.49
S 15+16+17	—	3.10	1.92	1.98	2.43
S 18	0.73	0.35	0.49	0.28	*0.56*
S 19	0.60	0.50	0.57	0.42	0.55
S 20	0.61	0.80	0.59	0.47	*0.67*
S 21	0.34	0.12	0.45	0.29	0.40

[a] The average copy numbers were determined from the amount of radioactivity recovered after cutting out the spots and burning the samples in a Packard sample oxidizer. The ribosomes had been labeled with ^{3}H and ^{14}C. The details of the procedure will be published [32].

[b] Analyses of ribosomes prepared according to Kurland [20] in our laboratory (D and N) are compared with data published by Kurland (K) [21] and by Weber (W) [64].

[c] Isolated 30 S particles that are unable to combine with active 50 S subunits even after heating.

[d] Data from 2-D PGE performed on tight (aA) couples.

teins S20 and L26 are apparently identical and occur in both subunits because during dissociation this protein can go either way [36].

Other analyses of vacant couples suggest that all 30 S subunit proteins are present in whole copy numbers with the exception of S1, S14, S19 and S21. This result would be compatible with the assembly map of Mizushima and Nomura [37] in which these proteins occupy peripheral positions. These findings together with analyses of the 50 S subunit suggest that Kurland's riddle of the fractional proteins is gradually vanishing, as our techniques of analysis and of preparing active ribosomes are improving.

Table 3. Effect of antibody binding on subunit association and GTP hydrolysis[a]

Antibodies (Fab fragments against ribosomal proteins)	% 70 S Ribosomes formed	% GTP (^{32}P) hydrolyzed
None	100	100
non-immune serum	111	106
L 15	107	102
S 5	100	110
L 14	34	25
L 19	12	$<$ 1
L 23	21	10
L 27	40	34
S 9	42[b]	30
S 11	1	3
L 7	102	$<$ 1

[a] Data from [34].

[b] At higher Fab titers subunit association could be completely prevented (J. Highland, personal communication).

11. Mechanism and Control of Initiation: A Model[5]

To conclude my survey, I am returning to the mechanism of initiation because many of the apparent inconsistencies that have confused and continue to confuse this issue can now be discussed and explained on the basis of the structural qualities of the ribosome.

The detailed molecular understanding of initiation is essential in order to answer central questions like: what makes a messenger translatable? Or, what possible translational controls are operating at the ribosomal level? It is convenient to analyze the recognition process between ribosomes and mRNA in terms of (a) the structural dynamics of the ribosome, (b) the structure of mRNA, and (c) the role of initiation factors.

---

[5] The model and the experiments on which it is based have been described previously: for details, consult Ref. [7, 10].

A general model derived from our experiments with tight ribosome couples and R17 RNA is shown in Fig. 25. From our studies on the structural dynamics of the ribosome discussed in the foregoing sections, we conclude that the ribosomes most closely resembling those active in the cell are tight couples, nearly all of which are potentially competent in initiation. At our standard conditions of 5 mM Mg^{2+} and 50 mM NH_4^+, nearly all active particles are in the associated state as indicated by the dissociation constant

$$D_e = k_1/k_2 = 10^{-5}\,(c_{70})_0 \tag{5}$$

and the rate of spontaneous dissociation is fairly slow:

$$k_1 = 3.5 \times 10^{-4}\,s^{-1}. \tag{6}$$

It follows that if initiation requires dissociation, a formidable energy barrier would have to be overcome and the process would be slow if it depended on the spontaneous dissociation rate k_1.

Our theory postulates two basic mechanisms of chain initiation: the first mechanism requires dissociation of the ribosome into subunits and participation of initiation factors IF-2 and IF-3, whereas the second mechanism involves direct attachment of 70 S particles to the messenger without dissociation and without a requirement for initiation factors. The operation of the two mechanisms is illustrated in Fig. 25 with R17 as messenger. In its native state (shown on the left), extensive secondary structure keeps the initiation sites of the synthetase (P) and A-protein (A) closed and inaccessible to the ribosomes, for example by base pairing with sequences of other regions of the messenger [38]. In our example the location of these regions is chosen in such a way that after the start of protein synthesis at the coat protein cistron, ribosomes moving along the messenger open first the polymerase and then the A-protein cistron to translation.

Which of the two mechanisms applies is governed by two simple rules:

(i) Ribosomes will attach with *low efficiency* and *without dissociation* to any open stretch of mRNA that is at least 30 nucleotides long [16]. This is the general mechanism encountered in initiation with synthetic messengers such as poly U. By binding of the cognate tRNA-OH to the P site, a primitive initiation complex is formed and translation may start at any triplet with the

EF-T-mediated insertion of the aa-tRNA specified by the triplet in the A site. However, such starts are often abortive owing to the instability of these complexes during the first translocation [15]. Attachment to an AUG codon, whether in phase or out of phase, is favored in the presence of fMet-tRNA because it results in a

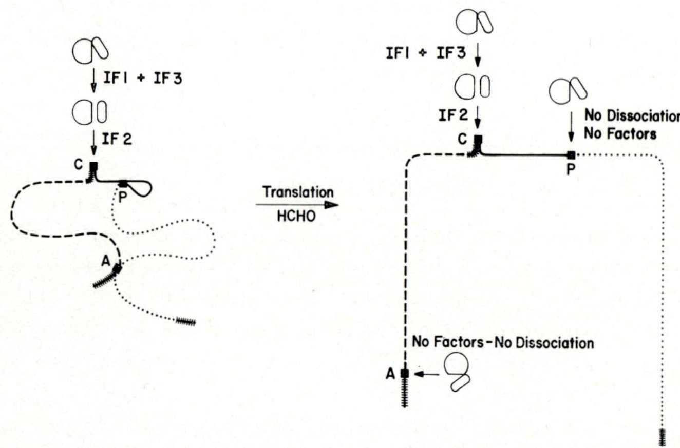

Fig. 25. Overall scheme for mechanism and control of initiation in the translation of RNA from R17 family. For explanation see text. From Nature New Biol. [7] with permission

more stable complex. Initiation at an AUG triplet exposed within a single-stranded region may occur by direct attachment of 70 S ribosomes to mRNA and does not require initiation factors.

(ii) Ribosomes attach preferentially to AUG starting triplets that are present at the end of a loop. This attachment requires both dissociation of 70 S ribosomes into subunits and the participation of IF-3. We may visualize that these loops are too bulky to be inserted into the cleft between the subunits of ribosome couples. On the other hand, 30 S subunits containing IF-3 and fMet-tRNA may recognize an AUG triplet at the end of a loop and become attached by opening up the loop in an action that might be compared to that of the σ-factor in opening up double-stranded DNA for transcription.

If these rules are applied to the translation of R17 messenger, the following picture emerges (left hand side of Fig. 25). Initiation of factor-dependent translation starts with the attachment of a 30 S subunit to the coat protein initiation site (C). Movement of the ribosomes along the messenger successively opens the synthetase (P) and A-protein (A) sites, both of which are not at the end of a loop and hence not recognized by IF-3. Initiation at these sites is

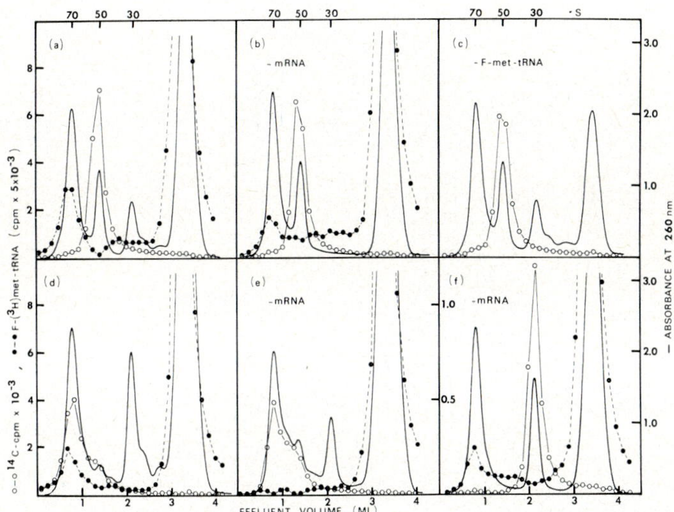

Fig. 26a—f. Formation of 70 S initiation complex without subunit exchange after unfolding R 17 RNA with formaldehyde. Pure ^{14}C-50 S subunits were incubated with pure tight couples, f-^{3}H-Met-tRNA and formaldehyde-treated R 17 RNA in the absence of initiation factors at 10 mM Mg^{2+} and 37 °C for 20 min. Controls (b, e, f) lacked R 17 RNA, in (c) fMet-tRNA was omitted; (f) was as (b) except that ^{14}C-30 S subunits were used to show absence of exchange under conditions of optimal peak separation. In (d) and (e) ^{14}C-50 S subunits were mixed with 30 S subunits instead of pure tight couples. The reaction mixtures were analyzed on sucrose gradients as described previously [7, 10]. From Nature New Biol. [7], with permission

less efficient and occurs by direct binding of 70 S ribosomes without initiation factors. Evidence for this mechanism comes from our experiments in which the secondary structure of the messenger was opened up by mild formaldehyde treatment [39]. We found that attachment of ribosomes to the four additional sites opened up by formaldehyde required neither initiation factors nor dissociation into subunits (Fig. 26).

Our model for the detailed mechanism of the factor-dependent initiation is shown in Fig. 27. At the Mg^{2+} concentration optimal for factor-dependent initiation (3 mM), vacant tight couples are present mostly in the associated state, although small changes in the binding energy will produce relatively large shifts in the position of the equilibrium (Fig. 7). This labile state may be visualized as a dynamic opening and closing of the subunits as if they were joined by a hinge (a). Binding of IF-1 to the 30 S subunit of the 70 S ribosomes (b) enhances their rate of thermal dissociation (c).

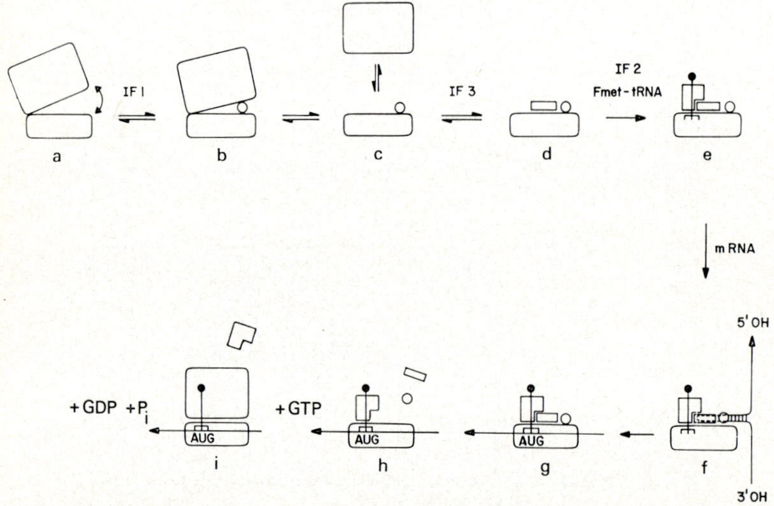

Fig. 27. Detailed model for mechanism of factor-dependent initiation at coat protein initiation site of R17 RNA. From Nature New Biol. [7], modified as indicated in text

The effect of IF-1 has been measured (i) by an increased rate of subunit exchange when added to pure tight couples in the presence of a radioactive subunit (Fig. 9), (ii) by an identical increase in the rate of initiation when added to an otherwise complete initiation mixture (Fig. 28), and (iii) by an increased rate of net dissociation in the presence of IF-3 (Fig. 29).

Binding of IF-3 to 30 S subunits (Fig. 27d) shifts the equilibrium toward partial dissociation. Although IF-3 produces net

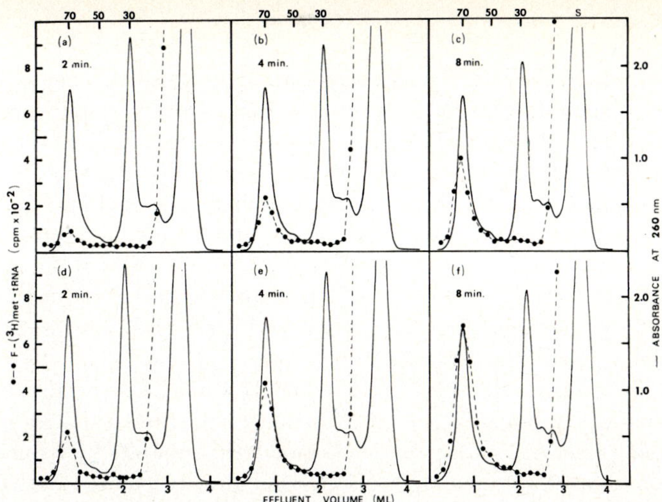

Fig. 28a—f. Rates of initiation in presence and absence of IF-1. Pure tight couples were incubated at 37 °C and 4.5 mM Mg^{2+} in the absence (a—c) or in the presence (d—f) of IF-1 plus all other components of the R17 initiation system. From [10]

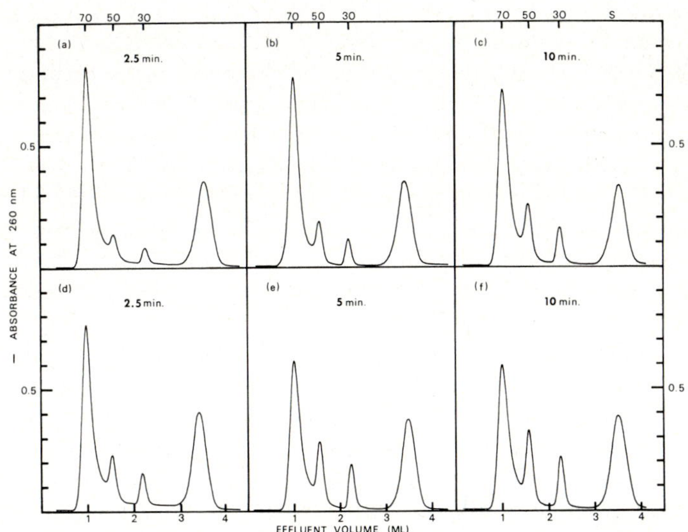

Fig. 29a—f. Initial rates of dissociation of pure tight couples in the presence of stoichiometric amounts of IF-3 (a—c) or IF-1 plus IF-3 (d—f) at 37 °C and 4.3 mM Mg^{2+}. From Nature New Biol. [7] with permission

dissociation, the rate at which dissociation occurs is the same as the rate of spontaneous dissociation k_1 measured in exchange experiments. Thus at 5 mM Mg^{2+} IF-3 acts merely as an anti-association factor (f. footnote 6), k_2, whereas IF-1 functions as a true dissociation factor by increasing k_1 (and probably decreasing k_2). Although the possibility that IF-3 is an anti-association factor has been considered by others [40, 41], it should be pointed out that valid experimental support for this claim requires rate measurements of exchange and dissociation with pure ribosome particles of defined structure.

The cooperative effect of IF-1 and IF-3 may be visualized as follows. IF-1 binds to the 30 S portion of the 70 S particles and lowers the energy of interaction between the subunits by reducing the area available for Mg^{2+} bridges. As already pointed out, the effect on dissociation of a slight shift in the effective Mg^{2+} concentration within the critical range of 3–5 mM would be greatly amplified because of the steep Mg^{2+} dependence of he dissociation curve below 3 mM (Fig. 7).

All steps so far are reversible and the equilibrium favors the vacant ribosome couples under physiological conditions. Subsequent attachment of the complex [IF-2 · fMet-tRNA] to the [30 S · IF-1 · IF-3] complex (Fig. 27e) strengthens the binding of IF-3 and hence shifts the equilibrium greatly toward formation of complex (e) that now acts as a trap for the messenger. Interaction with the messenger in the following step (f) involves recognition of the loop at the coat protein initiation site followed by the opening of the double-stranded loop as a result of the cooperative effects of IF-3 and fMet-tRNA (g). Initiation factors IF-1 and IF-3 now leave the 30 S subunit [42, 43] and the 50 S subunit joins the [30 S · IF-2 · fMet-tRNA] complex in a reaction that requires GTP and releases IF-2, possibly in form of a IF-2 · GDP complex (i). Chain elongation now begins with the binding of the first aminoacyl-tRNA followed by peptide bond formation.

In the original version, our model [7, 10] specified the release of IF-2 and IF-3 in form of a complex as reported by Groner and Revel [44]. However, since this finding has not been confirmed, it would seem more compatible with other evidence to assume that IF-2 is released later, during the joining reaction with the 50 S subunit.

Although initiation by way of an *obligatory* (30 S · mRNA · fMet-tRNA) complex [45] has become a generally accepted dogma, there never was sufficient evidence to support this claim. GUTHRIE and NOMURA [45] assumed that under the conditions of their experiments the 70 S complex would be stable in the absence of initiation. Hence, if initiation required dissociation and was carried out with density labeled 30 S*–50 S* couples in the presence of a very large excess of light 50 S subunits, fMet-tRNA should appear exclusively in the 50 S–30 S* hybrids.

Obviously, if subunit exchange was fast relative to the rate of initiation, exchange experiments cannot be used to resolve the question. In the paper cited, the control showing that no spontaneous exchange took place at 0° [45] is meaningless because even couples that exchange rapidly at 37° remain frozen at 0°. Furthermore, the light 50 S particles were not added in excess, and, consequently, a significant fraction of fMet label should have been associated with the dense 30 S*–50 S* couples. Even more serious is that their heavy 70 S ribosomes were contaminated with up to 50% of subunits. Hence, the total amount of fMet-tRNA (equivalent to about 4% of the input of heavy ribosomes) that was bound to "*hybrid 70 S particles*" may have arisen exclusively from the formation of initiation complexes by the added light 50 S with the contaminating heavy 30 S particles. Moreover, as pointed out by SUBRAMANIAN and DAVIS [25], the 60 S peak position is not sufficient evidence for the existence of the postulated hybrids because it coincides with that of vacant couples after pressure-induced dissociation.

If we accept for the moment that initiation can occur by way of a (30 S · fMet-tRNA · mRNA) complex, we have to distinguish between a mixed model in which initiation via 30 S complex competes with the direct conversion of vacant 70 S particles into an initiation complex and a model in which direct conversion is prohibited. For the case of initiation at the coat protein site of R17 RNA, we ruled out the mixed model by showing in exchange experiments with radioactive 50 S subunits (i) that each initiation corresponded to an exchange event and (ii) that stimulation of the rate of exchange by IF-1 resulted in an increase of the initiation rate by exactly the same factor (Fig. 8, 28, 30 and 31). Similarly, when exchange of tight couples was inhibited at 9.5 mM Mg^{2+}, no initiation was observed. Yet, at the same Mg^{2+} concentration loose couples could still initiate, presumably because exchange was possible (Fig. 17).

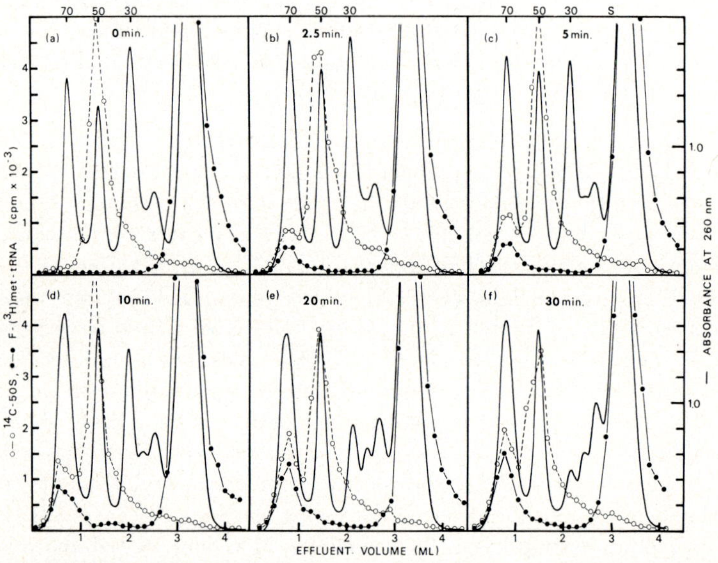

Fig. 30a—f. Time course of initiation and subunit exchange by pure tight couples in the presence of R17 RNA, f-^{3}H-Met-tRNA, ^{14}C-50 S subunits and crude initiation factors at 5 mM Mg^{2+} and 37 °C. Data from Nature New Biol. [7] with permission

Elucidation of the role of the initiation factors in the individual steps leading to the 70 S initiation complex as outlined in our model was greatly facilitated by our discovery that the so called 30 S initiation complex actually is a 40 S complex that can be well resolved from the unreacted components if the R17 messenger has not been degraded during incubation (Fig. 32). Formation of the 40 S complex was complete after 10 min at 37° and strictly dependent on IF-3 (compare Fig. 32d and e). While lack of a requirement for IF-1 (b) was expected, it was rather surprising that the 40 S complex formation was optimal in the absence of IF-2 as well (c). At 0° no complex was formed even with preactivated 30 S particles. The possibility that the 40 S peak was due to a 30 S dimer or some other artifact was ruled out by quantitative analysis of the complex formed with radioactive mRNA and fMet-tRNA. From the peak areas corresponding to absorbancy and radio-

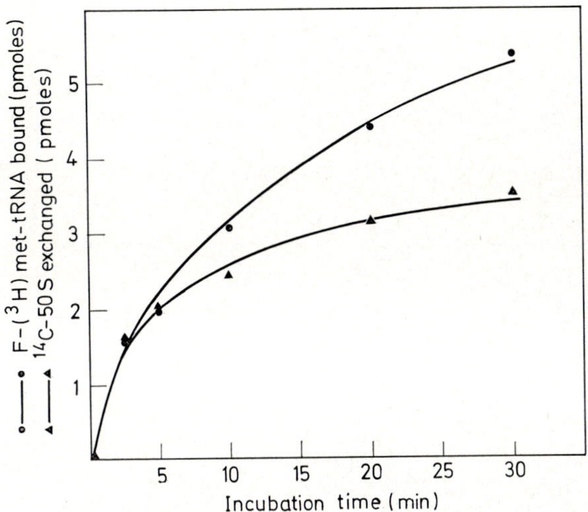

Fig. 31. Rate of subunit exchange and initiation complex formation as calculated from Fig. 30. While initially the rates are identical, the rate at which ^{14}C-50 S subunits appear in the initiation complex falls off more rapidly because the ^{14}C-50 S subunit pool is diluted progressively with non-radioactive particles originating from the initial exchange. Consequently, less than one ^{14}C-50 S particle is exchanged per initiation event. From [10]

activity, we computed that the complex contained per fMet-tRNA exactly one 30 S subunit and one intact R17 amB$_2$RNA. Moreover, addition of 50 S subunits in the cold converted the 40 S to a 76 S complex (not to a 110 S dimer) in a reaction that required GTP but no factors (Fig. 33a, d). The 76s complex was functional because the chain could be extended quantitatively to the amber triplet (coat protein hexapeptide) [10] and fMet could be released within 2 min by puromycin (Fig. 33b). It is interesting that the reaction of fMet with puromycin does not release the messenger, as evident from the persistence of the 76 S peak.

The appearance of a 76 S peak that is resolved from the 70 S peak of vacant couples at 15 mM Mg^{2+} in the gradient is observed only if the complex contains intact phage messenger. In the pres-

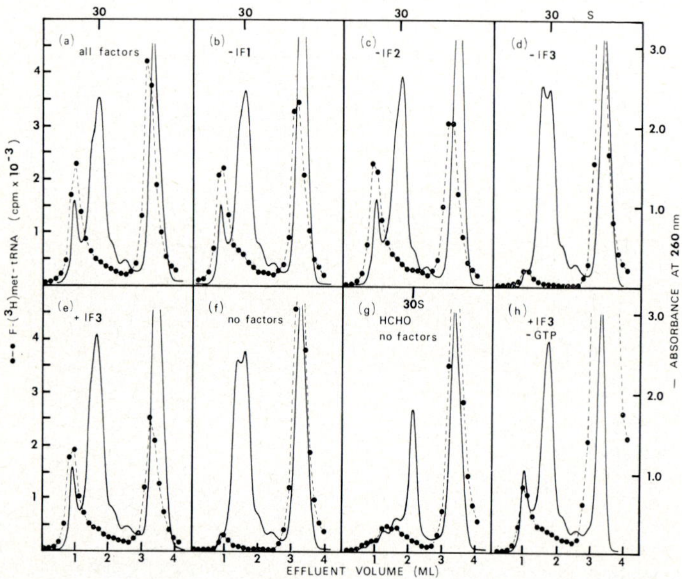

Fig. 32a—h. Requirement of initiation factors for formation of 40 S initiation complex with native and formaldehyde-treated R17 RNA. The incubation mixtures contained all the purified components for initiation, except as indicated in the individual panels. In the experiment corresponding to panel (e) IF-3 was the only initiation factor added. In (g) the messenger had been treated with formaldehyde. In (h) both GTP and IF 1 plus IF 2 were omitted and a tRNA-mixture free of GTP, charged with ^{3}H-Met and formylated was used, in all others formylated pure fMet-tRNA from Oak Ridge was the only tRNA added. In (g) the centrifugation time was reduced from 1.7 to 1.2 h to display the polysomes resulting from the attachment of several 30 S subunits to the R17 RNA messenger. From Nature New Biol. [7] with permission

ence of crude initiation factors, the messenger is partially degraded and the heterogeneous size of bound mRNA gives rise to a population of initiation complexes sedimenting anywhere between 70 S and 76 S.

Although it has long been known that a (30 S · mRNA · fMet-tRNA) complex can be formed, it has never been established rigorously that such a complex is an intermediate in initiation. Our experiments prove unequivocally that the 40 S complex is a precursor of the 76 S complex because it is obtained by addition of 50 S subunits and GTP under conditions (0°, absence of IF-2) that prohibit its formation from the individual components. That the 40 S complex is the precursor is also reflected in the fact that in

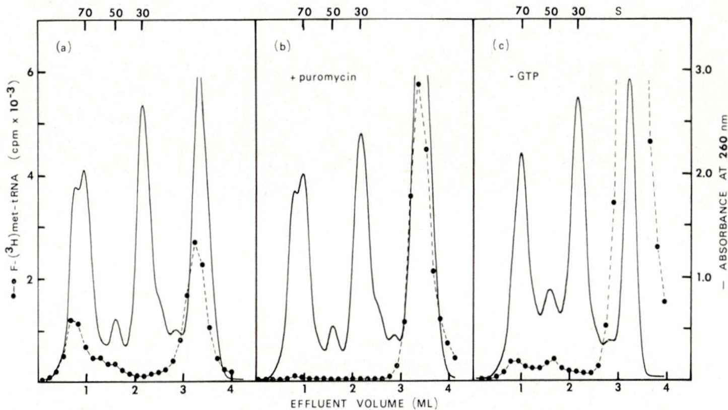

Fig. 33a—c. Effect of GTP on conversion of the 40 S initiation complex to the 76 S complex and release of fMet with puromycin. In (a) and (b) unwashed 50 S subunits were added in the cold to a 40 S initiation complex prepared in the absence of IF-1 and IF-2 as shown in Fig. 32e. (b) was made 1 mM with respect to puromycin and incubated for additional 2 min. In (c) unwashed 50 S subunits were added in the cold to a 40 S initiation complex prepared in the absence of GTP, IF-1 and IF-2 (Fig. 32h). The sucrose gradients contained 15 mM Mg^{2+}. Data from [10]

the cold the yield of 76 S complex corresponds to the much lower yield of the 40 S complex, despite the presence of an excess of competent subunits and the other components necessary for initiation. The 40 S complex is considerably less stable than the 76 S complex: it requires 15 mM Mg^{2+} in the gradient for stability and the yield of 40 S complex never exceeds about 30% of the 30 S particles, evidently because this corresponds to the equilibrium concentration.

The feature of our model that provoked a great deal of consternation was the proposal that fMet-tRNA binding precedes rather than follows the binding of the messenger [7, 10]. The key to our argument is the role of IF-2. As we have seen, IF-2 is needed neither for the formation of the 40 S complex nor for its conversion

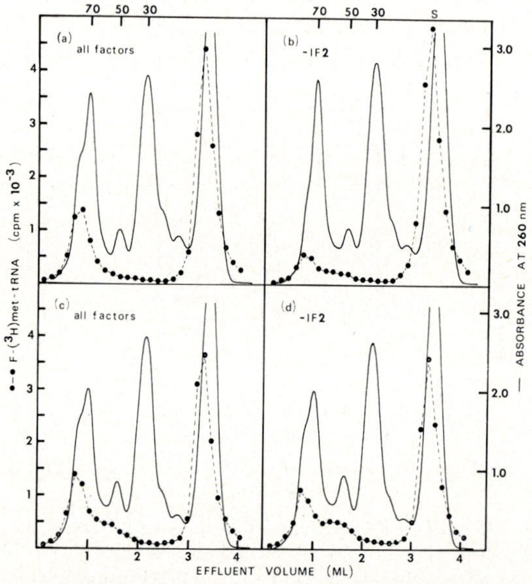

Fig. 34a—d. Role of IF-2 in initiation. Unwashed 50 S subunits were added either before (a, b) or after (c, d) incubation to reaction mixtures like those in Fig. 32a. In (b) and (d) the same experiment was carried out in the absence of IF-2. Sucrose gradient analysis (15 mM Mg^{2+}) after incubation for 15 min at 37 °C. Data from [10]

to the 76 S complex. It turned out, however, that IF-2 was required if 50 S subunits were present in the initiation mixture from the beginning (Fig. 34). In other words, in the absence of IF-2, 50 S subunits inhibit the formation of the 40 S complex by siphoning off the 30 S particles into vacant couples. This is a consequence of the fact that in tight couples the affinity of the 50 S-a particle for a 30 S-a subunit at 5 mM Mg^{2+} is several orders of magnitude greater than the affinity of IF-3 and hence rapidly displaces IF-3 from the 30 S subunit.[6] Hence, the only plausible explanation for the effect of IF-2 would be to postulate that attachment of IF-2, most likely in combination with fMet-tRNA, greatly enhances the binding of IF-3 (and thereby its own binding), either directly by sharing a common binding site or indirectly by allosteric induction. In short, IF-2 would change IF-3 from a poor into an efficient anti-association factor and thus shift the quilibrium toward formation of the 40 S complex. Since IF-2 by itself does not further enhance the IF-1/IF-3 promoted dissociation [10], we suggest that it acts in cooperation with fMet-tRNA, in agreement with earlier notions of the role of IF-2 as a binding factor for fMet-tRNA [47–49]. Binding of mRNA in the last step would further stabilize the complex which before addition of mRNA is demonstrable on millipore filters but does not survive centrifugation through a sucrose gradient, unless fixed with glutaraldehyde (M. Noll and H. Noll, to be published).

The existence of free 30 S subunits containing all three initiation factors (or at least IF-2 and IF-3) is easy to demonstrate: addition to an excess of pure vacant couples in the presence of mRNA and fMet-tRNA results in the conversion of all couples into 76 S initiation complexes by recycling of the factors [10]. The point of contention then is: starting with a [30 S · IF-1, 2, 3] complex, is fMet-tRNA or mRNA added first? Assuming mRNA was added first, the argument remains essentially the same. In order to explain why IF-2 relieves the inhibition by 50 S subunits, we have to postulate that it enhances the binding of IF-3. Since we know that IF-2 fails to do that by itself, at least to a sufficient degree,

[6] Examining the following two equilibria

$$70\,\mathrm{S} \underset{k_2}{\overset{k_1}{\rightleftharpoons}} 50\,\mathrm{S} + 30\,\mathrm{S} \text{ and } 30\,\mathrm{S}\cdot\text{IF-3} \underset{k_4}{\overset{k_3}{\rightleftharpoons}} 30\,\mathrm{S} + \text{IF-3},$$

we found

$$k_1/k_2 = (c_{50})_e\,(c_{30})_e/(c_{70})_e = 10^{-5}(c_{70})_0$$
$$k_3/k_4 = (c_{30})_e\,(c_F)_e/(c^F{}_{30})_e = 0.9\,(c_{30})_0;$$

where $(c^F{}_{30})_e$, $(c_F)_e$ denote the equilibrium concentrations of the [30S · IF-3] complex, of free IF-3 etc.

The terms on the right indicate the dissociation obtained when equimolar amounts of subunits or 30 S particles and IF-3 are mixed together. k_3/k_4 was calculated from published data on the binding of IF-3 to 30 S subunits [46].

we are again forced to invoke a cooperative effect, induced in this case by mRNA. This obviously does not make much sense because an (IF-2 + mRNA)-enhanced binding of IF-3 would not increase the proportion of 30 S particles capable of trapping messenger. It would be like saying you need to catch a mouse to set a mouse trap. Moreover, once the messenger is bound, there is no further need to prevent the joining of 50 S subunits since fMet-tRNA is readily bound to 70 S couples without dissociation as in the case of initiation with formaldehyde-treated R17 RNA. By contrast, in our model the absence of IF-2 would allow 50 S subunits to couple rapidly with 30 S particles by displacing IF-1 and IF-3.

There are other arguments of a more fundamental nature against the messenger-primed and in favor of our fMet-tRNA-primed model. Both IF-2 and fMet-tRNA have binding sites to interact with each other and with the 30 S subunit. The cooperative nature of these interactions should favor the formation of a complex even in the absence of mRNA. A similar case, of course, could be made for mRNA and IF-3. Yet, if in the first step the messenger were bound correctly to the 30 S subunit with the aid of IF-3, it is difficult to see why the subsequent positioning of fMet-tRNA by codon-anticodon interaction and binding to the 30 S subunit should require IF-2. Indeed, our experiments show that in the absence of 50 S subunits such a requirement does not exist. No matter how we turn the problem around, the conclusion seems inescapable that only models in which the fMet-tRNA precedes mRNA binding can explain adequately the conditional requirement for IF-2 as well as the reversal by IF-2 of the inhibitory effect of the 50 S subunits.

We have tried to determine the order of fMet-tRNA and mRNA addition from kinetic experiments. We found that the rate of complex formation was not shortened by preincubation of mixtures from which either messenger or fMet-tRNA had been omitted. Unfortunately, from this result we may only conclude that the second reaction is rate limiting but it does not tell us whether mRNA or fMet-tRNA binding is the rate limiting second reaction.

The last feature of our model that is at odds with previous notions is the role of GTP. Our experiments imply that GTP is required for an efficient joining of the 50 S subunit rather than for the binding of fMet-tRNA. For this reason, it would make more sense if GTP was binding directly to the 50 S subunit than by way of a (IF-2 · fMet-tRNA · GTP) complex. During the joining reaction fMet-tRNA might act as the trigger for the GTPase of the 50 S subunit. Hydrolysis of GTP would set off the "accomodation" [50] of fMet-tRNA to the P site with release of IF-2 and GDP.

One of the most crucial aspects of our model, if it can be generalized, is the central significance attributed to the secondary and tertiary structure of mRNA. According to this view, viral messengers (if not all messengers) are tightly coiled structures that are not accessible to the ribosomes except at one initiation site known to contain AUG at the end of a double-stranded loop [38]. Our proposal that it is the task of IF-3 to recognize and open up this loop has recently been taken up by KAEMPFER who claims evidence for such a mechanism in the eukaryotic system [51]. Serious experimental complications are likely to arise if masking of non-initiator AUG triplets by the tertiary structure of the messenger is the only safeguard against false starts. Thus, the exposure of additional initiator or non-initiator AUG triplets resulting from fragmentation of mRNA in the incubation mixture [52] might often make the interpretation of the results very difficult. For example, what might appear to be the effect of the addition of specific protein factors on translation might merely reflect the action of traces of RNase contained in the test fractions. Inefficient translation systems are particularly susceptible to this sort of artifact.

Our model resolves the conflict posed by the general claim that dissociation is a necessary condition of initiation [45] and the finding in the eukaryotic system that initiation of translation with poly U is possible at Mg^{2+} concentrations that prevent subunit exchange [53]. In experiments with formaldehyde-treated R17 RNA we have shown that initiation without dissociation is not a peculiarity of synthetic polynucleotides lacking AUG triplets but rather a consequence of unfolding the 3-D structure of mRNA. It should be noted, however, that initiation by direct attachment of ribosome couples to mRNA does not require initiation factors and is much less efficient than factor-dependent initiation via 30 S subunit. It is doubtful that initiation by direct attachment plays a significant role *in vivo*. A somewhat modified version of this mechanism may be important in the sequential translation of the genes of polycistronic messengers. The observation that R factor releases the polypeptide chain without releasing the ribosome from mRNA [7, 10] suggests that the ribosome, after loss of the uncharged tRNA in the P-site, can slide across the intercistronic region until it hits the AUG triplet at the beginning of the next cistron and is fixed by the binding of fMet-tRNA.

Initiation by direct attachment without factors probably reflects an earlier stage in evolution. It would make sense to postulate that initiator tRNA, initiation factors and the tertiary structure of mRNA all evolved together, replacing the inprecise and much less efficient primitive mechanism [27] of chain initiation. Because of the structural complexity of the ribosome, these later evolutionary modifications had to be added without changing the basic design of the ribosome core. It is not surprising therefore that under proper conditions the ribosome can be induced to function in its primitive mode.

How has our model fared since its publication [7, 10] more than a year ago? It is gratifying that most of its essential predictions have in the meantime been confirmed by direct experiments in a number of laboratories. Among these predictions are the cooperative nature of the binding of IF-1, 2, and 3 to the 30 S subunit [54, 55] and the enhancement by IF-1 of the IF-3-dependent dissociation [57]. Moreover, it has been confirmed that the GTPase activity, contrary to earlier views [58–60], is not associated with IF-2 but requires proteins L7 and L12 of the 50 S subunit [61]. Finally, the most startling feature of our model, namely the stipulation that binding of initiator tRNA to the small subunit precedes that of the messenger, has been dramatically demonstrated in the eukaryotic system by the isolation of a stable complex [62, 63] corresponding to the hypothetical intermediate (e) in our model (Fig. 27). Some of the remaining differences of interpretation, such as the question of the role of IF-2, may be related to differences in ribosome structure. For example, in loose couples that appear to predominate in the ribosome preparations used by most investigators, the greatly weakened interaction between the subunits would dispense with the need for IF-2 to prevent the premature association of subunits. On the other hand, the stabilizing effect of IF-2 on fMet-tRNA binding might become so critical with these ribosomes that initiation cannot proceed without this factor. We are confident that many of these seemingly conflicting results will vanish as more active and structurally better defined ribosomes are coming into general use. We are particularly hopeful that much new insight might be gained if we finally learn more about the mysterious trauma that converts tight into loose couples.

Acknowledgments

The authors are indebted to Dr. D. NOVELLI and the National Institutes of Health for a gift of purified $tRNA_f$ and to Mr. K. MÜLLER for preparing IF-3. We are particularly grateful to JANICE M. PETIT for cheerful help in typing the manuscript.

This work was supported by Research Grant No. P381F from the American Cancer Society and Grant No. 5R01CA11797 from the U. S. Public Health Service.

References

1. TISSIERES, A., WATSON, J. D.: Nature (Lond.) **182**, 778 (1958).
2. MANGIAROTTI, G., SCHLESSINGER, D.: J. molec. Biol. **29**, 395 (1967).
3. KAEMPFER, R., MESELSON, M., RASKAS, H. J.: J. molec. Biol. **31**, 277 (1968).
4. DAVIS, B. D.: Nature (Lond.) **231**, 153 (1971).
5. NOLL, M., HAPKE, B., SCHREIER, M. H., NOLL, H.: J. molec. Biol. **75**, 281 (1973).
6. LENGYEL, P., SÖLL, D.: Bact. Rev. **33**, 264 (1969).
7. NOLL, M., NOLL, H.: Nature (Lond.) New Biol. **238**, 225 (1972).
8. ZITOMER, R. S., FLAKS, J. G.: J. molec. Biol. **71**, 263 (1972).
9. HAPKE, B., NOLL, H.: Submitted for publication.
10. NOLL, M.: Thesis, Northwestern University, 1972.
11. NOLL, M., HAPKE, B., NOLL, H.: J. molec. Biol. **80** (1973)
12. NOLL, M., NOLL, H.: Submitted for publication.
13. NOLL, H.: Vierteljahresschr. Naturforsch. Gesell. Zürich **116**, 377 (1971).
14. SPIRIN, A. S.: In: Currents in Modern Biology, Vol. 2, p. 115. Amsterdam: North-Holland Publ. Co. 1968.
15. SCHREIER, M. H., NOLL, H.: Proc. nat. Acad. Sci. (Wash.) **68**, 805 (1971).
16. STEITZ, J. A.: Nature (Lond.) **224**, 957 (1969).
17. HOUSMAN, D., GILLESPIE, D., LODISH, H. F.: J. molec. Biol. **65**, 163 (1972).
18. DAHLBERG, A. E., DINGMAN, C. W., PEACOCK, A. C.: J. molec. Biol. **41**, 139 (1969).
19. VAN DIGGELEN, O. P., OOSTROM, H., BOSCH, L.: FEBS Letters **19**, 115 (1971).
20. KURLAND, C. G.: J. molec. Biol. **18**, 90 (1966).
21. KURLAND, C. G.: Ann. Rev. Biochem. **41**, 377 (1972).
22. NOLL, H.: In: Symp. Fundamental Cancer Research; Univ. Texas M.D. Anderson Hospital and Tumor Inst., Houston, Texas, p. 67. Baltimore: William & Wilkens Co. 1965.

23. Herzog, H., Ghysen, A., Bollen, A.: FEBS Letters **15**, 291 (1971).
24. Spirin, A. S.: FEBS Letters **14**, 349 (1971).
25. Subramanian, A. R., Davis, B. D.: Proc. nat. Acad. Sci. (Wash.) **68**, 2543 (1971).
26. Zamir, A., Miskin, R., Elson, D.: J. molec. Biol. **60**, 347 (1971).
27. Schreier, M. H., Noll, H.: Nature (Lond.) **227**, 128 (1970).
28. Lodish, H. F.: Nature (Lond.) **226**, 705 (1970).
29. Chuang, D. M., Simpson, M. V.: Proc. nat. Acad. Sci. (Wash.) **68**, 1474 (1971).
30. Kaempfer, R.: Nature **228**, 534 (1970).
31. Garcia-Patrone, M., Gonzalez, N. S., Algranati, I. D.: Proc. nat. Acad. Sci. (Wash.) **68**, 2822 (1971).
32. Van Dieijen, G., Noll, H.: to be published.
33. Marsh, R. D., Parmeggianai, A.: Proc. nat. Acad. Sci. (Wash.) **70**, 151 (1973).
34. Highland, J. H., Ochsner, E., Gordon, J., Bodley, J., Hasenbank, R., Stoeffler, G.: submitted for publication.
35. Wittmann, H. G., Stoeffler, G.: In: Bosch, L.: The Mechanism of Protein Synthesis and its Regulation, p. 295. Amsterdam: North-Holland 1972.
36. Wittman, H. G.: personal communication.
37. Mizushima, S., Nomura, M.: Nature (Lond.) **226**, 1214 (1970).
38. Min Jou, W., Haegeman, G., Ysebaert, M., Fiers, W.: Nature (Lond.) **237**, 82 (1972).
39. Lodish, H. F.: J. molec. Biol. **50**, 689 (1970).
40. Kaempfer, R.: Proc. nat. Acad. Sci. (Wash.) **68**, 2458 (1971).
41. Kaempfer, R.: J. molec. Biol. **71**, 583 (1972).
42. Hershey, J. W. B., Dewey, K. F., Thach, R. E.: Nature (Lond.) **222**, 944 (1969).
43. Vermeer, C., de Kievit, R. J., van Alphen, W. J., Bosch, L.: FEBS Letters **31**, 273 (1973).
44. Groner, Y., Revel, M.: Nature (Lond.) **219**, 232 (1968).
45. Guthrie, C., Nomura, M.: Nature (Lond.) **219**, 232 (1968).
46. Sabol, S., Ochoa, S.: Nature (Lond.) New Biol. **234**, 233 (1971).
47. Salas, M., Hille, M. B., Last, J. A., Wahba, A. J., Ochoa, S.: Proc. nat. Acad. Sci. (Wash.) **57**, 387 (1967).
48. Rudland, P. S., Whybrow, W. A., Clark, B. F. C.: Nature (Lond.) New Biol. **231**, 76 (1971).
49. Lockwood, A. H., Chakraborty, P. R., Maitra, U.: Proc. nat. Acad. Sci. (Wash.) **68**, 3122 (1971).
50. Thach, S. S., Thach, R. E.: Proc. nat. Acad. Sci. (Wash.) **68**, 1791 (1971).
51. Kaempfer, R., Kaufman, J.: Proc. nat. Acad. Sci. (Wash.) **70**, 1222 (1973).
52. Gesteland, R. F., Spahr, P. F.: Cold Spr. Harb. Symp. quant. Biol. **34**, 717 (1969).
53. Falvey, A. K., Staehelin, T.: J. molec. Biol. **53**, 21 (1970).

54. LOCKWOOD, A. H., SARHAR, P., MAITRA, U.: Proc. nat. Acad. Sci. (Wash.) **69**, 3602 (1972).
55. HERSHEY, J.: personal communication.
56. VAN DIEIJEN, G., NOLL, M., NOLL, H.: To be published.
57. GRUNBERG-MANAGO, M.: personal communication.
58. KOLAKOFSKY, D., DEWEY, K. F., HERSHEY, J. W. B., THACH, R. E.: Proc. nat. Acad. Sci. (Wash.) **61**, 1066 (1968).
59. CHAE, Y.-B., MAZUMDER, R., OCHOA, S.: Proc. nat. Acad. Sci. (Wash.) **62**, 1181 (1969).
60. LELONG, J. C., GRUNBERG-MANAGO, M., DONDON, J., GROS, D., GROS, F.: Nature (Lond.) **226**, 505 (1970).
61. HIGHLAND, J. H., BODLEY, J. W., GORDON, J., HASENBANK, R., STÖFLER, G.: Proc. nat. Acad. Sci. (Wash.) **70**, 147 (1973).
62. SCHREIER, M. H., STAEHELIN, T.: Nature (Lond.) New Biol. **242**, 35 (1973).
63. DARNBROUGH, C., LEGON, S., HUNT, T., JACKSON, R. J.: J. molec. Biol. **76**, 379 (1973).
64. WEBER, H. J.: Molec. gen. Genet. **119**, 233 (1972).
65. HAPKE, B., NOLL, M., NOLL, H.: Submitted for publication.

Discussion

B. MACH (Geneva): Thank you very much, Dr. NOLL, for your talk, not only for the interesting results, but also because you brought some ethics into the world of ribosomes. At least I have finally understood what fidelity of translation is. May I now open the discussion?

E. FUCHS (Heidelberg): How do you keep the tight couples?

H. NOLL (Evanston): One way to keep the tight couples is to have the Mg^{2+} concentration rather high. The loose couples are never formed, as I showed in my second last slide, unless you take the ribosomes apart into the subunits.

STARLINGER: If the loose couples have lost some proteins, how can you reactivate them? By heating?

H. NOLL: They cannot be reactivated by heat. I had to hurry a little bit in that section; the phenomenon is the following. The 30 S subunit is very sturdy, and even when you break the couples apart, the 30 S subunits have a tendency to assume the B conformation. By heating, you can quantitatively reconvert them to the A form. The really delicate partner is the 50 S ribosome subunit. It decomposes readily in a reaction we have not yet understood, and first goes to the B form capable of forming loose couples, and then to the C form which is totally inactive.

Translation II

Isoacceptor tRNA's

M. Staehelin

Research Department Pharmaceuticals Division, CIBA-GEIGY Ltd., Basle, Switzerland

The term "isoacceptor tRNA" is used for chemically different species of tRNA which are acylated by the same amino acid. It is clear that the degeneracy of the genetic code requires the presence of isoacceptor tRNA's. Analysis of tRNA by counter current distribution, MAK columns, BD-cellulose columns and reversed phase chromatography has shown, however, that the number of isoacceptor tRNA's is greater than would be predicted from the genetic code.

This opens the question whether functional roles other than recognizing the various codons can be ascribed to isoacceptor tRNA's An attempt will be made to discuss the structural variations of isoacceptor tRNA's in relation to known functional differences.

We were interested in the chemical differences between isoacceptor tRNA's that recognize different anticodons. For this purpose serine tRNA from rat liver seemed to be a good choice, since serine is the only amino acid with 6 codons differing in all three bases. In the reversed phase chromatographic system 4 (RPC4) rat liver serine tRNA is fractionated into three different species with coding specificities for UCU, UCC, UCA (I), UCG (II) and AGU, AGC (III) respectively. The tRNA profile looks quite different according to which chromatographic system is used. In the RPC 5 system serine tRNA II and III are not well separated. On the other hand, a new tRNA appears after serine tRNA I. This tRNA has also coding properties for UCA [1]. Thus, there are at least four different serine tRNA's in rat liver. We have purified and sequenced the species I and III. The other two tRNA's are not yet fully sequenced. We know, however, the anticodon structure and this clearly indicates that they are chemically different entities. Serine tRNA III differs from serine tRNA I by altogether 22 out of 86 nucleotides. These include an alteration of the three bases of the anticodon, 8 changes of base pairs in double helical regions, one alteration of a G-C into a G-U pair and two different modifications of major to minor nucleosides. These changes cover more or less the alterations which have been found in all isoacceptor tRNA's [2].

The anticodon region of the tRNA's is of special interest. From the degeneracy of the genetic code it is obvious that for some tRNA's isoacceptor species have to exist with different base compositions in the anticodon. In view of the fact that some tRNA's recognize different codons, Crick [3] had originally proposed the wobble hypothesis which postulated that a G-U pair can be functional in the third position of the codon and inosine in the anticodon can recognize U, C and A. With this in mind, it is interesting to look at reported sequences of tRNA's (Table 1). In *E. coli* only unmodified G and C residues have been found in the anticodon. In addition inosine, N-acetylcytidine, uridine-oxyacetic acid, methyl-amino-methyl-2-thiouridine and the nucleoside Q which appears to be modified guanosine occur at the 5′ site of the anticodon [4]. In eukaryotic cells (Table 1) several minor bases have also been found. One interesting finding is that of two thio-U derivatives which have been found in lysine and glutamic tRNA of rat liver [5], both tRNA's have been found to contain two species, one recognizing A and one recognizing G in the terminal position of the codon.

Apparently modification of the nucleotides in the anticodon is very important for codon recognition. In addition to the three bases of the anticodon, the base next to the 3′ end of the anticodon is also modified in a specific way. This modification seems, however, to be related only to the nature of the first base of the codon. Thus, in *E. coli* all tRNA's recognizing codons with U as the first base contain methylthioisopentenyladenosine, whereas those having A as the first base contain N-[9-(β-D-ribofuranosyl) purine-6-ylcarbamoyl]threonine [4]. Isoacceptors which differ only in recognition of the last base have, in the cases reported so far, always identical base modifications at the nucleotide next to the anticodon. The presence of this modified nucleotide next to the anticodon seems to be very important for coding recognition. The lack of methylthioisopentenyladenosine in tyrosine suppressor tRNA was

Table 1. Sequences of the anticodons and adjoining bases of some tRNA's

A. Prokaryotes							
Phe				Tyr			
GAA	ms^{2}i^{6}A			G*UA	ms^{2}i^{6}A		
		Ser					
		UacGA	ms^{2}i^{6}A				
						Trp	
						CCA	ms^{2}i^{6}A
Leu				His		Arg	
GAG	G*			QUG	m^{2}A	ICG	m^{2}A
				Gln			
Leu				s^{2}UUG	m^{2}A		
CAG	G*			CUG	m^{2}A		
Ile				Asn		Ser	
GAU	t^{6}A			QUU	t^{6}A	GCU	t^{6}A
Met f							
CAU	A						
ac^{4}CAU	t^{6}A						
Val				Asp		Gly	
GAC	A			QUC	m^{2}A	GCC	A
				Glu			
UacAC	m^{6}A			s^{2}UUC	m^{2}A	UCC	C
				CUG	m^{2}A		
B. Eukaryotes							
Phe		Ser		Tyr			
GmAA	Y	IGA	i^{6}A	GΨA	i^{6}A		
Leu						Trp	
m^{5}CAAm^{1}G		CGA	i^{6}A			CmCA	A
						Arg	
						ICG	A
Ile						Ser	
IAU	t^{6}A					GCU	mt^{6}A
				Lys		Arg	
						mcm^{5}UCU	t^{6}A
Met f				m^{5}s^{2}UUU	t^{6}A		
CAU	t^{6}A			CUU	t^{6}A		
Val		Ala		Asp			
IAC	A	IGC	m^{1}I	GUC	m^{1}G		
				Glu			
				m^{5}s^{2}UUC			

s^{2}U = mnm^{5}s^{2}U.

found to be correlated with an impaired function of the tRNA to bind to ribosomes [6]. Information about the role of the double-stranded regions comes mainly from mutant tRNA's [7, 8]. Mutants of suppressor $tRNA^{Tyr}$ with single base changes in the double-stranded regions of the clover leaf have led to deficiency in the suppressor function. Single base substitutions in the amino acid arm have led to temperature sensitive tRNA's. Mutation in the anticodon arm has led to tRNA with increased K_m for the synthetase and mutation in the DHU arm and the extra arm has led to a loss of function. Revertants by all those mutants leading to new but different base pairs have shown those tRNA's to be more or less fully functional.

The G-U base pairs have an interesting feature. Although a G-U base pair is apparently fully functional in the wobble position and G-U base pairs have been found in the double-stranded regions of many tRNA's, one case has been reported where the change from an A-U to a G-U base pair has led to an alteration in the function of tRNA. This is a $tRNA^{Trp}$ from *E. coli* where a A to G alteration in the stem of the DHU arm has led to a suppressor tRNA for UGA [9]. Apparently a small distortion of the molecule has altered it to the extent that it now binds to a suppressor codon. One important fact about mutations with incorrect base pairs is also that the altered tRNA is only made in very small amounts and that it lacks methylation of guanosine to O-methylguanosine. This seems to indicate that there is a defect of maturation from the precursor to the final tRNA. Surprisingly this defect cannot be overcome only by a reversion to correct base pairs but also by second mutation at other sites, *e.g.* in the DHU arm. Thus, tRNA's lacking correct base pairs can also be functional to some extent.

Another important site in the tRNA molecule which has become apparent by mutations of the suppressor tRNA's is the fourth base from the 3′ end. This was already implicated as an important recognition site by ROE and DUDOCK [10]. Mutations in suppressor $tRNA^{Tyr}$ have actually shown that a single mutation from A to G can cause mischarging [11, 12], *i.e.* the tRNA now accepts glutamine instead of tyrosine.

One of the sites where tRNA's differ most strikingly in structure and function is the TψC sequence which is common to all tRNA's and is involved in elongation in protein synthesis. This sequence is lacking in two types of tRNA with different functions, namely the methionine initiator tRNA from yeast [13] and the glycine tRNA's [14] from staphylococci involved in peptido-glycan synthesis. This suggests that the role for this region is binding of tRNA to the ribosomes.

Role of Isoacceptor tRNA's in Translational Control

Is there any evidence that isoacceptor tRNA's are involved in translational control? Evidence for this comes from various lines. Gilbert and Anderson [15] have developed an *in vitro* system for haemoglobin synthesis which was dependent on the addition of tRNA. In the presence of limiting amounts of tRNA only half as many α-chains were formed as β-chains. This would indicate that some tRNA's might be more rate limiting in the synthesis of the α- than of the β-chain. The system appears, however, quite complicated in that Rudloff and Hilse [16], for instance, have shown that of the two lysine tRNA's, the one coding for AAG is preferentially incorporated into the α-chain. With a modified system more effective in haemoglobin synthesis, however, Woodward and Herbert [17] have shown that in each chain 10 lysines are incorporated from the species recognizing AAG and 2 from the species recognizing AAA.

A second suggestion comes from the tRNA modification under T-even phage infection in *E. coli*. Under T4 infection 8 new species of tRNA are synthetized [18] and the appearance of changed chromatographic profiles for the following amino acids has been reported [19]: leucine, glycine, arginine, proline, isoleucine. All these tRNA's are coded for by the phage since the tRNA's hybridize with phage DNA. The most significant change, however, is the breakdown of one specific leucine tRNA of the host [20]. This occurs by a break of one inter-nucleotide bond in the extra loop in a tRNA which has the anticodon for CUG. The phage-coded leucine tRNA which is synthetized instead has a different anticodon [21]. It responds to UUA, a codon which is rather poorly recognized by *E. coli* tRNA [18]. Since the CUG-recognizing species is surprisingly the most abundant species in uninfected *E. coli*, this seems to suggest that the reading of a different message might require the presence of different tRNA's. The extent to which a tRNA is present in the cell does not necessarily indicate to what extent it is used. KANO-SUEOKA and SUEOKA [22] have examined the presence of the various leucine tRNA's in different fractions of the cell. Although the species recognizing CUG was the most abundant in the supernatant, it was present in polysomes to a much lesser extent than other leucine tRNA's. In mammalian virus infections the appearance of new species of tRNA have also been reported [16], but, unfortunately, coding properties have not been examined. The third line of evidence comes from specific changes of tRNA in tissues under different growth conditions. In chicken and rooster liver, treatment with oestrogens induces the formation of phosvitin, a protein of the egg, very rich in phosphoserine. MÄENPÄÄ and BERNFIELD [25] have shown that there is a marked increase in the synthesis of only two particular serine tRNA's. Thus, a specific function seems to be related to the appearance of a specific tRNA. Similarly, in granuloma tissue which shows increased collagen biosynthesis LANKS and WEINSTEIN [26] have shown an increase in the amount of proline tRNA.

Effect of tRNA Modification on Other Functions

Two cases have been reported where tRNA's function in a sense other than protein synthesis and modification of minor bases has altered this function. The first case is that of the histidine tRNA in *Salmonella typhimurium*. The studies of AMES and coworkers have shown that it is the charged histidyl tRNA which is responsible for the regulation of the histidine operon [27]. The operon involves 9 structural genes for proteins. Genetic analysis has shown that there are 6 different loci which can be affected to produce derepressed constitutive mutants. The function of some of the 6 genes has become clear in recent years. It is obvious that one of the genes is the operon where the repressor-operator region can be so changed as to prevent repression. But since it is not histidine but histidyl tRNA which is the repressor, it is also understandable that genes related to the formation of histidyl tRNA are involved. Thus, derepression occurs when modified synthetases are produced which have a much higher K_m for either histidine or tRNA. If the K_m of histidine is raised, repression is still obtained at very high histidine concentrations whereas in mutants in which the K_m for tRNA is raised no repression by histidine is possible anymore.

Gene R is apparently the structural gene for tRNA mutants producing a reduced amount (about 50 %) of histidine tRNA with the same sequence as the normal tRNA. The reduced amount produced is reminiscent of the mutants obtained by SMITH and ANDERSON [8] where some mutants with single base changes in the anticodon arm also led to reduced amounts of tRNA and where it was believed that the mutation might have affected the maturation from the longer tRNA precursor to tRNA.

An especially interesting class are the T mutants. They have the full complement of tRNA but they lack the two ψ residues near the anticodon [28]. The mutation apparently arises from a lack of the enzyme converting U into ψ in this region. ALLAUDEEN, YANG and SÖLL [29] have sequenced the leucine tRNA from the same strain and have shown that it also lacks the two pseudouridine residues near the anticodon. The ψ residue in the TψC loop was always present, suggesting that two different enzymes are involved in converting U to ψ at the two sites.

The lack of modification of U to ψ only effects the function of histidyl tRNA in repression. The protein synthetic function of the tRNA is hardly affected by the mutation and the mutants grow quite well with only a slightly increased doubling time. The tRNA has the same affinity for the synthetase and is fully charged *in vivo*. Thus, in the case of histidine tRNA of *Salmonella typhimurium*, tRNA modification might not primarily involve protein synthesis, but rather, other functions in the cell.

An intriguing case of regulatory function in isoacceptor tRNA's is that of the tyrosine tRNA's in *Drosophila melanogaster*. JACOBSON [30] has observed that in the vermillion mutant which is deficient in tryptophan pyrrolase, the enzyme activity can be restored by treatment with RNAase. The restored enzyme activity can again be inhibited by tRNA from either the wild type or the mutant. Analysis of the individual tRNA's fractionated by reversed phase chromatography indicated that two tyrosine tRNA's could be separated and that one of them was responsible for the inhibition of the enzyme. Apparently the structure of the enzyme is so altered in the mutant that, upon interaction with tyrosine tRNA, the activity is lost. The mutation can be suppressed by the Su $(s)^2$ gene. Twardzik et al. [31] have analyzed the tRNA's in Su $(s)^2$ mutants and the wild type. In mutants homozygous for the Su $(s)^2$ gene, only one tyrosine tRNA species was found and the peak of the other corresponding to the inhibitory tRNA was absent. This indicated that the suppressor activity is due to the lack of formation of the inhibitory species. Recently White et al. [32] have examined the tRNA from wild type and vermillion *Drosophila*. They have found that $tRNA^{Tyr}$, $tRNA^{His}$, $tRNA^{Asp}$ and $tRNA^{Asn}$ were all altered. The alteration of $tRNA^{Asn}$ could be ascribed to the modification in the nucleotide Q which in many tRNA's is present in the anticodon. The evidence available suggests that during the different growth stages alteration of the tRNA's take place and apparently there are quantitative differences in the extent of these modifications in the wild type and the vermillion mutant. The inhibitory activity of one species suggests again, as in the case of histidine tRNA in *Salmonella*, that tRNA can be involved in regulatory functions and that modifications of minor bases which have no effect on protein synthesis can drastically alter these regulatory functions. Thus, it seems possible that tRNA modification might play a role in other functions of tRNA separate from its role in protein synthesis.

Discussion, see p. 331.

References

1. Müller, P., Wehrli, W., Staehelin, M.: Biochemistry **10**, 1885 (1971).
2. Svensson, I.: Modification in transfer RNA, a compilation. Uppsala, Jan. 1973.
3. Crick, F. H. C.: J. molec. Biol. **19**, 548 (1966).
4. Nishimura, S.: In: Davidson, J. N., Waldo, E., Cohn (Eds.): Progress in Nucleic Acid Research and Molecular Biology, Vol. 12. New York-London: Academic Press 1972.
5. Kimura-Harada, F., Saneyoshi, M., Nishimura, S.: FEBS Letters **13**, (1971).
6. Gefter, M. L., Russell, R. L.: J. molec. Biol. **39**, 145 (1969).
7. Smith, J. D., Barnett, R., Brenner, S., Russell, R. L.: J. molec. Biol. **54**, 1 (1970).
8. Anderson, K. W., Smith, J. D.: J. molec. Biol. **69**, 349 (1972).
9. Hirsh, D.: J. molec. Biol. **58**, 439 (1971).
10. Roe, B., Dudock, B.: Biochem. Biophys. Res. Commun. **49**, 399 (1972).
11. Shimura, Y., Aono, H., Ozeki, H., Sarabhal, A., Lamfrom, H., Abelson, J.: FEBS Letters **22** (1972).
12. Hooper, M. L., Russell, R. L., Smith, J. D.: FEBS Letters **22** (1972).

13. SIMSEK, M., RAJBHANDARY, U. L.: Biochem. Biophys. Res. Commun. **49**, 508 (1972).
14. ROBERTS, R. L.: Nature (Lond.) New Biol. **237**, 44 (1972).
15. GILBERT, J. M., ANDERSON, W. F.: J. biol. Chem. **245**, 2342 (1970).
16. RUDLOFF, E., HILSE, K.: Europ. J. Biochem. **24**, 313 (1971).
17. WOODWARD, W. R., HERBERT, E.: Science **177**, 1197 (1972).
18. SCHERBERG, N. H., WEISS, S. B.: Proc. nat. Acad. Sci. (Wash.) **69**, 1114 (1972).
19. SCHERBERG, N. H., WEISS, S. B.: Proc. nat. Acad. Sci. (Wash.) **67**, 1164 (1970).
20. YUDELEVICH, A.: J. molec. Biol. **60**, 21 (1971).
21. PINKERTON, T. C., PADDOCK, G., ABELSON, J.: Nature (Lond.) New Biol. **240**, 88 (1972).
22. KANO-SUEOKA, T., SUEOKA, N.: Proc. nat. Acad. Sci. (Wash.) **62**, 1229 (1969).
23. CLARKSON, S. G., RUNNER, M. N.: Biochim. biophys. Acta (Amst.) **238**, 498 (1971).
24. SEKIYA, T., KIN-ICHIRO ODA: Virology **47**, 168 (1972).
25. MÄENPÄÄ, P. H., BERNFIELD, M. R.: Biochemistry **8**, 4926 (1969).
26. LANKS, K. W., WEINSTEIN, I. B.: Biochem. Biophys. Res. Commun. **40**, 708 (1970).
27. BRENNER, M., AMES, B. N.: In: Metabolic pathways, 3rd Ed., Vol. V. Metabolic Regulation.
28. SINGER, C. E., SMITH, G. R., CORTESE, R., AMES, B. N.: Nature (Lond.) New Biol. **238**, 72 (1972).
29. ALLAUDEEN, H. S., YANG, S. K., SÖLL, D.: FEBS Letters **28** (1972).
30. JACOBSON, K. B.: Nature (Lond.) New Biol. **231**, 18 (1971).
31. TWARDZIK, D. R., GRELL, E. H., JACOBSON, K. B.: J. molec. Biol. **57**, 231 (1971).
32. WHITE, B. N., TENER, G. M., HOLDEN, J., SUZUKI, D. T.: J. molec. Biol. **74**, 635 (1973).

Functional and Modulated Adaptation of tRNA to Fibroin Biosynthesis in the Silk Gland of *Bombyx mori* L.

J. P. Garel

Centre de Neurochimie C.N.R.S., Strasbourg, France

and

A. Fournier and J. Daillie

Laboratoire de Biologie Générale, Université Claude Bernard, Villeurbanne, France

With 6 Figures

Writing about tRNA isoacceptors in the silk gland of *B. mori* demands a review of this lepidopter. The silkworm is a very interesting system for the molecular biologist at the end of its larval life, the fifth instar, when the silkworm is 5–8 cm long and the silk glands have grown to occupy an appreciable amount of the body space.

Silk is produced in silk glands which are a part of tubes lying on each side of the larva. Each gland is made of three parts. The *posterior part* – also called *secretor* – is specialized for the synthesis of the main silk protein, *fibroin*. At the end of the fifth instar, fibroin accounts for at least 70% of the proteins synthesized by this part of the gland. The fibroin moves forward into the lumen of the middle part which is a storage *reservoir*. Moreover, the reservoir also synthesizes the other silk protein, *sericin*. The anterior part is only a duct ended by the spinneret.

So we can retain a first feature of the silk gland: its morphological compartments producing different preponderant proteins.

The second advantageous feature of the silk gland is the peculiar amino acid composition of fibroin and sericin. Four amino acids (gly, ala, ser, tyr) account for 93% of the total amino acids of fibroin. Sericin is rich in serine (37 %), glycine (15 %) and aspartic acid (15 %).

Finally, we will point out a third feature. During the fifth instar we can distinguish two physiological phases.

– A growth phase (from the first to the fifth day) during which the silk gland triples its weight and produces its own enzymes, the major part of rRNA and begins to synthesize silk protein mRNA's [1].

– A secretory phase during the next four days (5th–8th day) accompanies massive silk production.

When we consider a highly differentiated cell system such as the silk gland from the point of view of tRNA function, we have to ask some questions.

– Is the tRNA pool a rate limiting factor of protein synthesis since translation is directly dependent on the availability of acylated tRNA?

– What can occur at the tRNA level when a cell starts the synthesis of massive amounts of a protein with a peculiar amino acid composition?

There are only two possibilities: either the rate of cytosolic acylation-polysomal deacylation will increase or the composition of the tRNA pool will change in order to maintain the concentration at optimum levels. In the latter case, the tRNA population is *adjusted* or *adapted*.

The answer has been given by the silk gland (Fig. 1): we have studied quantitative changes in tRNA's extracted from a post-microsomal supernatant of the posterior part. The dotted line represents the secretion of silk. The tRNA pool can be divided into two groups which are independently regulated.

– The lower curve represents thirteen tRNA's which are used mainly for the synthesis of protein other than fibroin. Their three-fold increase can be related to the normal growth of the silk gland.

– The upper curve represents the four tRNA's which are the carriers of the four major amino acids of fibroin: gly, ala, ser, tyr. Their fifteen-fold increase from the first to the eighth day leads to their remarkable preponderance at the end of the secretory phase. At this time they account for two thirds of the total supernatant tRNA pool.

This enrichment can be followed for each tRNA species (Fig. 2). tRNA amount is given in percent of total tRNA which can be acylated *in vitro*. In the post-microsomal supernatant we observe a constant enrichment of the four preponderant tRNA's so that their ratio at the eighth day can be correlated with the ratio of the

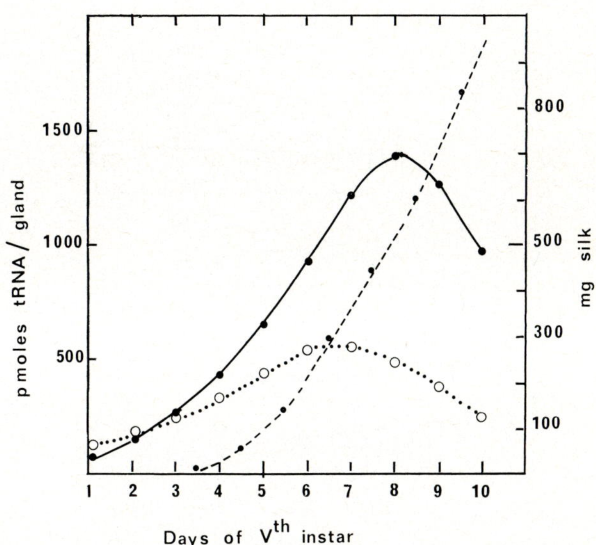

Fig. 1. tRNA distribution during the fifth instar in the posterior part of the silk gland of *Bombyx mori* L. tRNA's were extracted and acylated *in vitro*. ● ——— ● the total amount of 4 tRNA's (ala, gly, ser, tyr) corresponding to the four major amino acids of fibroin. ○ · · · · · · ○ the distribution of the other tRNA species. ● - - - - ● the production of silk proteins (in mg/gland)

four major amino acids of fibroin. On the contrary, the normal concentrations on the second day for other tRNA species decrease significantly and attain very low values during the secretory phase.

Fig. 3 shows the linear correlation between the amino acid composition of fibroin and sericin and the composition of secretor and reservoir tRNA pools [2, 3, 9, 12]. In Fig. 4 the correlation is shown from literature data [4–6] for immunoglobulin, rabbit and mouse hemoglobins, and crystalline.

This functionally adapted situation of supernatant tRNA describes a steady state between free tRNA's and tRNA's associated with polysomes. Daillie et al. have shown that membrane-bound polysomes are exclusive sites for fibroin synthesis [7]. The distribution of tRNA's collected in pellets of membrane bound polysomes obviously reflects the codon frequency of fibroin amino acids. For free polysomes the tRNA distribution is reversed: $tRNA^{ser}$ and $tRNA^{lys}$ are the most abundant [8].

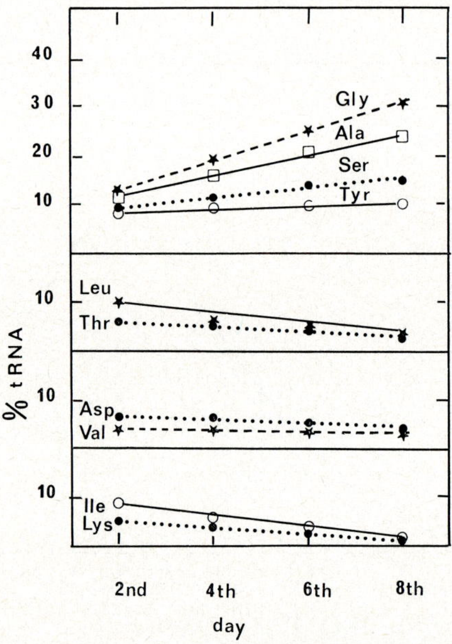

Fig. 2. Evolution of some species of tRNA during the fifth instar in the posterior part of the silk gland

Our previous results are based on *in vitro* heterologous and/or homologous acylations which are not the best tools for quantitative analysis. An objection can be raised: they do not reflect the *in vivo* actual state. We have shown with the highly reproducible technique of countercurrent distribution (C.C.D.) that an evolution takes place between the second and the eighth day [9]. These changes are related to changes in the tRNA distribution, concerning particularly the polar species of the left peak (Fig. 5a). These species can be indentified (Fig. 5b) as $tRNA^{ala}$ (two isoacceptors difficult to separate) and $tRNA^{gly}$ for the left peak, for the middle

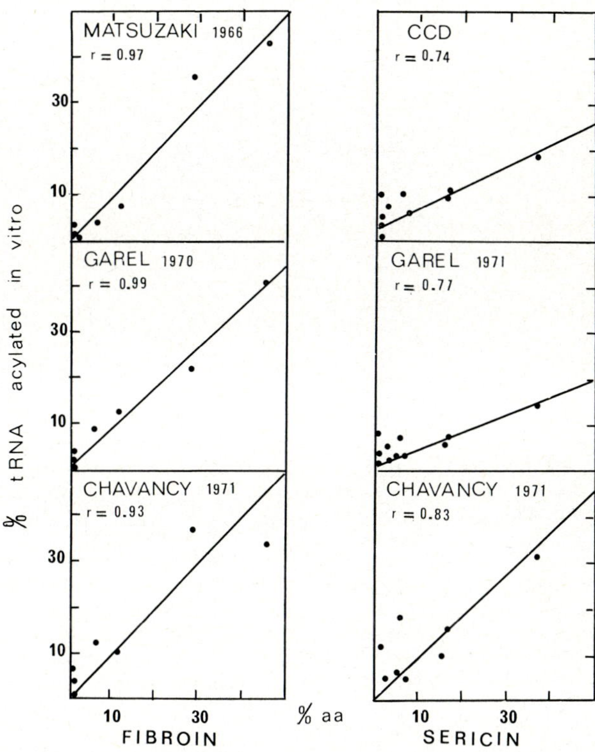

Fig. 3. Correlation between *in vitro*-acylated tRNA in the posterior (left) and middle (right) parts of *Bombyx mori* L. silk gland and amino acid composition of fibroin and sericin according to Chavancy et al. [2], Garel et al. [9—12], Matsuzaki [3]; r indicates the correlation coefficient

peak only $tRNA^{gly}$ and for the right peak $tRNA^{ser}$. The noteworthy situation of a functional adaptation of tRNA's in the posterior part of silk gland is clear when comparing a C.C.D. of a yeast tRNA population (upper curve) and tRNA's from *B. mori* on the eighth day.

An adaptative mechanism adjusts the level of tRNA isoacceptors to that required by the codon repartition of a given mRNA, here fibroin mRNA. We have good reasons to think that isoacceptor

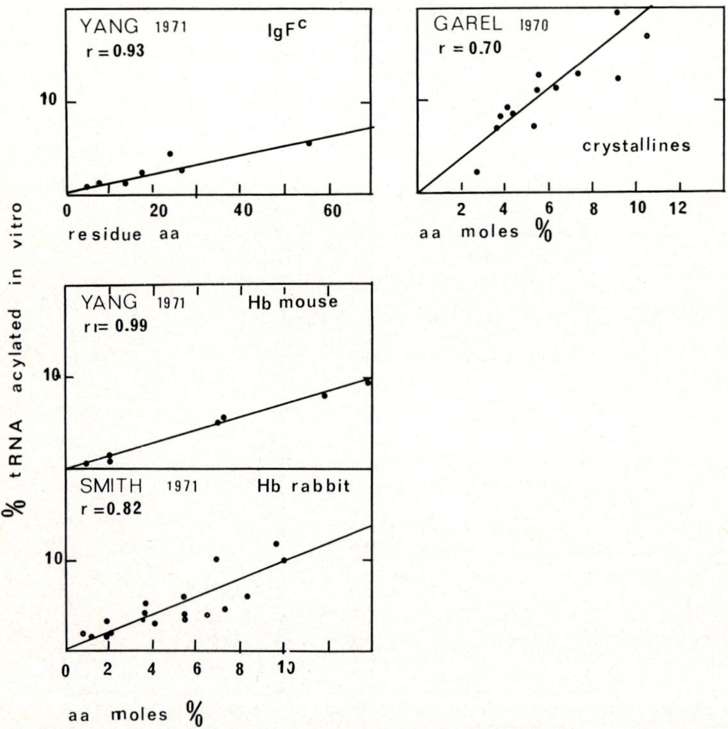

Fig. 4. Correlation between tRNA population and the amino acid composition of rabbit hemoglobin — Smith and MacNamara [5] —, mice hemoglobin and H chain of IgF — Yang [4] —, crystalline from the internal cortex of calf lens — Garel et al. [6] — A 10% background noise due to a residual protein synthesis other than hemoglobin was substracted from the values of acylated tRNA given by Smith and McNamara

distribution of $tRNA^{gly}$ species at the end of the secretory phase must be adjusted to glycine codons (GGN) of fibroin mRNA. This simple hypothesis can be checked under one condition: glycine codon frequency of fibroin mRNA must be significantly different from a random distribution of the four glycine codons occuring in all others mRNA's.

In order to find an answer we have analyzed by means of various chromatographic procedures (adsorption and partition) radioactive elution profiles of [^{14}C]gly-tRNA^{gly} extracted on the fourth

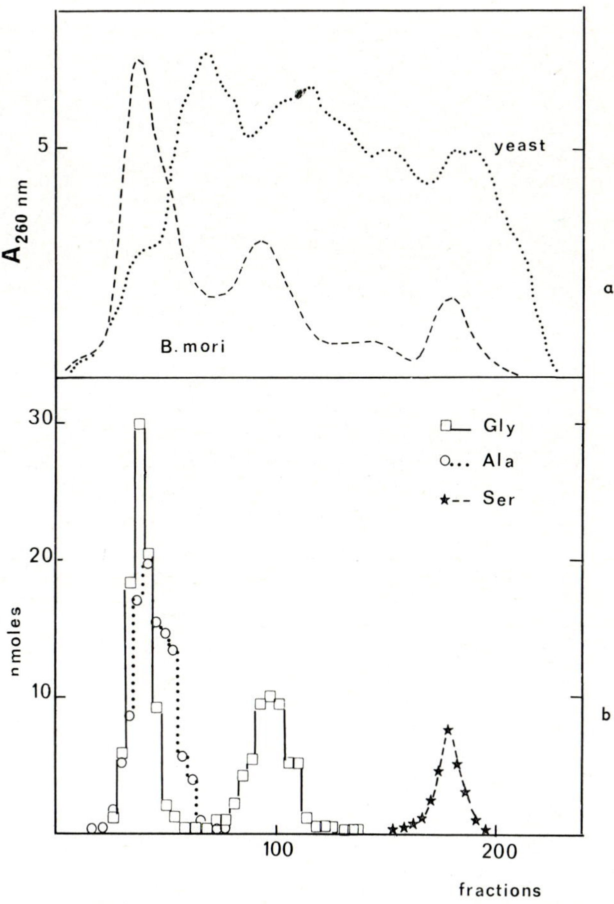

Fig. 5. Countercurrent distribution of tRNA from baker's yeast and from the posterior part of the silk gland of *B. mori* L. (8th day of the 5th instar). (a) optical densities at 260 nm (*B. mori*and yeast). (b) acylation of fractions obtained by CCD (*B. mori* 8th day Vth instar)

day (Fig. 6, dotted line) and on the eighth day (broken line).

– The distribution of the isoacceptor species is different. The ratio peak 1/peak 2 = 1.1 on the fourth day against 0.6 on the eighth day.

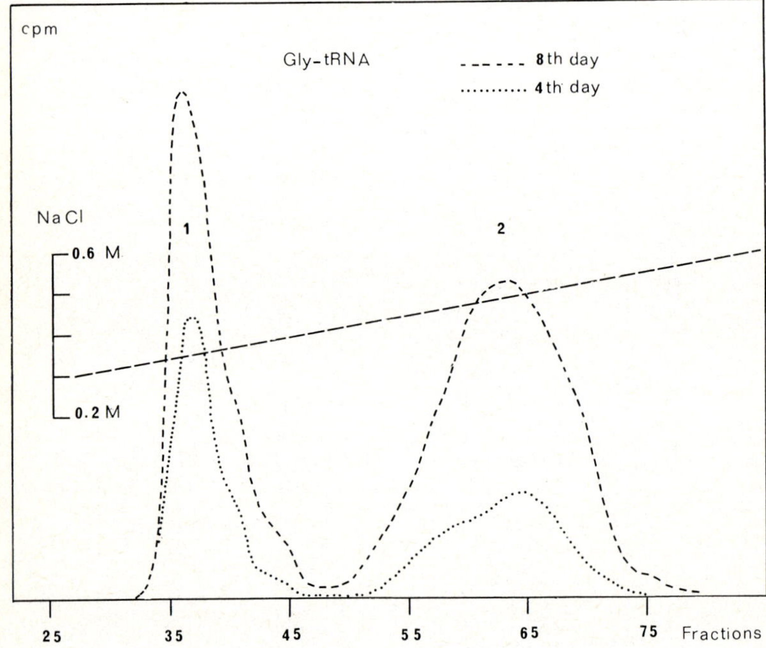

Fig. 6. Chromatography on RPC (microscale) of [^{14}C]gly-tRNA^{gly} ..... 4th day Vth instar, ------- 8th day Vth instar

– The increase of tRNA^{gly} during the fifth instar affect differentially their isoacceptors. In our opinion this differential increase of isoacceptor, from a factor 2 for isoacceptor 1 and 3.5 for isoacceptor 2, must be related to the glycine codon frequency of fibroin mRNA. Unfortunately we did not have the opportunity until now to check directly the codon response of the two tRNA^{gly} isoacceptors. However, their behaviour by C.C.D. allows us to presume that isoacceptor 1 could recognize GGA and isoacceptor 2, GGU [10]. Brown and Suzuki [11] have isolated fibroin mRNA and studied its structure. From their results it can be deduced that glycine is coded by two codons: GGA and GGU with a ratio 3 : 5.

From our results on the silk gland one can show that tRNA metabolism is selective and related to the presence of the mRNA to be translated. We suggest defining this phenomenon as "*modulated tRNA biosynthesis*". It implies an indirect coupling at the transcriptional level between mRNA and tRNA cistrons.

References

1. Daillie, J.: C.R. Acad. Sci. (Paris) **261**, 4872 (1965).
2. Chavancy, G., Daillie, J., Garel, J.P.: Biochimie **53**, 1187 (1971).
3. Matsuzaki, K.: Biochim. Biophys. Acta (Amst.) **114**, 222 (1966).
4. Yang, W.K.: Cancer Res. **31**, 639 (1971).
5. Smith, D.W.E., McNamara, A.L.: Science **171**, 577, 1040 (1971).
6. Garel, J.P., Virmaux, N., Mandel, P.: Bull. Soc. Chim. Biol. **52**, 987 (1970).
7. Daillie, J., Grasset, L., Prudhomme, J.C., Beck, J.P., Ebel, J.P.: FEBS Letters **13**, 321 (1970).
8. Fournier, A., Garel, J.P., Daillie, J.: In preparation.
9. Garel, J.P., Mandel, P., Chavancy, G., Daillie, J.: Biochimie **53**, 1195 (1971).
10. Garel, J.P., Mandel, P.: J. Chromatog. **71**, 203 (1972).
11. Suzuki, Y., Brown, D.D.: J. molec. Biol. **63**, 409 (1972).
12. Garel, J.P., Mandel, P., Chavancy, G., Daillie, J.: FEBS Letters **7**, 327 (1970).

Discussion

H. Noll (Evanston): What is the explanation for reports indicating that translation of messengers in a certain system, *e.g.*, globin synthesis in a mammalian ribosome system, is much more efficient with the homologous tRNA than heterologous preparations, *e.g.*, bacterial tRNA?

M. Staehelin (Basel): Two effects might be responsible for the low efficacy of prokaryotic tRNA's in eukaryotic protein synthetizing systems. It is known that some bacterial tRNA's are only very poorly aminoacylated with eukaryotic synthetases (Doctor, B.P., Mudd, J.A.: J. Biol. Chem. **238**, 3677 (1963). The other might be that bacterial initiator tRNA does not interact with mammalian initiator factors.

H. Kersten (Erlangen): Some modifications of a tRNA could be necessary to protect certain sequences in a tRNA loop from endonucleolytic attack during the maturation process. We know that tRNA's from mycoplasma do not contain modified nucleosides. Is something known whether the tRNA's from mycoplasma are already made in the correct size?

M. Staehelin: The only precursor I know of is that of the phage-induced tyrosine tRNA [Altman, S., Smith, J. D.: Nature New Biology **233**, 35 (1971)].

That tRNA's from mutants which produce lesser amounts of tRNA contain little O-methylguanosine whereas revertants contain a full complement could indicate that your suggestion seems a distinct possibility.

The fact that precursor tRNA accumulates in these mutants, however, would point out that normally tRNA maturation proceeds faster than degradation by nucleases.

Weil: The idea that tRNA's might be involved in the regulation of translation is derived from what is known about the mechanisms of suppression where a new tRNA molecule allows the translation of a message that contains a nonsense codon which could not be read in the normal cell, but can be read by the new modified (suppressor) tRNA in the mutant cell.

H. Noll: There must be a recognition sites for RNA on the ribosomes. Therefore one could expect that a mammalian tRNA works better with mammalian ribosomes than with the *E. coli* ribosomes. What is your experience in this respect?

M. Staehelin: The fact is that recognition of codons on messengers by mammalian tRNA normally work better with *E. coli* ribosomes than with mammalian ribosomes. Why this is so, we don't know.

Matthaei (Göttingen): *In vitro*, we have shown that AA-tRNA binding appears to be a rate-limiting step in both enzymatic binding of Phe-tRNA and poly-Phe synthesis (Biochim. Biophys. Acta **281**, 113, 1972). Polypeptide synthesis coded by natural mRNA is likewise competitively inhibited by elongator AA-tRNA species (Biochim. Biophys. Acta **299**, 468, 1973). It seems desirable to have concentrations of the AA-tRNA species adapted to the actual demand for translation of the mRNA's present because AA-tRNA species that are not required or in a relative excess would be inhibitory. In addition to an optimal supply of substrate, activity of more than one AA-tRNA binding site per ribosome might increase the velocity of mRNA translation. It would be good, therefore, to determine the number of active sites on silk fibroin producing ribosomes. Do you know how many tRNA molecules per ribosome the silk glandular tissue contains?

J.P. Garel (Strasbourg): About 10.

Matthaei: This number is similar to the one in fast-growing *E. coli* cells. Since tRNA species are adapted to demand in both their number and concentrations, each of the ribosomes could probably bind up to three molecules of AA-tRNA without interfering with the simultaneous substrate requirements of other ribosomes.

Drews: Can you isolate your messenger RNA and can you do all this *in vitro*, and then check the functional significance of your various isoacceptors?

J.P. Garel: The group of Day in Lyon is now doing some experiments along these lines, but I don't know their results.

C.A. Thomas (Boston): Can you tell us about the existence of mutations, or the hope for such mutations, affecting fibroin structure?

J.P. Garel: Yes, there probably exist mutants in the fibroin, but they can be only detected after cocoon spinning, not before. This is a rather limiting factor.

Functional Characterization of Five Initiation Factors for Mammalian Protein Synthesis

Max H. Schreier and Theophil Staehelin

Basle Institute for Immunology, Basle, Switzerland

With 7 Figures

Although there has been rapid progress in the separation, purification and partial functional characterization of several protein factors involved in mammalian polypeptide chain initiation from rabbit reticulocytes [1—9], the mechanism of eukaryotic chain initiation is still not clear. In our earlier work [10] we used well-defined ribosomes (purified 60 S and 40 S subunits) whose structural and functional integrity was shown by their capacity to participate quantitatively and efficiently in the *in vitro* translation of rabbit globin mRNA.

We found that each globin mRNA molecule was translated several times in our system and that each ribosome synthesized, on the average, several globin chains. Figs. 1 and 2 demonstrate these conclusions. In Fig. 1, mRNA was titrated from 0.25 to 1 mRNA molecule per ribosome. Reaction mixtures (0.4 ml) containing 2 OD_{260} units of mouse liver ribosome subunits and all other components for globin synthesis were incubated at 30 °C for 15 min. The chilled mixtures were analysed in 5 ml concave exponential sucrose gradients with 0.1 M KCl, 0.005 M magnesium acetate, 0.02 M Tris, pH 7.6. Fig. 1A shows the controls without mRNA. In the absence of initiation factors almost all ribosomes form 80 S couples (dashed curve). After incubation, in the presence of saturating amounts of initiation factors, about 40% of the ribosomes are dissociated into free subunits (solid curve) suggesting the presence of a dissociation or anti-association factor. The incubation mixtures in B through F contained initiation factors and increasing amounts of mRNA: 2 μg (B), 3 μg (C), 4 μg (D), 5 μg

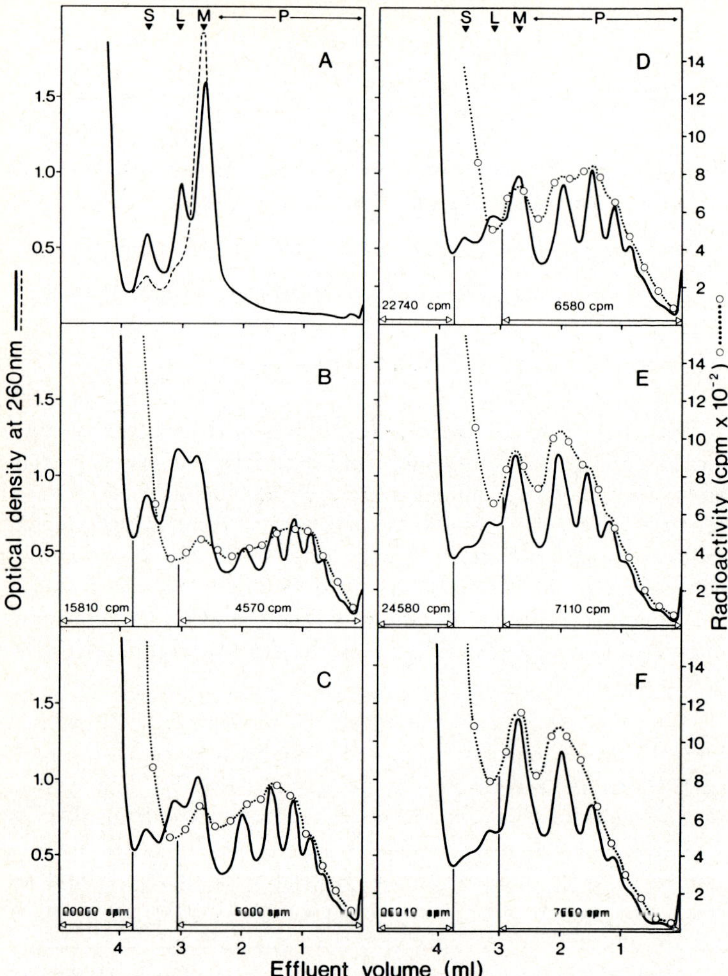

Fig. 1A—F. 9 S mRNA-dependence of polysome distribution during the steady state of *in vitro* globin synthesis. 0.4 ml reaction mixtures contained 2 A_{240} units mouse liver ribosome subunits. The samples displayed in (*A*) were incubated in the absence of mRNA and in the absence of initiation factors (dashed curve) or in the presence of initiation factors (solid curve). Samples shown in (*B*) to (*F*) had initiation factors, and 9 S mRNA as follows: 2 µg (*B*), 3 µg (*C*), 4 µg (*D*), 5 µg (*E*), and 8 µg (*F*). Incubation was at 30 °C for 15 min. The chilled reaction mixtures were centrifuged in 5 ml concave exponential 0.3—1.5 M sucrose density gradients. Only half the input (0.2 ml reaction mixture) was analysed in the gradients shown in (*A*). Fifteen 0.25 ml

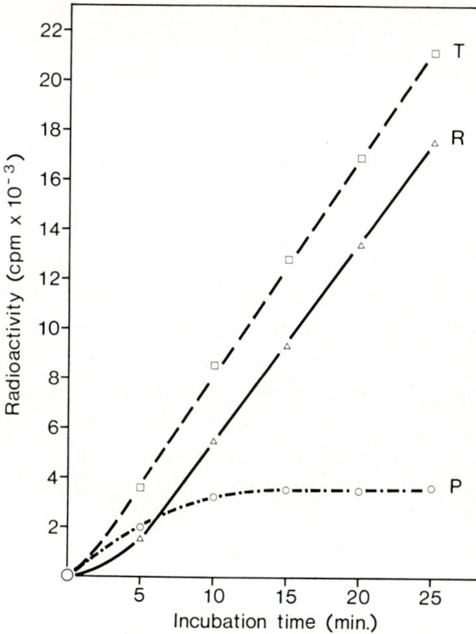

Fig. 2. Kinetics of *in vitro* globin synthesis. The distribution of polysome-plus monosome-bound and released polypeptide chains as a function of incubation time at 30 °C was analysed in sucrose density gradients as described in the legend to Fig. 1. 0.2 ml reaction mixtures contained 1.0 A_{260} unit mouse liver ribosome subunits, 2.5 μg 9 S mRNA (ribosome to mRNA ratio of sample in Fig. 1, panel (e)) and 45 μg of a mixture of further separated and purified initiation factors. Incubation was at 30 °C for the lengths of time indicated on the abscissa. Each gradient was collected into two fractions only, namely, one containing polysomes and monosomes and one containing the remaining upper part of the gradient. After addition of 0.5 mg bovine serum albumin as carrier protein to the ribosome fraction, both fractions were precipitated with 10% trichloroacetic acid, heated to 90 °C and processed on filter papers for scintillation counting. Curve *P*, polysome- + monosome-bound labeled protein; curve *R*, released labeled protein; curve *T*, total (i.e. $P + C$) hot acid-precipitable radioactivity recovered per gradient. (This Figure is from [10])

---

fractions were collected and mixed with scintillation liquid for direct counting. The upper 1.5 ml of each gradient was precipitated with 10% trichloroacetic acid, heated at 90 °C and collected and washed on glass fiber filters for counting. The hot acid-precipitable radioactivity from the top of the gradient and the sum of the ribosome-bound radioactivity are indicated in each panel. *S* small subunit; *L* large subunit; *M* monomer; *P* polysomes. (This figure is from [10])

(E) and 8 μg (F). This corresponds to one mRNA molecule per four ribosomes in (A) and to one mRNA molecule per one ribosome in (F). As expected, the polysome distribution shifts from large polysomes to smaller polysomes and more protein-synthesizing monomers as the mRNA concentration increases.

Fig. 2 shows the kinetics of synthesis determined from sucrose gradient analyses. The mRNA input was the same as in Fig. 1 E, *i.e.*, one mRNA molecule per 1.5 ribosomes. The OD_{260} profile of the gradients showed a strong increase of polysomes from 1–10 min incubation with a shift to larger polysomes. Between 15 and 25 min incubation, the polysome pattern remained unchanged [identical to that of Fig. 1 E]. The gradients were collected into two fractions only, one including all polysomes-and monosomes, the second consisting of the upper part of the gradient. Both fractions were assayed for hot acid-insoluble radioactivity. The curve *P* in Fig. 2 represents polysome-and monosome-bound radioactivity, curve *R* released protein and curve *T* the sum of the two. The latter corresponds within 2 % to the radioactivity directly precipitated from duplicate incubation mixtures. The duplicate assays showed further that the synthesis is linear to 35 min and then levels off. This experiment demonstrates that the system at 30 °C is in a steady state of ribosome recycling from about 10 min to about 35 min of incubation. From the rate of globin chain release and the number of ribosomes engaged in synthesis in the steady state, one can calculate the average time needed for the synthesis of a complete chain. This was between 7 and 8 min in this experiment, although the first finished chains were released after about 3 min.

Now we have separated and purified five protein factors (IF-$E_{1,2,3,4,5}$, E = eukaryote) from rabbit reticulocytes required for the initiation of globin synthesis by purified ribosome subunits in the presence of exogenous mRNA, initiator Met-$tRNA_f$, GTP and ATP. Two independent tests were used to show the requirement for initiation of the different factors and to sugest their mode of action.

The first assay is a polypeptide synthesizing system consisting of the following purified and defined components; 40 S + 60 S ribosomal subunits, globin mRNA, purified tRNA precharged with all 20 amino acids, purified elongation factors EF-T_1 and EF-T_2, GTP and ATP, and initiation factors. This system differs from a poly U-dependent phenylalanine polymerization system only by its additional requirement for all 20 aminoacyl-tRNA's, ATP and the five initiation factors. The optimal free Mg-ion concentration (not including Mg bound by ATP and GTP) of this system is 1.5 mM as compared to about 7 mM in the poly U-system.

Table 1. Globin synthesis with purified components

	[^{14}C] amino acids polymerized, c.p.m.
Complete system	14910
Minus mRNA	980
Minus ATP	740
Minus GTP	1050
Minus IF-E_1	4710
Minus IF-E_2	3300
Minus IF-E_3	630
Minus IF-E_4	690
Minus IF-E_5	4820

0.1 ml of the complete reaction mixture contained 0.5 A_{260} units 40S + 60S mouse liver ribosome subunits; 2 μg globin mRNA, 1.5 A_{260} units rat liver tRNA charged with 14 (^{14}C]-amino acids (average spec. act. = 70—80 mCi/mmol) and 6 [^{12}C]-amino acids; 3.5 μg IF-E_1, 6 μg IF-E_2, 7 μg IF-E_3, 5 μg IF-E_4, 2.8 μg IF-E_5, 35 μg partially purified elongation factor EF-T_1, 2 μg EF-T_2; 0.5 mM ATP-Mg^{++}, 0.1 mM GTP-Mg^{++}, 1.5 mM $MgCl_2$, 120 mM KCl, 1 mM DTT, 30 mM Tris HCl pH 7.6. Incubation was for 20 min at 37 °C followed by 15% TCA precipitation at 90° and filtration onto glass fiber filters for liquid scintillation counting as described in [10].

The dependence of polypeptide synthesis on the individual components is shown in Table 1.

The initiation factors 3 and 4 are absolutely required. The same is probably true for factor 2. IF-E_1 is a rate factor. It stimulates synthesis and shifts the Mg optimum from less than 1 mM to 1.5 mM. At 1.5 mm Mg, the stimulation by IF-E_1 is three- to four-fold. IF-E_5 is also not absolutely required but stimulates synthesis about three- to five-fold. The most striking finding in this purified system of natural chain initiation and amino acid incorporation is the absolute requirement for ATP in addition to GTP.

The second assay measures initiation complex formation between ribosomes, initiator Met-tRNA$_f$ and mRNA analysed in sucrose density gradients. In this assay system we find the initial mRNA-independent binding of initiator Met-tRNA to the 40 S ribosome subunit [11]. This [40 S · Met-RNA$_f$] complex formation occurs with IF-E$_2$, GTP and IF-E$_3$. IF-E$_2$, like *E. coli* IF$_2$, recognizes and combines with the initiator tRNA in the presence of

Table 2. The GTP-dependent formation of a complex between purified IF-E$_2$ and initiator Met-tRNA$_f$

Factor input	[^{3}H]Met-tRNA$_f$ on millipore filter (c.p.m.) + GTP	— GTP	× Stimulation by GTP
IF-E$_2$ (4 μg)	13405	782	17.1
IF-E$_3$ (11 μg)	33	15	—
IF-E$_2$ (4 μg) + IF-E$_3$ (11 μg)	13040	845	15.4

Reaction mixtures of 0.1 ml contained 20 pmol [^{3}H]Met-tRNA$_5$, factors as indicated, and 0.01 μmol GTP where indicated. The ionic conditions were 0.07 M KCl, 1.5 mM MgCl$_2$, 0.02 M Tris HCl, pH 7.6, and 1 mM DTT. After 8 min incubation at 30 °C, 3 ml of ice cold buffer was added, and the samples were filtered through millipore filters and washed with 9 ml of cold buffer. The dried filters were counted in toluol-PPO-POP scintillation liquid. 1 pmol [^{3}H]Met-tRNA$_0$ had approximately 2500 c.p.m. A blank of 112 c.p.m. (assay without factors) was subtracted from each value. (This Table is from [11]$_5$)

GTP. This complex is retained on millipore filters in the absence of ribosomes and template (Table 2). IF-E$_3$ catalyses the mRNA-independent binding reaction of Met-tRNA$_f$ to the 40 S subunit.

Figure 3 shows a series of experiments demonstrating the interdependent roles of IF-E$_2$ and IF-E$_3$ in the [40 S · Met-tRNA$_f$] initiation complex formation. IF-E$_3$ alone does not bind initiator Met-tRNA$_f$ to ribosomes in the presence of the artificial template poly (A, U, G) (or globin mRNA, not shown) and GTP (Fig. 3A). Purified IF-E$_2$, however, promotes very efficiently the GTP- and poly (A, U, G)-dependent Met-tRNA$_f$ binding to 40 S subunits (Fig. 3B). We can calculate that all subunits in the 40 S peak, that is, about 50% of those present in the assay, had bound a [^{3}H]Met-tRNA$_f$. When the artificial template poly (A, U, G) was replaced by the natural globin mRNA (3 pmol 9 S mRNA per pmol 40 S subunits), almost no Met-tRNA$_f$ was bound (Fig. 3C). Addition of IF-E$_3$ to IF-E$_2$ and globin mRNA results, however, in

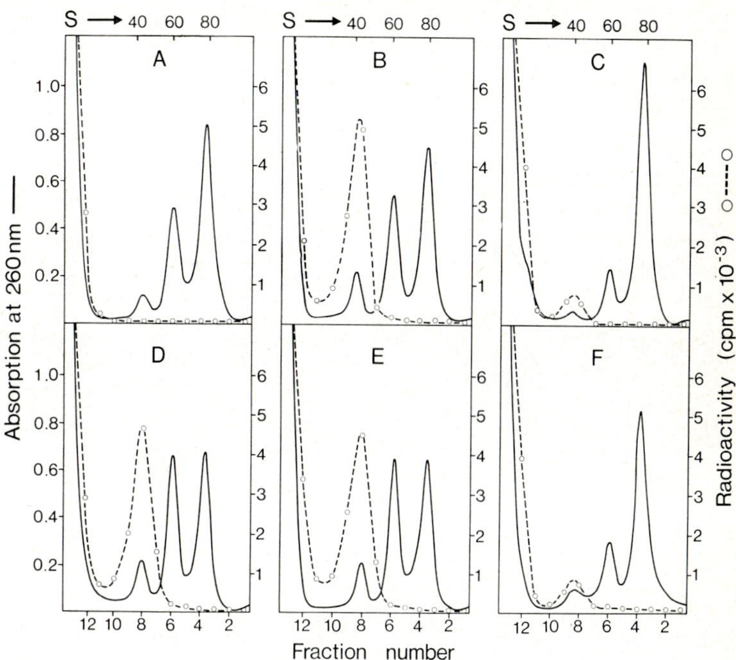

Fig. 3A—F. [Met-tRNA$_f$ · 40 S) ribosome subunit initiation complex formation dependent on IF-E_2 and poly (A, U, G) or IF-E_3. 0.1 ml reaction mixtures contained 0.13 A_{240} units (8 pmol) mouse liver 40 S subunits, 20 pmol [^{3}H]Met-tRNA$_f$ (from guinea-pig liver), 0.04 μmol GTP, 7 μmol KCl, 0.43 μmol $MgCl_2$, 3 μmol Tris HCl, pH 7.6, 0.1 μmol DTT and the following additions: *A* 5 μg poly (A, U, G), and 6 μg IF-E_3; *B* 5 μg poly (A, U, G), and 4 μg IF-E_2; *C* 4 μg IF-E_2 and 6 μg rabbit globin mRNA; *D* 4 μg IF-E_2, 6 μg IF-E_3, and 6 μg globin mRNA; *E* 4 μg IF-E_2 and 6 μg IF-E_3, no template; *F* 4 μg IF-E_2, 6 μg globin mRNA, and 6 μg IF-E_3 treated with N-ethylmaleimide. (IF-E_2 was from rabbit reticulocytes, and IF-E_3 from mouse Krebs ascites cells.) Incubation was at 36 °C for 20 min, then 0.33 A_{260} units of mouse liver 60 S subunits were added and incubation was continued for 10 min at 30 °C to allow ribosome subunit association. The samples were analysed on 4 ml convex exponential 10—28% sucrose gradients containing 0.07 M KCl, 0.008 M $MgCl_2$, 0.02 M Tris HCl, pH 7.6. Centrifugation was at 56,000 r.p.m. for 70 min, at 4 °C in a 'Beckman SW 56 Ti rotor'. The gradients were analysed and counted as described in ref. 10. 1 pmol [^{3}H]Met-tRNA ~ 2500 c.p.m. (This Figure is from [11])

efficient 40 S initiation complex formation (Fig. 3D). The most plausible interpretation of these results would be that IF-E_2 can bind Met-$tRNA_f$ to 40 S subunits in the presence of a suitable template, namely the artificial poly (A, U, G), whereas natural mRNA becomes only an effective template for Met-$tRNA_f$ in the

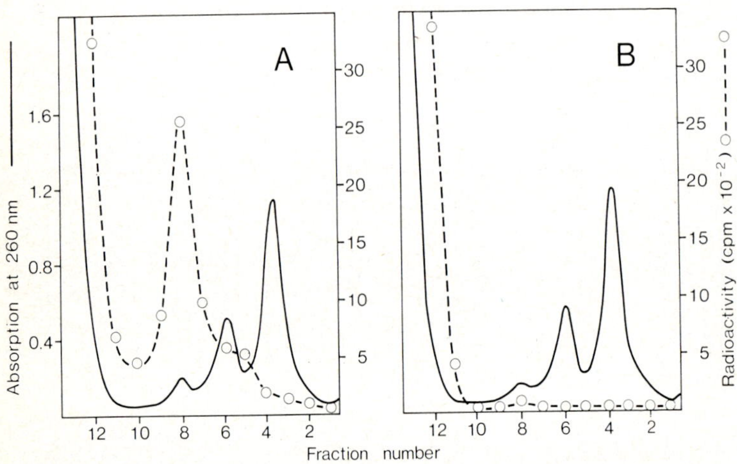

Fig. 4A and B. Specificity of IF-E_2 and IF-E_3 for binding of initiator Met-$tRNA_f$ versus Met-$tRNA_M$. Incubation mixtures (0.1 ml; ions as in Fig. 3) contained partially purified IF-E_2 (10 μg) purified IF-E_3 (6 μg), GTP, and globin mRNA and (*A*) 20 pmol [^{3}H]Met-$tRNA_f$ or (*B*) 20 pmol [^{3}H] Met-$tRNA_M$. Both 40 S and 60 S ribosome subunits were present throughout the 20 min of incubation at 36 °C. Sucrose gradient centrifugation and analysis were as described for Fig. 3. Specific activity of both [^{3}H] Met-tRNAs was 2,500 c.p.m. $pmol^{-1}$. (This Fig. is taken from [11])

presence of IF-E_2 and IF-E_3, thus suggesting that IF-E_3 is required for binding of natural mRNA. When mRNA was omitted, however, Met-$tRNA_f$ binding to 40 S subunits was just as efficient with IF-E_2 and IF-E_3 (Fig. 3E). This rather surprising result clearly demonstrates that IF-E_3 promotes the template-independent binding of initiator Met-$tRNA_f$, presumably complexed with IF-E_2 and GTP, to 40 S ribosome subunits. The experiment shown in Fig. 3F was identical to that in Fig. 3D, except that IF-E_3 was treated with the sulphydryl group blocking reagent N-ethyl-maleimide, demonstrating the sensitivity of IF-E_3 to this reagent. IF-E_2 is also inactivated by N-ethyl-maleimide (result not shown, essentially as in Fig. 3F). It is important to note that in the presence of IF-E_2 and IF-E_3 alone, no [80 S · Met-tRNA$_f$] initiation complexes are formed.

The IF-E_2 and IF-E_3 dependent initiation complex formation is specific for the initiator Met-tRNA$_f$ (Fig. 4A), whereas the Met-tRNA$_M$ used for methionine transfer into internal positions of protein is not recognized by these initiation factors (Fig. 4B).

Figure 5 demonstrates the function of IF-E_5 and the IF-E_4 dependent stimulation of Met-tRNA$_f$ binding by mRNA and vice versa. IF-E_5 promotes the joining of the 40 S initiation complex with the 60 S subunit to form an 80 S complex even in the absence of mRNA and IF-E_4 (Fig. 5A and B). ATP (present in all assays of Fig. 5) and IF-E_4 have no significant effect on Met-tRNA$_f$ binding in the absence of mRNA (Fig. 5C). Also mRNA, in the absence of IF-E_4, does not stimulate Met-tRNA$_f$ binding in the absence of IF-E_4 (Fig. 5B). However, in the presence of IF-E_4 *and* ATP, mRNA strongly stimulates the 80 S initiation complex formation (Fig. 5D). The stimulation of Met-tRNA$_f$ binding by mRNA requires both ATP and IF-E_4; and the stimulation by IF-E_4 requires both mRNA and ATP. Therefore, it seems that IF-E_4 together with ATP effects binding of natural mRNA to the [ribosome · Met-tRNA$_f$] complex.

The interdependent and cooperative functions of IF-E_4, ATP, IF-E_1 and also IF-E_5 are demonstrated in more detail in Fig. 6.

IF-$E_{2,3,4,5}$, without factor 1, promote quite efficiently a mRNA- and ATP-independent, apparently fortuitous [80 S · Met-tRNA$_f$] complex (Fig. 6A). The addition of globin mRNA decreases the Met-tRNA$_f$ binding to 80 S ribosomes (Fig. 6B). ATP in addition to mRNA (Fig. 6C) stimulates Met-tRNA$_f$ binding again to at least the level without mRNA. The addition of IF-E_1 to all the other factors (Fig. 6D, E, F) strongly influences Met-tRNA$_f$ binding to ribosomes. In the absence of mRNA and ATP (Fig. 6D), IF-E_1 markedly inhibits binding to 80 S ribosomes. This inhibition is less pronounced in the presence of ATP (○······○). In the presence of mRNA and absence of ATP, IF-E_1 seems to inhibit the efficient conversion of the [40 S · Met-tRNA$_f$] complex into the 80 S complex (Fig. 6E). Addition of mRNA plus ATP (Fig. 6F) gives a very strong stimulation of Met-tRNA$_f$ binding with almost com-

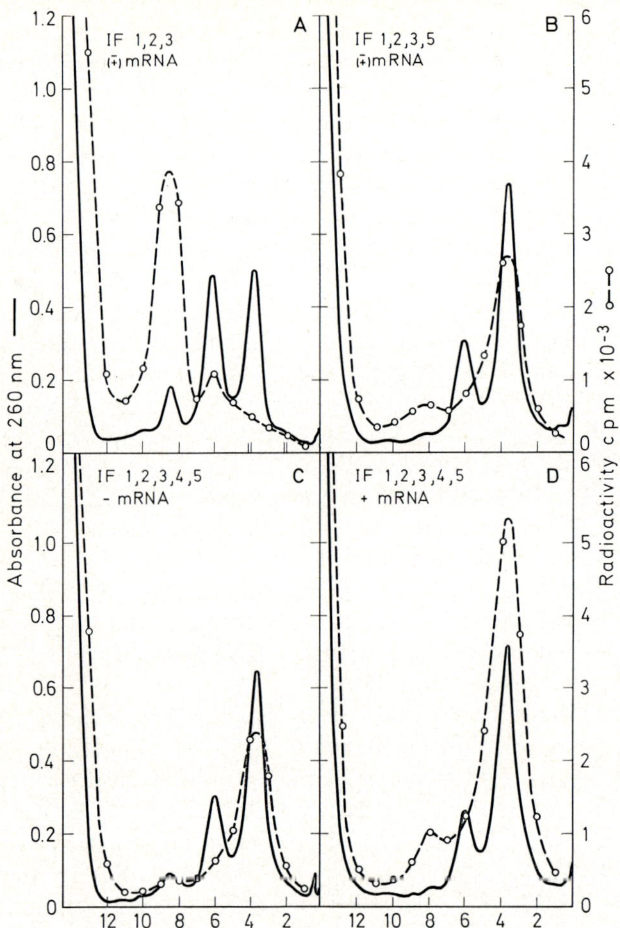

Fig. 5A—D. Effects of IF-E_4, IF-E_4 and globin mRNA on initiation complex formation: 0.1 ml incubation mixtures contained 20 pmol [^{3}H]Met-tRNA$_f$ and the ionic conditions, ATP, GTP and 9 pmol mixed mouse liver ribosome subunits as described for globin synthesis in Table 1; initiation factors (amounts as in Table 1) and globin mRNA (4 μg) were present as indicated for each panel. Incubation was for 10 min at 37 °C. The chilled reaction mixtures were diluted with 0.1 ml of gradient buffer followed by sucrose density gradient centrifugation and analysis as described for Fig. 3. 1 pmol [^{3}H]Met-tRNA ~ 2700 c.p.m.

plete conversion into 80 S initiation complex. From the specific activity of the Met-tRNA and the area of the 80 S peak, we can calculate that about 90% of the 80 S ribosomes have bound a Met-tRNA$_f$.

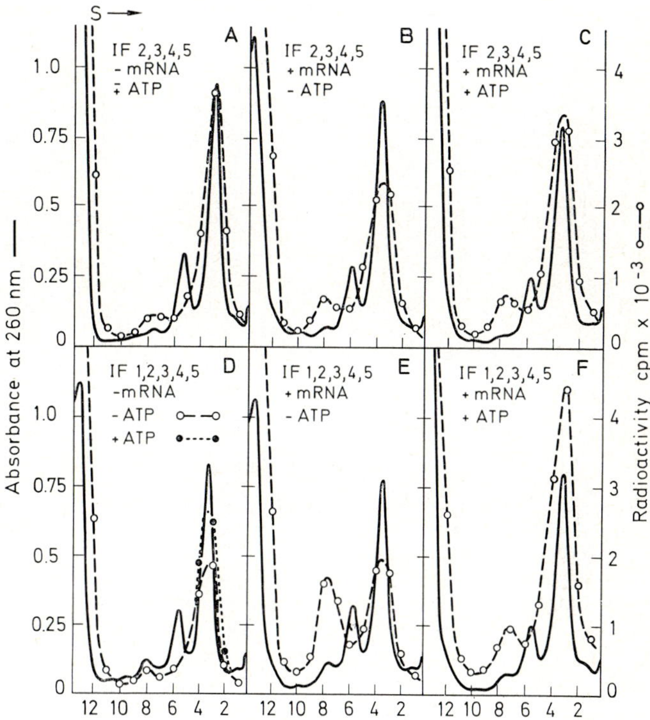

Fig. 6A—F. Effects of IF-E$_1$, ATP and mRNA on initiation complex formation: incubation mixtures of 0.1 ml were identical to those of Fig. 5 except for ATP, mRNA and initiation factors which are indicated on each panel. Incubation, sucrose gradient centrifugation and analysis were as described for Fig. 3 and 5

Finally, we tested to what extent the initiator Met-tRNA$_f$ is bound in the proper P-site of the 80 S ribosome measured by reaction with puromycin. The results in Fig. 7 demonstrate that the Met-tRNA$_f$ is fully (*i.e.* $>$ 80%) reactive with puromycin (Fig. 7C) only in the presence of globin mRNA with all initiation factors,

GTP *and* ATP. In the absence of mRNA (Fig. 7A and B), only about 45% of the bound Met-$tRNA_f$ reacts with puromycin. The slight stimulatory effect of ATP in the absence of mRNA (Fig. 7B, also Fig. 6D) might be due to traces of globin mRNA present in the IF-E_5 preparation.

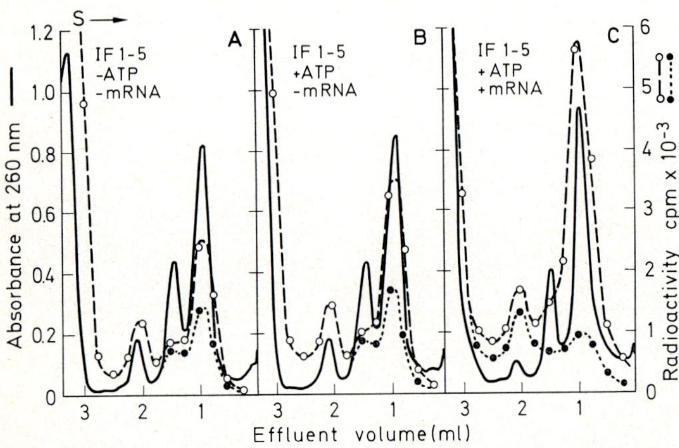

Fig. 7A—C. Effect of mRNA + ATP on the puromycin reactivity of Met-$tRNA_f$ in the 80 S initiation complex. Duplicate incubation mixtures of 0.1 ml containing GTP, all the initiation factors, and ATP and mRNA as indicated were incubated for 5 min at 37 °C. Then puromycin (2 mM final concentration) was added to one of each duplicate and incubation was continued for 5 more minutes; sucrose gradient centrifugation and analysis were as described for Fig. 3. ○ ---- ○ = control, ● .... ● = puromycin treated

Discussion

The mechanism for mammalian polypeptide chain initiation most consistent with our results requires the initial selection of the initiator Met-$tRNA_f$ by IF-E_2 for the formation of a [Met-$tRNA_f$ · IF-E_2 · GTP] complex. This complex is bound independently of mRNA to the 40 S ribosome subunit by IF-E_3 [11]. Subsequently, mRNA binding and 60 S subunit joining is promoted by the cooperative action of IF-E_4, ATP, IF-E_1 and IF-E_5. Any stimulation

[1] See contribution by GRUMMT in discussion.

of Met-tRNA$_f$ binding by natural mRNA requires IF-E$_4$ and ATP, and vice versa. This strongly suggests that IF-E$_4$ and ATP are involved in natural mRNA binding. Thus, binding of mRNA probably stabilizes the Met-tRNA$_f$ on the ribosome, making an initially reversible binding irreversible and shifting the equilibrium to the ribosome complex. IF-E$_5$ promotes the joining of the 40 S complex with the 60 S subunit. IF-E$_1$ inhibits the [80 S · Met-tRNA$_f$] complex formation in the absence of mRNA binding, possibly by destabilizing Met-tRNA$_f$ on the 40 S subunit during the IF-E$_5$-promoted subunit joining in the absence of mRNA binding. However, in the presence of mRNA binding, *i.e.*, of ATP and IF-E$_4$ together with mRNA and the other factors, IF-E$_1$ significantly stimulates Met-tRNA$_f$ binding.

The artificial manipulation of the *in vitro* system by omitting single factors, mRNA or ATP strongly suggest distinct roles of the individual components in the assembly process of the [mRNA · Met-tRNA$_f$ · 80 S ribosome] initiation complex. Of particular interest is IF-E$_1$ which seems to act like a policeman in the complicated process by blocking or inhibiting the "forbidden" formation of [Met-tRNA$_f$ · 80 S ribosome] complexes in the absence of concomitant mRNA binding but stimulating this complex formation in the presence of mRNA binding. Therefore, in the presence of (excess) IF-E$_1$, the [Met-tRNA$_f$ · 80 S] complex formation can be shown to depend strongly not only on IF-E$_5$ (subunit joining factor) but also on mRNA, ATP and IF-E$_4$ (not shown). In the undisturbed *in vivo* situation, we visualize the mRNA binding and 60 S subunit joining to be promoted by a strongly cooperative action of IF-E$_{1/4/5}$ and ATP.

The experiments presented here suggest that at least 5 protein factors are involved in mammalian polypeptide chain initiation. The most outstanding and novel features of the mechanism suggested by our work are (i) the mRNA independent binding of the initiator Met-tRNA$_f$ to the 40 S ribosome subunit promoted by the two initiation factors IF-E$_2$ and IF-E$_3$ and (ii) the absolute requirement for ATP in addition to GTP for initiation. ATP is presumably needed for mRNA binding. This latter finding might relate to an ATPase activity recently found in both eukaryotic and prokaryotic ribosomes (Grummt and Erdmann, personal communication), and it strongly supports the observation of Marcus

[12 a, b] who described on ATP requirement for the binding of a plant viral RNA to wheat germ ribosomes. IF-E_3 used for the experiments in Table 1 and Figs. 5, 6, 7 had not been purified with the final step of glycerol gradient centrifugation [11]. It seems now that an additional component besides IF-E_3 required for initiation (IF-E_4) is present in the cruder IF-E_3 preparations.

References

1. SHAFRITZ, D. A., ANDERSON, W. F.: Nature (Lond.) **227**, 918 (1970).
2. PRICHARD, P. M., GILBERT, J. M., SHAFRITZ. D. A., ANDERSON, W. F.: Nature (Lond.) **226**, 511 (1970).
3. SHAFRITZ, D. A., LAYCOCK, D. G., ANDERSON, W. F.: Proc. nat. Acad. Sci. (Wash.) **68**, 496 (1971).
4. CRYSTAL, R. G., SHAFRITZ, D. A., PRICHARD, P. H., ANDERSON, W. F.: Proc. nat. Acad. Sci. (Wash.) **68**, 1810 (1971).
5. SHAFRITZ, D. A., LAYCOCK, D. G., CRYSTAL, R. G., ANDERSON, W. F.: Proc. nat. Acad. Sci. (Wash.) **68**, 2246 (1971).
6. PRICHARD, P. M., PICCIANO, D. J., LAYCOCK, D. G., ANDERSON, W. F.: Proc. nat. Acad. Sci. (Wash.) **68**, 2752 (1971).
7. CRYSTAL, R. G., ANDERSON, W. F.: Proc. nat. Acad. Sci. (Wash.) **69**, 106 (1972).
8. SHAFRITZ, D. A., PRICHARD, P. M., GILBERT, J. M., MERRICK, W. C., ANDERSON, W. F.: Proc. nat. Acad. Sci. (Wash.) **69**, 983 (1972).
9. CRYSTAL, R. G., NIENHUIS, A. W., PRICHARD, P. M., PICCIANO, D., ELSON, N. A., MERRICK, W. C., GRAF, H., SHAFRITZ, D. A., LAYCOCK, D. G., LAST, J. A., ANDERSON, W. F.: FEBS Letters **24**, 310 (1972).
10. SCHREIER, M. H., STAEHELIN, T.: J. molec. Biol. **73**, 329 (1973).
11. SCHREIER, M. H., STAEHELIN, T.: Nature (Lond.) New Biol. **242**, 35 (1973).

12a. MARCUS, A.: J. biol. Chem. **245**, 955 (1970).
12b. MARCUS, A.: J. biol. Chem. **245**, 962 (1970).

Discussion

H. NOLL (Evanston): I have two questions. 1. In the experiment where you combined f met with F2 alone in the presence of poly AUG, can this binding also be obtained with a AUG triplet alone? 2. In the case of this 40 S and 80 S initiation complex which releases about 40% of the puromycin, can the messenger be inserted afterwards?

TH. STAEHELIN (Basel): The answer to both questions is very simple: I don't know. We have not used AUG triplets. And for the second point, the 30 S initiation complex puzzles us. It is not certain if the messenger must bind at the 30 S complex; there is evidence suggesting that it *can* bind there by its stabilizing action. But it might be able to enter at the 80 S stage. We just don't know at the time.

F. Grummt: Did you find an ATPase activity in your ATP-dependent IF4 reaction or could AMP-PCP substitute for ATP? We have isolated an ATPase activity from the rat liver ribosomes which has a very high Km for ATP of about 10^9—10^8 M and a Ki for GTP of about 1 m M.

Th. Staehelin: We have not looked yet for an ATPase activity, but in view of our results we expected that a ribosome-dependent ATPase activity should exist.

F. Grummt: A short remark to your factor IF5 which makes coupling of 60 S to the 40 S subunit initiation complex. You mentioned that your IF 5 is similar to Anderson's factor M_{2A}. We could isolate a factor with the same elution properties as M_{2A} from rat liver cytosol. This factor was shown to be active in coupling small and large subunits in the initiation step of protein synthesis. So it seems sure to us that IF5 and M_{2A} of Anderson are the same factors.

Th. Staehelin: We have found IF-E_5 activity in both guinea pig liver and Krebs ascites tumor cells. From the latter source, this factor has been partially purified and is very similar to the reticulocyte factor (Schreier and Lewis, unpublished results).

J. Drews: How do your factors IF1—5 relate to the factors M_1, M_{2A-B} and M_3 described by Anderson and to the initiation factor isolated by Ochoa's group from the supernatant of arteria saliva?

Th. Staehelin: It seems quite clear that our factor IF-E_3 is identical with Anderson's M_3. Our factor IF-E_4 is most likely Anderson's M_{2B} and our IF-E_5 is Anderson's M_{2A}. Our IF-E_2 is not identical with any of Anderson's factors. However, it is likely that Anderson's M_1 is related to IF-E_2, since it seems to have a rather similar function. Our IF-E_1 does not seem to correspond to any of Anderson's factors, but it might well be the equivalent of one of several stimulatory factors Anderson has observed but not further characterized (personal communication by W.F. Anderson).

Hilz (Hamburg): I have a comment to the lecture of Dr. Mach with respect to the trick to get rid of ribonuclease and to isolate undegraded mRNA by the application of proteinase K. I would like to point out that originally we have introduced proteinase K for the isolation of polysome-associated mRNA about two years ago.